Objective
Horticulture

Objective
Horticulture

Neeraj Pratap Singh

Executive (Horticulture) (Planning and Architecture)
Sahara India Commercial Corporation Limited
Sahara India Centre, 9th Floor, 2 Kapoorthala Complex
Aliganj, Lucknow 226 024 (U.P.) India

CBS

CBS Publishers & Distributors Pvt. Ltd.

New Delhi • Bengaluru • Chennai • Kochi • Kolkata • Mumbai
Hyderabad • Uttarakhand • Nagpur • Patna • Pune • Jharkhand

Objective Horticulture

ISBN: 978-93-85915-16-1

First CBS Reprint: 2016
Reprint: 2017, 2019

Published by **Satish Kumar Jain** and produced by **Varun Jain** for
CBS Publishers & Distributors Pvt. Ltd.,
4819/XI Prahlad Street, 24 Ansari Road, Daryaganj, New Delhi - 110002
delhi@cbspd.com, cbspubs@airtelmail.in • www.cbspd.com
Ph.: 23289259, 23266861, 23266867 • Fax: 011-23243014

Corporate Office: 204 FIE, Industrial Area, Patparganj, Delhi - 110 092
Ph: 49344934 • Fax: 011-49344935
E-mail: publishing@cbspd.com • publicity@cbspd.com

Branches:
• *Bengaluru:* 2975, 17th Cross, K.R. Road, Bansankari 2nd Stage,
  Bengaluru - 70 • Ph: +91-80-26771678/79 • Fax: +91-80-26771680
  E-mail: cbsbng@gmail.com, bangalore@cbspd.com
• *Chennai:* No. 7, Subbaraya Street, Shenoy Nagar, Chennai - 600030
  Ph: +91-44-26681266, 26680620 • Fax: +91-44-42032115
  E-mail: chennai@cbspd.com
• *Kochi:* Ashana House, 39/1904, A.M. Thomas Road, Valanjambalam,
  Ernakulum, Kochi • Ph: +91-484-4059061-65
  Fax: +91-484-4059065 • E-mail: cochin@cbspd.com
• *Kolkata:* 6-B, Ground Floor, Rameshwar Shaw Road, Kolkata - 700014
  Ph: +91-33-22891126/7/8 • E-mail: kolkata@cbspd.com
• *Mumbai:* 83-C, Dr. E. Moses Road, Worli, Mumbai - 400018
  Ph: +91-9833017933, 022-24902340/41 • E-mail: mumbai@cbspd.com

Representatives:

• Bhubaneswar	0-9911037372	• Hyderabad	0-9885175004	• Jharkhand	0-9811541605
• Nagpur	0-9021734563	• Patna	0-9334159340	• Pune	0-9623451994
• Uttarakhand	0-9716462459	• Dhaka (Bangladesh)	01912-003485		

Printed at:
J.S. Offset Printers, Delhi (India)

Preface

We are living in an era of Competition. This is visible everywhere. In educational institutions too, the competitive exams are no exception to this. Most of the Universities, Institutes, Organisations and Government staff selection Agencies have made competitive examinations the basis for choosing candidates. The questions asked in these exams are of objective type, large in number and of varied nature. They require quick answering within a specified period of time to secure high merit and assured success. A subject like horticulture encompasses many branches viz., Pomology, Olericulture, Floriculture, Spices & Condiments, Medicinal & Aromatic plants, Plantation crops, Post-harvest Management Technology and Mushroom culture, etc.. The aspirants have to scan through a number of textbooks, journals, magazines and periodicals to cover each topic of syllabus thoroughly. This is so because there is hardly any high quality book on horticulture that covers the entire syllabus of this subject for competitive exams. Keeping this in mind, an earnest effort has been made to write this book named 'OBJECTIVE HORTICULTURE' covering various branches of horticulture.

The book has been written keeping in view the pattern of different horticulture/ agriculture competitive exams viz. JRF, SRF, ARS, ASRB, NET, SET, Pre PG (Ag.) & Ph.D. (Horticulture) and UPSC.

The book is divided into 20 Steps. (Sets/Modules) with 4000 questions. Each Step is not only enriched with all the relevant information but also trains the reader to answer the questions in the most systematic manner in minimum possible time. The book brings a unique opportunity to take an early leap and win the deciding edge in competitions.

Yes, Go on to groom yourself for the ultimate success step by step.

The path from dream to success does exist; certainly.

May you have the vision to find it,

The courage to get on to it, and

The perseverance to follow it.

May you have a successful journey that sees your dreams come true...

Hanuman Jayanti   **Neeraj Pratap Singh**

Think it over

Ten Lessons

If you seek salvation and perfect wisdom, you should be aware of ten failures. **The first.** If, having been taught the truth, you do not follow it, you are like someone returning emtpy-handed from a land rich in precious gems. **The second.** If, having known the way, you turn to the other, you are like a moth flying into a flame. **The third.** If you live with a wise teacher, and yet remain ignorant, you are like someone dying of thirst on the shore of a lake. **The fourth.** If you flout the moral percepts you have been given, you are like a sick person who refuses to take medicine. **The fifth.** If you preach religion, but do not practise it, you are like a parrot. **The sixth.** If you give to the poor money which you have obtained by robbery and deceit, you are like lightning striking the surface of water. **The seventh.** If you pretend to be patient, you are like a cat stalking a mouse. **The eight.** If you do good merely to obtain respect, you are exchanging a precious gem for a piece of goat's dung. **The ninth.** If you are clever in expounding religion, but have had no spiritual experiences, you are like a rich man who has lost the key to his treasury. **The tenth.** If you try to explain teachings which you do not understand yourself, you are like a blind man leading the blind.

"Unfortunately some people fail to see the great values of life because of sadly twisted thought processes. A chief reason I write about having faith and practising positive thinking and finally winning over every form of defeat and adversity is that I believe in life and love it. I want to encourage others to believe in it and love it too"

- Neeraj Pratap Singh

Contents

Keys

A biology teacher was teaching his students how a caterpillar turns into a butterfly. He told the students that in the next couple of hours, the butterfly would struggle to come out of the cocoon, but no one should help the butterfly. Then he left.

The students were waiting and it happened. The butterfly struggled to get out of the cocoon, and one of the students took pity on it and decided to help the butterfly out of the cocoon against the advice of his teacher. He broke the cocoon to help the butterfly so it didn't have to struggle anymore. But shortly afterwards the butterfly died.

When the teacher returned, he was told what happened. He explained to this student that by helping the butterfly, he had actually killed it because it is a law of nature that the struggle to come out of the cocoon actually helps develop and strengthen its wings. The boy had deprived the butterfly of its struggle and the butterfly died.

Apply this same principle to our lives. Nothing worthwhile in life comes without a struggle. Trials in life can be tragedies or triumphs, depending on how we handle them.

Script your success story

[Triumphs don't come Without Effort]

Step-One

The Loser is always part of the problem.
The Winner is always part of the answer.

Multiple Choice

1. Lye peeling is done by
    A] 1% caustic + cool   B] 5% caustic soda +cool
    C] 1% Caustic + hot   D] 5% caustic soda + hot

2. Spongy tissue is due to
    A] Pathogen   B] Physiological factor
    C] Both a & b   D] None

3. Mass selection is essentially a-
    A] Hybridization technique B] Population improvement
    C] Clonal selection   D] Mutation breeding

4. 'Essential oil' is extracted from_____fruit-
    A] Lemon   B] Mangosteen
    C] Olive   D] Almond

5. Daria cultivation is followed in which crop
    A] Tomato   B] Root crops
    C] Cucurbits   D] Bulb crops

6. Axillary bearing is found in-
    A] Loquat   B] Bael
    C] Phalsa   D] Litchi

7. Diploid apogamy is found in
    A] *Allium spp*   B] *Perithelium spp*
    C] Tobacco   D] Rice

8. Dioecious form of sex is found in-
    A] Citrus   B] Aonla
    C] Papaya   D] Litchi

9. Maximum acreage under cocoa is found in-
    A] Karnataka   B] Andhra Pradesh
    C] Kerala   D] Tamil Nadu

10. Maximum acreage under coffee is found in-
    A] India   B] Mexico
    C] Brazil   D] Britain

Bulb dealer (jobber)
An individual of firm who stores, ships and distributes bulbs.

Bulb grower (producer)
An individual or firm who propagates bulbs for wholesale purpose.

Sodium hypochlirite (Clorox)
Common household bleach, used for commercial sterilization of greenhouse benches and tools.

Quintal
A measure of weight, 100 kilograms.

Acorn
The fruit of oaks, a thick walled nut with a woody cup-like base.

Diffuse
Loosely or widely spreading, an open form.

NAA
Naphthalene acetic acid, and auxin used as a rooting hormone.

Stub
A short blunt projection, as a blunt projecting stem, branch or root.

Axillary
Borne in an axil, as a bud or a flower.

Obscure
Not easily made out, as applied to buds, bundle scars, etc.

Compressed
Flattened from the sides, as applied to nodes or buds.

Orbicular
circular or globed, as in the shape of a leaf

11. Sweet orange is generally trained on-
A] Single stem
B] Multiple stem
C] Two branches
D] None of above

12. *Rauwolfia srpentina* is harvested after
A] 6 months
B] 12-15 months
C] 24 months
D] 36 months

13. 'Pusa Swarnima' is hybrid variety of-
A] Radish
B] Yam
C] Turnip
D] Carrot

14. Metal which is important constituent of chlorophyll is
A] Fe
B] Cu
C] N_2
D] Mg

15. 'Auxin' production is inhibited by-
A] 'Cu' deficiency
B] 'P' deficiency
C] 'Mg' deficiency
D] 'Mn' deficiency

16. 'Auxin' production is inhibited by-
A] Boron deficiency
B] Zinc deficiency
C] Molybdenum deficiency
D] Potash deficiency

17. Maximum acerage under cooconut is in-
A] Andhra Pradesh
B] Tamil Nadu
C] Maharashtra
D] Kerala

18. Which fruit is known as 'King of temperate fruits'-
A] Almond
B] Apple
C] Peach
D] Walnut

19. Kamala Retreat at Kanpur is laid on the pattern of-
A] Mughal style
B] Japanese style
C] English style
D] Formal style

20. RRII headquarter is located at
A] Kasargod
B] Calicut
C] Kottayan
D] Bangalore

21. Which product is made by yeast fermentation
A] Apple cider
B] Mango pulp
C] Citrus nector
D] All

22. Cassava (Tapioca) is mainly grown in-
A] Kerala
B] Bihar
C] Andhra Pradesh
D] Assam

23. Sugarcane production is maximum in which state
    A] TN   B] Maharashtra
    C] UP   D] MP

24. Vitamin C is maximum in
    A] Guava   B] Aonla
    C] Papaya   D] Mango

25. Which instrument is most helpful for preparing a 'colour scheme'-
    A] Colour box   B] Colour wheel
    C] Colour scale   D] Refractometer

26. Which flower is most suitable for drying-
    A] Poppy   B] Acroclinum
    C] Tuberose   D] Cosmos

27. Which annual is most susceptible to frost-
    A] Hollyhock   B] Aster
    C] Pot marigold   D] Salvia

28. Micro-irrigation system is much efficient for-
    A] Grape   B] Banana
    C] Coconut   D] All of them

29. Black Tip of mango is much prominent in-
    A] Ripening fruits   B] Mature fruits
    C] 6-8 weeks old fruits   D] Fertilization stage

30. Sapota (Chiku) is native of-
    A] India   B] China
    C] Iran   D] Mexico

31. Regular pruning is needed in
    A] Peach   B] Strawberry
    C] Guava   D] Sapota

32. Which of the following is not true-
    A] Mango-Drupe   B] Guava-Berry
    C] Pear-Pome   D] Apple-Pepo

33. Synptonimal complex is observed in
    A] During miosis I   B] During miosis II
    C] On completion of fertilization   D] All

Lignin
A complex substance which, associated with cellulose, causes the thickening of plant cell-walls and so forms wood.

Anterior
On the front side, away from the axis, toward the subtending bract.

Layering
A method of propagation, by which a branch of a plant is rooted while still attached to the plant by securing it to the soil with a piece of wire or other means..

Pendulous
hanging or drooping, can refer to plant form or parts, like branchlets or inflorescences

34. Heterostyly condition is observed in
    A] Tomato   B] Brinjal
    C] Potato   D] Chilli

35. Which annual is of Indian origin
    A] Balsam   B] zinnia
    C] Phlox   D] None

36. Japanese persimmon is originated in-
    A] China   B] Japan
    C] Malaya   D] S. Africa

37. 'Dashehari' mango is planted at a distance of 10 m, but 'Amrapalli' is planted at distance of-
    A] 15 m   B] 12 m
    C] 5 m   D] 2.8 m

38. *In-situ* grafting is practiced in-
    A] Laboratory   B] Planting site
    C] Test tube   D] Aerial parts

39. Novalonic acid is precursor of-
    A] GA & ABA   B] IAA & IBA
    C] Ethylene & CCC D] Cytokinin & ABA

40. Which variety of rose is preferred as 'cut flower-
    A] Sonia   B] Montezuma
    C] Super Star   D] None of these

41. Acetic acid is present in
    A] Nectar   B] Sweet chutney
    C] Vinegar   D] Apple cider

42. Guava jelly is preserved by
    A] Pectin   B] Sugar
    C] Na benzoate   D] None

43. Sweet potato is primarily a crop of-
    A] Tropical region B] Temperate region
    C] Arid region   D] None

44. Purple blotch is problem in
    A] Okra   B] Tomato
    C] Onion   D] Turmeric

45. The commercial tartaric acid is produced from-
    A] Citrus   B] Grape
    C] Loquat   D] Tamarindus

46. Bush pepper is produced from-
    A] Runners   B] Laterals
    C] Top shoots   D] Hanging shoots

47. *Polargonium gravedens* is propagated by
    A] Hardwood cutting   B] Semihardwood cutting
    C] Herbaceous cutting D] Leaf cutting

48. Water melon is propagated by-
    A] Seed   B] Cutting
    C] Tuber   D] None

49. Large cardamom is propagated through-
    A] Seed   B] Cutting
    C] Layer   D] Sucker

50. Which organization is related to agricultural marketing
    A] NAPHED   B] APEDA
    C] NABARD   D] All

51. Fruit which is rich in pectin & acid is
    A] Jamun   B] Apple
    C] Pear   D] Pomegranate

52. Which vegetable is rich source of Calcium
    A] Amarenthus   B] Cowpea
    C] Potato   D] Brinjal

53. Which fruit is rich source of fat
    A] Amla   B] Bael
    C] Avocado   D] All

54. Which vegetable is rich source of Vit. C (>100 mg/100g)
    A] Tomato   B] Cabbage
    C] Turnip   D] Radish

55. Which crop is richest source of oil
    A] Groundnut   B] Castor
    C] Sesamum   D] Oil palm

56. Rice Gandhibug is scientifically known as
    A] *Oresalia oryzae*   B] *Leptocorisa acutea*
    C] *Benesia taba*   D] None

Sorbitol
A white crystalline alcohol, found in certain fruits and berries and manufactured by the catalytic hydrogeneration of sucrose; used as a sweetner and in the manufacture of ascorbic acid and synthetic resins.

Glabrous
Smooth.

Spatulate
Spoon shaped.

Subtending
Standing below.

Tumid
Swollen

Filimentous
Thread-like.

Comose
Tufted with hairs.

Voluable
Twining.

Torsion
Twisting

57. Rossette formation is seen due to deficiency of
A] Mn   B] Zn
C] Cu   D] Mo

58. Epigial germination is seen in
A] Cashew   B] Rice
C] Jackfruit   D] Citrus

59. Karnal bunt is serious disease of
A] Apple   B] Wheat
C] Mango   D] Tomato

60. Phomopsis blight is serious disease of
A] Tomato   B] Brinjal
C] Chilli   D] Onion

61. 'Shalimar Bagh' is situated in-
A] H.P.   B] J&K
C] Assam   D] Rajasthan

62. The ratio of NPK is starter solution should be-
A] 2:1:1   B] 1:2:1
C] 1:1:2   D] 1:2:2

63. Which annual is substitute for 'dry flowers'-
A] Holly hock B] Ice plant
C] Helichrysum D] Corn flower

64. Which plant is suitable for Bonsai making
A] Pinus roxburghi   B] Hibicus
C] Ixora   D] All

65. Which month is suitable for pruning in peach
A] Early November   B] Late December
C] Late November   D] Early December

66. *Ficus lyrata* is the botanical name of-
A] Banyan   B] Banzo tree
C] Rubber plant D] Pilkhan

67. *Allium porrum* is the botanical name of-
A] Shallot   B] Leek
C] New Zealand spinach D] Asparagus

68. *Laginaria siceraria* is the botanical name of-
A] Bitter gourd   B] Pumpkin
C] Cucumber   D] Bottle gourd

69. *Cyanopsis tetragonolobus* is the botanical name of-
    A] Cluster bean   B] Winged bean
    C] Kidney bean   D] Hyacinth bean

70. *Ipomoea batata* is the botanical name of-
    A] Lily   B] Sweet Sultan
    C] Carrot   D] Sweet Potato

71. *Lagerstroemia spaciosa* is the botanical name of-
    A] Sawani   B] Pride of India
    C] Kadamb   D] Bottle brush

72. Curcuma longa is the botanical name of-
    A] Turmeric   B] Ginger
    C] Tuberose   D] Nutmeg

73. 'Palash' is the common name for
    A] *Plumeria alba*   B] *Butea monosperma*
    C] *Cassia fistula*   D] *Bauhinia vehil*

74. Indian ginseng is the common name of
    A] *Azadiracta indica*   B] *Rauwolfia serpentaria*
    C] *Hemidesmus indicus*   D] None

75. Black stem is the disease of fruit crop
    A] Apple   B] Mango
    C] Peach   D] Litchi

76. Bitter pit is the disorder of
    A] Apple   B] Pineapple
    C] Ber   D] None

77. Which of the following is the fruit of family Anacardiaceae-
    A] Walnut   B] Phalsa
    C] Carambola   D] Cashewnut

78. Die back is the major disease of which flower crop
    A] Rose   B] Chrysanthe mum
    C] Carnation   D] Marigold

79. Calyx splitting is the major problem in
    A] Orchid   B] Rose
    C] Carnation   D] Marigold

Distichous
Two ranked.

Abortive
Undeveloped;
rudimentary.

Ceriferous
Waxy.

Flaccid
Wilted.

Spicate
With spikes.

Bristly
With stiff hairs.

Fastigiate
With upright
branches.

Apetalous
Without petals.

Incised
intermediate
between toothed
(dentate or
serrate) and lobed,
being a sharply
inward cut leaf
(the inward cuts
are called
incisions)

Oleuropein
Bitter glycoside of olive (*Olea europaea*).

Supplementary illumination (lighting)
A means of supplying additional light (usually in the middle of the dark period) to lengthen the photoperiod, used to keep short day plants in vegetative phase or to give additional light to the plants.

Buttressed
With supporting wings.

Inequilateral
With unequal sides.

Ton
A unit of weight, generally equal to 2,000 lb. Short ton : 2,000 lb, long ton 2,240 lb.

80. Which variety of grape is the most suitable for 'raisin' making-
A] Perlette   B] Thompson Seedless
C] Anab-e-Shahi   D] Black Prince

81. Which method of planting is the most suitable for an area located near a city-
A] Square   B] Contour
C] Hexagonal   D] Rectangular

82. Which group of crops is the most suitable for intercropping in new orchard-
A] Short duration fodders B] Short duration fruits crops
C] Poplar & Eucalyptus   D] Short duration legume crops

83. Which region is the most suitable for litchi cultivation in U.P.-
A] Lucknow   B] Saharanpur
C] Bundelkhand   D] Varanasi

84. Which crop is the most susceptible to water logging-
A] Banana   B] Pomegranate
C] Papaya   D] Jack fruit

85. *Feijoa sellowiana* is the scientific name of-
A] Bread fruit   B] Pineapple guava
C] Pineapple   D] Rambutan

86. *Artocarpus lakoocha* is the scientific name of-
A] Barhal   B] Kathal
C] Chiku   D] Kamrakh

87. *Citrus maxima* is the scientific name of-
A] Mosambi   B] Pummelo
C] Mandarin   D] Sweet orange

88. Root wilt is the serious disease of
A] Arecanut   B] Coconut
C] Cashew   D] Cocoa

89. Yellow dwarf is the serious disease of
A] Coconut   B] Arecanut
C] Oil palm   D] Cashew

90. Spongy tissue is the serious problem in
A] Alphonso   B] Pairi
C] Deshehari   D] Neelam

91. 'Breeder seed' is the source of-
    A] Certified seeds   B] Nucleus seeds
    C] Foundation seeds   D] Registered seeds

92. Arka abir is the var of
    A] Brinjal   B] Cucurbit
    C] Tomato   D] Chilli

93. Pusa deepti is the var of
    A] Brinjal   B] Pumpkin
    C] Okra   D] Chilli

94. Arka navneet is the var of
    A] Brinjal   B] Tomato
    C] Okra   D] None

Matching

95. Match thc following :-
    A] Hazelnut   i]   Anacardiaceae
    B] Chestnut   ii] Corylaceae
    C] Pistachionut   iii] Oleaceae
    D] Pecan   iv] Fagaceae
    E] Olive   v]   Juglandaceae

96. Match the followilig :-
    A] Rosaceae   i]   Gooseberry
    B] Actinidiaceae   ii] Cherry
    C] Ericaceae   iii] Walnut
    D] Saxifragaceae   iv] Cranberry
    E] Juglandaceae   v]   Kiwiberry

97. Match the following-
    A] Quince   i]   Nut
    B] Sapota   ii] Pome
    C] Mulberry   iii] Sorosis
    D] Rambutan   iv] Capsule
    E] Carambola   v]   Berry

98. Match the following .-
    A] Big ovary   i]   Pseudo short styled flowers
    B] Medium size ovary   ii] Cross pollination
    C] Rudimentary ovary   iii] Medium styled flowers
    D] Long styled brinjal   iv] Self-pollination
    E] Medium & True short   v]   Long styled flowers
       styled brinjal

Chromatin
A complex between DNA, histones and other proteins and chromosomal RNA which make the chromosomes.

Stolon
along with runners, horizontal stems involved in vegetative propagation; a horizontal stem (usually just above, at, or below ground level) that roots at its tip and gives rise to a new plant; similar to runners

Millerandage
A disorder of grape, in which the ovary persists but the seed remain small or do not attain the usual size.

Dormant

Also called as resting, inactive, quiescent. Applied to buds when they are not actively growing and to plants when they are not in leaf.

Armed

With pines (barberry, Japanese quince) or prickles (brambles). Leaves that are pungent at the tip or around the margin (holly), are not included in this limited definition.

Disarticulating

Falling away by abscission, leaving a clean cut scar, as with most leaves, many flowers, some twig tips, etc.

Ciliate

Hairy on the margin, like the eyelids.

99. Match the following :-
    A] *Annona squamosa*   i]   Sour sop
    B] *Annona reticulata*   ii]   Sweet sop
    C] *Annona muricata*   iii] Bullock's heart
    D] *Annona cherimola*   iv]   Atemoya
    E] *Annona atemoya*   v]   Lakshaman Phal

100. Match the following :-
    A] Chlorophyll   i]   Pumpkin
    B] Carotenoids   ii]   Potato
    C] Anthocyanins   iii] Red beet
    D] Flavonoids   iv]   Cabbage
    E] Betalains   v]   Apple

101. Match the following :-
    A] Curcumin   i]   Bougainvillea
    B] Lahar   ii]   Rose
    C] Thimma   iii] Potato
    D] Y2-K   iv]   Turmeric
    E] Browning   v]   Chrysanthemum

102. Match the following :-
    A] Dioecious papaya   i]   Responsible for maleness papaya
    B] Gynodioeclous papaya ii]   Madhu Bindu, Washington, Pusa Giant
    C] M_1 (dominant gene) iii] Responsible for 'femaleness' in papaya
    D] M_2 (dominant gene) iv]   Pusa Majesty, Coorg Honey Dew; CO-3
    E] m (recessive gene)   v]   Responsible for 'hermaphroditism'

103. Match the following : -
    A] Elephant foot yam   i]   Alliaceae
    B] Asparagus   ii]   Chenopodiaceae
    C] Leek   iii] Convolvulaceae
    D] Yam   iv]   Malvaceae
    E] Spinach   v]   Araceae
    F] Lettuce   vi]   Compositae
    G] Sweet potato   vii] Dioscoreaceae
    H] Knolkhol   viii] Liliaceae
    I] Okra   ix]   Euphorbiaceae
    J] Tapioca   x]   Cruciferae

104. Match the following :-
     A] Herbaceous cutting   i] Bryophyllum
     B] Hard wood cutting   ii] Jack fruit
     C] Leaf cutting   iii] Guava
     D] Seed propagation   iv] Litchi
     E] Veneer grafting   v] Grape
     F] Mound layering   vi] Ber
     G] Marcottage   vii] Rose
     H] Ring budding   viii] Jamun
     I] Shield budding   ix] Chrysanthemum
     J] Polyembryonic seed   x] Mango

105. Match the following :-
     A] Mulching   i] Tomato
     B] Evergreen   ii] Brinjal
     C] Radial Cracking   iii] Chilli
     D] Epilachna beetle   iv] Olive
     E] Male sterile line   v] Strawberry
        (G-2)

106. Match the following :-
     A] Mushroom (PG)   i] Mint
     B] Menthol   ii] Ginger
     C] Nutmeg   iii] Seed spice
     D] Rhizome   iv] Edible fungus
     E] Fennel   v] Tree spice

107. Match the following :-
     A] October   i] Grape
     B] Dec.January   ii] Coconut
     C] River bed   iii] Muskmelon
     D] Aridzone   iv] Potato
     E] Sea coast   v] Phalsa

108. Match the following :-
     A] Powdery mildew   i] *Pythium* sp.
     B] Colocasia   ii] Guava
     C] Die-back   iii] Pomegranate
     D] Granulation   iv] Cucurbits
     E] Fruit cracking   v] Citrus

Nageire
A style of Ikebana, meaning literally "thrown in" where tall upright vases are used for making the arrangements and flowers with sufficent stem length are held in the vases by bar fixtures.

Jam
A preserved fruit food, mixture of fruit and sugar cooked to the consistency of a gel, firm enough to hold fruit tissues. Contains fruit pulp and hence not clear.

Woolly
Having long, soft, more or less matted or tangled hairs; like wool.

Rosin
A product of resin, obtained when turpentine (from pine trees) is distilled.

Break
New lateral shoot, often developed following removal of apical dominance by pinching.

Bract
a modified leaf, often small and associated with leaves or inflorescences, but sometimes large and showy

Milky sap
Whitish in color, often thicker than water.

Achene
A dry, indehiscent, one-seeded fruit.

Catkin
A scaly, cylindrical, pendulous, unisexual spike of flowers e.g. mulberry, birch, acalypha, etc.

109. Match the following :-
A] Pusa Nanha   i] Selection from Honey Dew
B] Coorg Honey   ii] Hybrid of CO-2 x Sunrise Solo
C] CO-1   iii] Inbred selection of Ranchi
D] CO-3   iv] Mutation bred papaya
E] CO-4   v] Hybrid of CO-1 x Washington

110. Match the following :-
A] Self-barrenness   i] M-12
B] Earliest rootstock of Apple ii] Apricot
C] Tallest rootstock of Apple   iii] Rymer
D] Cooking Apple   iv] M-7
E] Indigenous Apple   v] Bramley's Seeding

111. Match the following :-
A] Self-compatible cvs   i] Lucknow Seedless (lemon), Clementine (mandarin)
B] Self-incompatible cvs   ii] Nagpur Santra (mandarin)
C] Male-sterile (No viable pollen) iii] Marsh (grape fruit), Eureka & Lisbon (lemon)
D] Female-sterile (No viable stigma) iv] Washington Naval, Satsuma
E] Pollinizers for Pant Lemon-1 v] Italian and Nepali oblong lemon

112. Match the following :-
A] Shield budding   i] Grape
B] Cutting   ii] Epicotyle grafting
C] Seed   iii] Mound layering
D] Guava   iv] Aonla
E] Mango   v] Jack fruit

113. Match the following :-
A] Sterilization   i] Raisin & Dates
B] Carbonation   ii] Canning
C] Dehydration   iii] Jam & Jelly
D] Fermentation   iv] Fruit juice beverages
E] Sugar addition   v] Vinegar

GLOSSARY

Carotenoids
A group of yellow, orange and orange-red fat-soluble pigments present in plant parts. These are either hydrocarbons or its derivatives and are composed of isoprene units. Carotenoids occur in different forms like, carotene, lycopene, lutein, violaxanthin, neoxanthin, capsanthin, bixin, xanthophyll, etc.

Trophy
An arrangement of colourful, potted plants which may be annual or herbaceous perennials.

Flower induction
An unobservable, preparatory step that occur prior to visible flower bud initiation.

---

114. Match the following :-
    A] Turmeric   i] Crosin
    B] Black pepper   ii] Bark Spice
    C] Saffron   iii] Panniyur-2
    D] Coriander   iv] Seed spice
    E] Cinnamo   v] Sudarshan

115. Match the following :-
    A] Gardening in India   i] W.B. Hayes
    B] Vegetable Crops of   ii] S.L. Jindal
       India
    C] Fruit Growing in India   iii] M.S. Randhawa
    D] Flowering Shrubs   iv] B.P. Pal
       in India
    E] Beautiful Climbers of   v] K.S. Yawalkar
       India

116. Match the following :-
    A] CITH   i] Bikaner
    B] ICMR   ii] Srinagar
    C] NRCAZH   iii] Calicut
    D] ICAR   iv] Lucknow
    E] ISRI   v] New Delhi

True / False

117. Apple is known as 'King of temperate fruits'.   T/F

118. Cashewnut & banana are rich source of Vit-B$_1$.   T/F

119. Dr. B.P. Pal was an eminent rose breeder.   T/F

120. Kanchan and Krishna are varieties of ber.   T/F

121. Spongy tissue of mango is a viral disease.   T/F

122. The fruit of potato is botanically berry.   T/F

123. Tip rot/*Koeli* is a disorder of mango.   T/F

124. Vegetables are alkaline in nature.   T/F

125. Agroson-GN is suitable for treating nursery beds.   T/F

126. C.M.R.S. is situated at Lucknow.   T/F

127. Coffee belongs to family Rubiaceae.   T/F

128. Fruits are rich source of vitamins.   T/F

Berry
A fleshy, indehiscent, pulpy, multi-seeded fruit resulting from a single pistil.

Stellate
Arranged like a star, radiating, as leaves surrounding a stem in a whorl.

Villous
Having long, soft, shaggy hairs that are not matted.

Tendril
The twisting, clinging, slender growth on many vines, which allows the plant to attach themselves to a support or trellis.

Spongy
Irregularly interrupted by small, sometimes scarcely distinguishable cavities; porous.

129. In high density orcharding (H.D.O.) of pomegranate, planting distance should be 5x5m. T/F
130. Mughal Gardens are replica of informal style. T/F
131. Phalsa requires severe pruning for optimum yield. T/F
132. Red Delicious is an important variety of plum. T/F
133. Self-incompatibility in mango cvs like Dashehari, Langra, Chausa, Bombay Green has made easier the work of hybridization. T/F
134. Sweet potato is a modified stem. T/F
135. 'Chakaiya' aonla is suitable for Rajasthan. T/F
136. In datepalm male & female flowers appear on same plant. T/F
137. Kaliyanpur Bundeli is a variety of colocasia. T/F
138. Papaya is susceptible to waterlogging. T/F
139. Pectic acid is essential for jelly making. T/F
140. Pruning in rose is done in Dec.-Jan. T/F
141. Sweetness (low acid) in peaches is controlled by dominant gene. T/F
142. Vellaikolumban is a mango root stock. T/F
143. Alphonso is a variety of banana. T/F
144. Hermaphodite or perfect flowers of mango are protogynous. T/F
145. Japanese garden is picturesque type of garden. T/F
146. Lettuce can be grown round the year in plains. T/F
147. Mango is known as 'king of trees'. T/F
148. Pomegranate is dioecious fruit plant. T/F
149. Powdery mildew is serious disease of grape. T/F
150. Tea belongs to family Theaceae. T/F
151. Tropical America is the origin place of tomato. T/F
152. Brown Turkey is root stock of fig. T/F
153. *Camellia japonica* tea is grown for ornamental purpose. T/F
154. Gibberellic acid and kinetin delays ripening process in mango. T/F

155. Nymphaeas are suitable for hanging basket.   T/F
156. Phylloxera is a root louse of grape.   T/F
157. Pineapple is propagated by cutting.   T/F
158. Plant population is higher in Quincunx method.   T/F
159. Pungency in onion is due to 'quercetin'.   T/F
160. Stock-Scion incompatibility is found in citrus.   T/F
161. Vindhya hills are ideally suited for apple growing.   T/F
162. Garden pea and grapes are preserved by dehydration.   T/F
163. Guava is commonly propagated by compound layering.   T/F
164. Intensity orcharding was first time practised in apple.   T/F
165. Kaolin is an antitranspirant.   T/F
166. Pithiness in radish is developed due to 'B' deficiency.   T/F

Fill in the Blanks

167. Mango and Citrus produces _____ seedlings.
168. Panama wilt is serious disease in _____
169. _____ produces seedless fruits.
170. Cabbage 'head' is modification of _____
171. Sweet potato is modification of _____
172. Muskmelon is _____ cucurbit.
173. Cucumber has _____ sex form.
174. Shield budding is practiced in _____
175. Roots of _____ are transplanted for seed production.
176. For seed production _____ of turnip are transplanted.
177. Guava is propagated by _____
178. Papaya plants are _____
179. _____ is important for setting of jelly.
180. _____ is important constituent of fruits of aonla.
181. Bael is rich source of _____
182. Alphonso is a variety of _____

183. Chrysanthemum is propagated by _____
184. Mosaic is a disease of _____
185. _____ is practised, where is problem of scion-stock incompatibility.
186. Coconut has _____ sex form.
187. _____ is important constituent of fruits of mango & papaya.
188. Spongy Tissue disorder is found in _____
189. Seed variation is greatly observed in _____
190. KMS is used for _____ of fruit juice.
191. Papain is a/an _____
192. Papain is obtained from _____
193. Sweet potato is propagated by _____
194. Cauliflower 'curd' is modified _____
195. _____ is commercial method of mango propagation.
196. _____ is the most dangerous disease of potato.
197. _____ is the richest source of Vit-C.
198. _____ leads in total fruit production.
199. Guava contains _____ Vit-C in 100g pulp.
200. _____ is generic name of grape.

There was a man who failed in business at the age of 21; was defeated in a legislative race at age 22; failed again in business at age 24; overcame the death of his sweetheart at age 26; had a nervous breakdown at age 27; lost a congressional race at age 34; lost a senatorial race at age 45; failed in an effort to become vice-president at age 47; lost a senatorial race at age 49; and was elected president of the United States at age 52. This man was Abraham Lincoln.

Would you call him a failure? He could have quit. But to Lincoln, defeat was a detour and not a dead end.

Script your success story

Failure is the **H**ighway to **S**uccess.

Step-Two

The Loser always has an excuse;
The Winner always has a program

Multiple Choice

1. India contributes % mango production in the world-
   A] 56   B] 80
   C] 20   D] 10

2. Maximum 1 % of oil is found in
   A] Groundnut   B] Mustard
   C] Linseed   D] Coconut

3. 'Hen & Chicken' is disorder of
   A] Banana   B] Custard Apple
   C] Grape   D] Strawberry

4. Bell pepper & chilli belong to
   A] Same spp   B] Same family
   C] Samgye genus   D] All of above

5. Beet spinach & European spinach belong to-
   A] Same family   B] Same genus
   C] Same spp   D] None

6. Animal cells & plant cells can be differentiated by-
   A] Conductivity   B] Shape
   C] Size   D] Presence or absence of
                                 cell-wall

7. Benzoic acid & SO_2 are added to preserve fruit & veg. product to check-
   A] Micro organism   B] Ethylene production
      activity
   C] Reduce acidity   D] None

8. The fruits & vegetables are known as
   A] Nutritive food   B] Energy food
   C] Protective food   D] None

9. Ikebana is
   A] Flower arrangement   B] Bonsai culture
   C] Soil culture   D] All

Stem topples
In tulips, the physiological disorder that is characterized by the collapse of a small portion of the inter-node of the flower stalk located just underneath the flower.

Crown
On a clumping plant, the point at which the main stem (or stems) and the roots join together.

Thicket
A number of shrubs, trees, growing close together.

Astroturf
A synthetic fibrous lawn, used in roof gardens as well as in stadia.

Recurved
Bent or curved backward, usually roundly or obtusely so.

Pickle
A preserve of fruit, vegetable in salt solution, with spices and condiments added. One of the oldest methods of food preservation.

Discoid
The same as chambered, when applied to pith.

Landscape plants
Those plants, which are used with the objective of beautifying our surrounding. They must serve certain functional, architectural or engineering uses.

Agriculture
An activity of man, which is primarily aimed at the production of food, fibre, fuel, etc. by optimum use of terrestrial resources.

10. Photophosphorylation is
A] Hill reaction
B] Formation of ATP in presence of light
C] Formation of NADP in presence of light
D] None

11. Xeriseaping is
A] Landscaping in arid areas
B] Use of xerophytic plants for indoor decoration
C] Soil conservation measures
D] All

12. Terrarium is
A] Plant grow on terraces
B] Transparent container for keeping flower
C] Solution used in flowers
D] Pot for raising plants

13. Isotope have
A] Same no. of protons
B] Same no. of neutrons
C] Same no. of electrons
D] None

14. Isotone have
A] Same no. of protons
B] Same no. of neutrons
C] Same no. of electrons
D] None

15. Cleistogamy favours
A] Self pollination
B] Cross pollination
C] Both
D] None

16. Herkogamy favours
A] Self pollination
B] Cross pollination
C] Mutation
D] Wide hybridization

17. Pruning involve
A] Thinning
B] Heading back
C] Both a & b
D] Bending

18. Scientists Recruitment Board (ASRB) was established in
A] 1969
B] 1973
C] 1975
D] 1980

19. Lemon (*C. limon*) flowers...............a year
A] Once
B] Twice
C] Thrice
D] Throughout the year-

20. Guava (L-49) is a chance seedling selection of-
    A] Behut Coconut   B] Apple Guava
    C] Allahabad Safeda   D] Arka Amulya

21. Polyantha (Modem Floribunda) rose was developed by-
    A] Guillot (1867)   B] Pernet Dutcher (1898)
    C] Poulsen (1912)   D] William Robinson (1820)

22. Ornamental banana (*Musa coccinea*) has the somatic chromosome numbers-
    A] $2n = 22$ B] $2n = 33$
    C] $2n = 20$ D] $2n = 44$

23. *Diospyros kaki* (persimon) belongs to family-
    A] Anacardiaceae   B] Dioscoreaceae
    C] Ebenaceae   D] Oxalidaceae

24. *Nephelium lappaceum* (Rambutan) belongs to family-
    A] Sapotaceae   B] Apocynaceae
    C] Sapindaceae   D] Oxalidaceae

25. First seedless (triploid) variety of watermelon was developed by-
    A] Dr. Kihara (1951)   B] Baily (1949)
    C] Freidlander (1977)   D] Heslop (1963)

26. Potassium metabisulphite (KMS) is used in form of-
    A] Gas   B] Liquid
    C] Solid   D] Any form

27. 'Stunted growth', 'yellowing' and 'leaf curling' are the symptoms of-
    A] Bacterial infection   B] Viral infection
    C] Fungal infection   D] Nutrient deficiency

28. 'Night blindness', a disease of eyes is caused due to deficiency of -
    A] Ascorbic acid   B] Retinol
    C] Riboflavin   D] Folic acid

29. *Bemisia tabaci*, a pest of cotton is commonly known as
    A] Boll worm   B] Hairy caterpillar
    C] White fly   D] Sticking moth

PGRs
Plant Growth Regulators, which modify plant physiological process.

Oblong
longer than broad, with the margins parallel except at the extreme basal and apical ends

Trichomes
A bristle or hair.

Strobile
A cone of pines.

Sporophyll
A spore bearing leaf.

Vespertinus
Appearing in the evening.

Stratified
Arranged in horizontal layers.

Conidia
Asexual fungus spores.

Basifixed
Attached by the base.

Asepsis
Being free from microorganisms.

Cauline
Belonging to the stem.

Dysgenic
Biologically defective or deficient.

Weeping
Dropping conspicuously, pendent.

FTE
Fritted Trace Element.

Spreading
Growing outward or horizontally.

Calcarate
Having a spur.

Twiggy
Having many divergent twigs.

30. In India , irrigated area is about
     A] 70 MH   B] 50 MH
     C] 30 MH   D] 80 MH

31. In North Indian conditions , loquat flowers-
     A] Once a year   B] Twice a year
     C] Thrice a year D] No any flush

32. After preservation , nutrients, vitamins & minerals are
     A] Less than before   B] More than before
        preservation   preservation
     C] Constant   D] None

33. For seed production purpose , onion is considered as-
     A] Annual plant   B] Biennial plant
     C] Perennial plant D] Flowering plant

34. For breaking bud dormancy , suitable chemical is-
     A] Cytokinin   B] Cycocel
     C] Ethylene   D] I.B.A

35. In a papaya orchid , the ratio of female and male plants should be-
     A] 25:10   B] 10:1
     C] 50:25   D] 1:10

36. In cherries , universal donors-Stella & Vista are grouped into-
     A] 'A' group   B] 'B' group
     C] 'AB' group   D] 'O' group

37. In tea , which of the following beverage substance is present
     A] Caffein   B] Thiobromin
     C] Piperine   D] None

38. In mango , which var is monoembryonic
     A] Neelam   B] Salem
     C] Goa   D] Olour

39. First H . T. rose was 'La France' which was developed by-
     A] Guillot (1867)   B] Dr. B.P. Pal (1962)
     C] Poulsen (1912)   D] Pernet Dutcher (1898)

40. In U.P ., grape fruit is commercially grown in-
     A] Eastern region B] Southern region
     C] Western region D] Central U.P.

41. Papaya contains .... I unit of Vitamin-A
     A] 2500 IU   B] 2020 IU
     C] 3000 IU   D] 4000 1U

42. First D .D.G. (Horti) in ICAR was-
     A] Dr. R.M. Pandey   B] Dr. S.P. Ghosh
     C] Dr. K.L. Chadha   D] Dr. G.L. Kaul

43. First D .G. of ICAR was-
     A] Dr. O.P. Gautam   B] Dr. B.P. Pal
     C] Dr. G.S. Randhawa   D] Dr. P.P. Trivedi

44. Which P .G.R. is used to induce femaleness in fruit-
     A] GA$_3$ B] IAA
     C] Cytokinin   D] Maleic Hydrazide

45. The Headquaters of N .H.B. is situated at-
     A] Abohar (Punjab)   B] Gurgaon (Haryana)
     C] Pusa (New Delhi)   D] Bangalore (Karnataka)

46. *Kochia scoparia* is a /an-
     A] Flowering annual   B] Fruiting perennial
     C] Foliage annual   D] Foliage perennial

47. *Victoria regia* is a /an-
     A] Oxyegenerator   B] Marginal aquatic
     C] Surface flowering aquatic   D] Foliage Plant

48. Which fruit crop is /are suitable for 'dry-land orcharding'-
     A] Aonla   B] Custard apple
     C] Pomegranate   D] All of above

49. High auxin : Cytokinin ratio in callus favours the-
     A] Shoot formation   B] Flowering
     C] Root formation   D] Embryo development

50. How many plants of 'Amarpali' mango can be planted in one hectare-
     A] 100   B] 400
     C] 1600   D] 1000

51. Capsicum varieties 'Arka Mohini, Arka Gaurav and Arka Basant' are bred by-
     A] Dr. G. Kalloo   B] Dr. J. V. Peter
     C] Dr. D.P. Singh   D] Dr. K.L. Chadha

GLOSSARY

HWT
Hot Water Treatment.

X
Indicates a hybrid.

Tomentulose
Microscopically tomentose or woolly.

Native
Original to an area.

Prominent
Projecting outward, conspicuous.

Trailing
Prostrate or not rooting.

Pilose
Shaggy with soft hairs.

Divergent
The same as spreading.

Sporulate
To produce spores.

Erect
Upright habit of growth.

Bracteate
With bracts. Bracted.

Venulose
With very fine hairs.

Drupe
Also called stone fruit. A fruit derived entirely from an ovary, one seeded, with an exocarp, fleshy mesocarp, and stony endodcarp as in peach, cherry and plum.

Template
A pattern or mould. Here refers to DNA which stores coded information and acts as a model or template from which information is taken by messenger RNA.

52. Which is considered as 'Bathroom' fruit-
A] Cherry   B] Banana
C] Mango   D] Litchi

53. To control 'biennial bearing' in mango, which is effective-
A] Paclobutrazol   B] Naphthalene Acetic Acid
C] Gibberellic acid   D] Ethylene

54. Which is 'C$_4$' horticultural plant-
A] Litchi   B] Mango
C] Date palm   D] Cherry

55. In India 'Epicotyl grafting' in mango was developed by-
A] Lynch (1940)   B] Bhan *et al* (1969)
C] S.K. Mukherjee (1960)   D] Sant Ram (1965)

56. Which is known as 'Garden city of India'-
A] Bangalore   B] Chandigarh
C] New Delhi   D] Mysore

57. The term 'Heaven of man' is used for gardesn
A] Japanese garden   B] Mugal garden
C] Italian garden   D] none

58. The main function of 'Hormones' is-
A] Stimulation   B] To Start
C] To Inhibit   D] All of above

59. 'Dholka' and 'Kandhari' are the varieties of-
A] Pomegranate   B] Aonla
C] Loquat   D] Jamun

60. 'Amrapali' and 'Mallika' varieties of mango were developed by-
A] Dr. B.P. Pal   B] Dr. R.N. Singh
C] Dr. N.S. Randhawa   D] Dr. K.L. Chadha

61. The scientific name of 'mango stem borer' is-
A] *Dacus dorsalis*   B] *Batocera rufomaculata*
C] *Inderbella* sp.   D] *Drosicha mangiferae*

62. The bitter compound of 'Neem' is-
A] Pyrethrum   B] Melicin
C] Solanin   D] Rotenone

63. The width of a 'nursery bed' should be-
    A] 30-60 cm   B] 100-120 cm
    C] 150-175 cm   D] 15-30 cm

64. Which is suitable for 'river bed cultivation'-
    A] Guava   B] Muskmelon
    C] Litchi   D] Banana

65. 'Panchmukhi' and 'Sahasra Mukhi' are the cultivars of-
    A] Sweet Potato   B] Elephant Yam
    C] Colocasia   D] Cassava

66. In India 'Sanjose Scale' in apple was first reported during-
    A] 1902   B] 1906
    C] 1912   D] 1932

67. The ideal concentration of 'sodium benzoate' for preserving fruit juices is-
    A] 0.01-0.05%   B] 0.06-0.10%
    C] 1.1-1.5%   D] 1.6-2.0%

68. 'Mangala' and 'Sumangala' are the varieties of-
    A] Rubber   B] Coffee
    C] Coconut   D] Arecanut

69. The commercial source of 'tartaric acid' is-
    A] Grapefruit   B] Grape
    C] Karonda   D] Sour Orange

70. Which is known as 'vinegar bacteria'-
    A] *Lactobacillus*   B] *Acetobactor*
    C] *Clostridium*   D] *Bacillus* sp.

71. Which is described as 'Wonder-tree'-
    A] Mango   B] Pipal
    C] Neem   D] Eucalyptus

72. For planting 1 hac bitter gourd how much seed rate is required-
    A] 10kg   B] 20kg
    C] 40kg   D] 5kg

73. To plant 1 hectare rice, nursery area should be
    A] .01H   B] .1H
    C] .5H   D] .2H

Syringing
Application of water. Sprayed on the foliage. Used to reduce transpication, control of insectpests, to reduced leaf temperature, or as a mean of watering.

Catechin
A tannin $(C_{15}H_{14}O_6.3H_2O)$ usually called 'Katha', extracted from the chips of heart wood of *Acacia catechu* (Khair).

Carbide
Binary compound of carbon ; a term loosely applied for calcium carbide (CaC_2), a compound used for enhancing ripening in fruits.

Laxative
A mild purgative.

Dryad
A shade plant

Carpoptosis
Abnormal fruit drop.

Epigeal
Above ground

Sanicle
An umbeliferous plant.

Verticillate
Arranged in whorls.

Anthotaxy
Arrangement of flowers.

Venation
Arrangement of veins.

Anthophorous
Bearing flowers.

Seminiferous
Bearing seed.

Formosus
Beautiful, handsome.

74. The temperature below - 150°C is called as-
A] Freezing   B] Cryogenic
C] Refrigeration   D] Sharp freezing

75. Vegetable of 20th century is-
A] Cluster bean   B] Amaranthus
C] Winged bean   D] Chikurmanis

76. Chromosome (2n) no. of rice is
A] 22   B] 24
C] 26   D] 28

77. A gully 3 m deep, 18 m or more width is
A] Very small   B] Small
C] Medium   D] None

78. 9: 3:3:1 ratio in F_1 hybrid denote
A] Characters are   B] There is 50% linkage
   independent to each other
C] There is more than   D] All
   50% linkage

79. 2, 4-D is used to control-
A] Flowering   B] Weeds
C] Vegetative growth D] Root initiation

80. More than 50% of area under grape in India is under the var. of -
A] Anab-e-shahi   B] Perlette
C] Thompson seedless D] Dilkhush

81. N- 53 is a variety of-
A] Onion   B] Garlic
C] Arvi   D] Okra

82. 39. 9 : 7 ratio of gene interaction is
A] complementary   B] Supplementary
C] Inhibiting   D] Polymorphic

83. Among the 8 Regional Fruit Research Stations established by ICAR, which were allotted to work on mango-
A] Kodur & Pune   B] Saharanpur & Sabour
C] Abohar & Chethali D] Mashobra & Kahikuchi

84. Which is a 'dwarf root stock' of pear-
    A] Quince-A   B] Quince-B
    C] Quince-C   D] Kainth

85. Which is a 'triploid' variety of guava-
    A] L-49   B] Arka Mridula
    C] Seedless Guava   D] Allahabad Safeda

86. 'Parthenocarpy' is a characteristic feature of-
    A] Guava   B] Mangosteen
    C] Banana   D] Loquat

87. 'Monocarpism' is a characteristic feature of-
    A] Jack fruit   B] Papaya
    C] Banana   D] Olive

88. Which is a chocolate colour rose variety-
    A] Shabnam   B] Mohini
    C] Sindoor   D] Arunima

89. Which is a commercial crop in 'coastal region' of India-
    A] Walnut   B] Date palm
    C] Cashewnut   D] Tea

90. 'Chakaiya' is a commercial variety of -
    A] Custard apple   B] Ber
    C] Aonla   D] Mandarin

91. 'Protandry' is a common characteristic of 'monoecious' plants. Such type of plant is-
    A] Date palm   B] Papaya
    C] Mango   D] Coconut

92. 'Anthracnose' is a common disease of-
    A] Muskmelon   B] Cucumber
    C] Tomato   D] Watermelon

93. 'Wintering' is a common practice in-
    A] Rose   B] Dahlia
    C] Apple   D] Chrysanthemum

94. 'Sterility' is a common problem in-
    A] Litchi   B] Guava
    C] Loquat   D] Grape

Glabrescent
Becoming smooth.

Floret
Technically a minute flower ; applied to the flowers of grasses and Composites.

Crisped
Wavy on the margin ; short and curly when applied to pubescence.

Globose
Shaped like a globe ; spherical.

Interunfruitfulness
Also called cross sterility. The inability of a variety of fruit to produce viable seed if only the pollen of another variety is used.

Mhos
Measurement of conductivity, which denotes amounts of soluble salts in a medium.

Notching

A partial ringing of a branch above a dormant lateral bud by removing a small narrow strip of bark just above and close to a dormant bud, as a result of which the bud is forced to grow either into a vegetative shoot or a mixed shoot containing both foliage and flowers according to the natural habit of growth of tree. Generally practised to encourage more seasonal growth in deciduous trees whose branches have a natural tendency to produce a few new growths from the lateral buds situated near the ends while the buds in the middle and the base remain dormant and do not grow.

Matching

95. Match the following :-

A]	'Ca' deficiency	i]	Water Core
B]	Low temp. (-1 to 20ºC) in storage	ii]	Core Flush/Brown Core
C]	Excess of 'B' & Low 'Ca'	iii]	Jonathan Spot
D]	Due to deficiency of 'Ca' & 'B'	iv]	Bitter Pit in apple
E]	High level of 'B'	v]	Cork Spot

96. Match the following :-

A]	'Cork spot' in apple	i]	'S' deficiency
B]	Tea yellow	ii]	'Cu' deficiency
C]	'Heart rot' of cabbage	iii]	'Zn' deficiency
D]	'Reclaimation' in citrus	iv]	'Ca' deficiency
E]	'Browning' in jack fruit	v]	'Catfaced' fruit in strawberry
F]	'Mottle leaf' in citrus	vi]	'Mo' deficiency
G]	'Blindness' in cauliflower	vii]	'B' deficiency
H]	'Mo' deficiency	viii]	'Splitting' of citrus
I]	'Ca' deficiency	ix]	Downward 'cupping' in radish
J]	'B' deficiency	x]	Very low temp. (-1ºC)

97. Match the following :-

A]	'Finger tip' disease	i]	*Vitis rotundifolia*
B]	Millerandage	ii]	Banana
C]	Dioecious grape	iii]	Patilo
D]	Phylloxera resistant root stock	iv]	*Vitis rupestris*
E]	Wilt resistant guava	v]	Grape
F]	Culinary guava	vi]	*S. xanthocarpum*
G]	Root knot tolerant brinjal	vii]	Chinese guava
H]	Phomopsis blight resistant brinjal	viii]	Mariana 2624
I]	Root stock of cherry	ix]	*Solanum elaegnifolium*
J]	Root stock of almond	x]	Mazzard

98. Match the following :-

A] 'Heterosis' for earliness   i] Cabbage

B] Self-incompatibility in kale   ii] 700 g/ha

C] 'Stumping' method   iii] Cabbage, cauliflower

D] 'Scooping'method   iv] 1000 g/ha

E] Seed production of 'Snowball'   v] Single dominant gene

F] Seed production of early Cauliflower   vi] 200-450 g/ha cauliflower

G] Seed rate of early cauliflower   vii] Cauliflower

H] Seed rate of late cauliflower   viii] Kullu, Kalimpong

I] Seed rate for cabbage   ix] 500 g/ha

J] Seed rate for kohlrabi   x] Plains of U.P. & Bihar

99. Match the following :-

A] 'Metsubre' in Taro   i] Overage seedlings

B] 'Pox & Scurf' in Sweet Potato   ii] Temperature fluctuation

C] 'Bolting' in Onion   iii] Imperfect fertilization

D] 'Zoning' in Beet   iv] Undecomposed manures

E] 'Forking' in Radish   v] High temperature

F] 'Cracked stem' in Celery   vi] 'Ca' deficiency

G] 'Bitterness' in Celery leaves   vii] 'B'deficiency

H] 'Hollow heart' in Watermelon   viii] High pH (>6.7)

I] 'Crook-neck' in cucumber   ix] Alkaloids

J] 'Bitterness' in Bitter gourd   x] Over maturity

Tomentose
With short matted hairs ; woolly.

Botanical name
The Latin or "scientific" name of a plant, usually composed of two words, the genus and the species.

TIBA
2,3 -6- triiobobenzoic acid- a synthetic plant growth regulator.

Arcuate
Arched, bent like a bow.

Scierophyte
A plant adopted to a climate of wet winter and hot dry summer characterized by thick, hard, usually small, evergreen leaves capable of functioning during winter and withstanding the dry summer.

Fountain
Any device by which a continuous supply of water is assured.

Tempering
Bringing a product to a desired moisture content or temperature for processing.

Schuurkassem
Dutch term for a dual purpose building used for both bulb storage and forcing.

Stamen
The male part of a flower composed of the anther and filament; the pollen bearing organ of a seed plant.

Antholyse
The failure of a flowering sized tulip bulb to initiate a flower after forming a complete

100. Match the following :-

A] $2n = 14$ chromosomes   i] Round melon, Muskmelon, Pointed gourd

B] $2n = 22$ "   ii] Spine gourd, Cho Cho, Litchi

C] $2n = 24$ "   iii] Cucumber, Custard Apple

D] $2n = 26$ "   iv] Radish, Papaya, Phalsa

E] $2n = 28$ "   v] Bottle gourd, Bitter gourd, Watermelon

F] $2n = 40$ "   vi] Loquat, Apple

G] $2n = 18$ "   vii] Pumpkin, Summer Squash, Winter Squash

H] $2n = 36$ "   viii] Sponge gourd, Ridge gourd, Fig

I] $2n = 16$ "   ix] Onion, Garlic, Lasoda

J] $2n = 34$ "   x] Bael, Date palm

101. Match the following :-

A] 45% portion of prepared fruit   i] Jelly

B] 45% portion of fruit juice   ii] Preserve

C] 45% portion of citrus fruits   iii] Marmalade

D] 25% fruit juice   iv] Jam

E] 55% portion of prepared fruit   v] Squash

102. Match the following :-

A] 700 ppm KMS   i] Artificially coloured products

B] 350 ppm KMS   ii] Squash, Cordial

C] 100 ppm KMS   iii] Naturally coloured products

E] Sodium benzoate   iv] Fruit juices

E] Patassium metabisulphite   v] R.T.S., Nectar

103. Match the following :-

A] 85°C tor 30 minutes   i] Sterilization temp. for fruit product

B] 88°C for 30 minutes   ii] U.H.T. sterilization

C] 100°C for 30 minutes   iii] Pasteurization temp. for non-acidic fruits

D] 116°C for 30 minutes   iv] Sterilization temp. for vegetables

E] 149°C for 1-2 seconds   v] Pateurization temp. for acidic fruits

104. Match the following :-

A] Abscissic Acid   i] Retardation of sprouting

B] B-Nine & Cycocel   ii] Cell elongation

C] MH-40   iii] Dormancy breaking

D] Acetylene   iii] Extending vase-life

E] Cytokinin   v] Dwarfing of plants

F] Gibberellic acid   vi] Cell enlargement

G] Auxin   vii] Cell division

H] Thio-Urea   viii] Senescence of leaves

I] H.Q.C-8   ix] Inhibition of root & shoot elongation

J] Ethylene   x] Fruit repening

105. Match the following :-

A] Adam's apple   i] *Syzygium jambos*

B] Elephant apple   ii] *Samradhi*

C] Rose apple   iii] *Manilkara kauki*

D] Monkey jack   iv] *Annona glabra*

E] Alligator's apple   v] *Dillenia indica*

F] Surinam cherry   vi] *Syzygium javanicum*

G] Water apple   vii] *Artocarpus lakoocha*

H] Export quality cashew   viii] *Caveri*

I] Soft seeded Anar   ix] *Syzygium uniflora*

J] Arecanut variety   x] *Priyanka*

Plicate
Folded, as in a folding fan, or approaching this condition.

Flora
Assemblage of plants of a given area, usually arranged in a systematic manner.

Rootstock
the portion of a grafted plant that provides the root; grafted plants typically consist of a scion, which develops into the shoot or crown and a rootstock that provides the root system; rootstocks may include a signicant length of stem, called standards, commonly used for weeping trees or shrubs

Albinism
Absence of chlorophyll in a green plant.

Conservatory
A structure, generally a greenhouse in which plants are conserved or displayed during difficult season.

Saran cloth
Material installed a greenhouse or field to reduce light intensity and radiant heat.

Essence
A substance that contains a high degree of essential oils or other stored products of plant.

Branchlet
A smaller division of a branch.

Lawn
Any open ground about a house or other building grown to fine grasses and maintained in good turf, especially for its aesthetic value.

106. Match the following :-

A] *Aegle mormelos*   i] Rambutan (Hairy litchi)
B] *Euryale ferox*   ii] Bael
C] *Nephelium lappaceum*   iii] Cape Goose berry (Jam fruit)
D] *Grewia subinaequalis*   iv] Makhana (Gorgon nut)
E] *Physalis peruviana*   v] Phalsa (Star apple)

107. Match the following :-

A] *Allamanda cathartica*   i] Purple flowers
B] *Pyrostegia venusta*   ii] Screening purpose
C] *Clerodendron splendens* iii] Yellow flowers
D] *Passiflora edulis*   iv] Blue-Mauve
E] *Porana peniculata*   v] Golden flowers
F] *Petrea volubilis*   vi] White, scented
G] *Jasminum grandiflorum*   vii] Crimson flowers
H] *Monstera deliciosa*   viii] White flowers
I] *Vernonia elaegnaefolia* ix] Annual climber
J] *Lathyrus odorata*   x] Foliage Climber

108. Match the following :-

A] Allicin   i] Aroma in orange
B] Isopentyl acetate ii] Aroma in apple
C] Citral   iii] Odour in garlic
D] Benzaldehyde   iv] Aroma in almond
E] 2, methyl butyrate v] Aroma in banana

109. Match the following :-

A] Alluvial soils   i] Coconut, Black-pepper
B] Arid soils   ii] Apple, Potato
C] Black soils   iii] Jamun, Asparagus
D] Red soils   iv] Litchi, Mulberry
E] Laterite soils   v] Guava, Jack fruit
F] Saline soils   vi] Muskmelon, Cucumber
G] Acidic soils   vii] Phalsa, Datepalm
H] Marshy soils   viii] Aonla, Ber
I] Tarai soils   ix] Citrus, Sapota
J] Sandy soils   x] Tea, Rubber

110. Match the following :-
     A] Allyl-propyle di-sulphite   i] Turmeric
     B] Capsaicin   ii] Bitter gourd
     C] Allicin   iii] Chilli
     D] Bromelin   iv] Aonla
     E] Curcumin   v] Onion
     F] Amyl acetate   vi] Grape
     G] Methyl salicylate   vii] Garlic
     H] Insulin   viii] Pineapple
     I] Essential oil   ix] Banana
     J] Ascorbic acid   x] Jasmine

111. Match the following :-
     A] *Aloe barbadeusis*   i] Liliaceae
     B] *Lawsonia inermis*   ii] Piperaceae
     C] *Piper longum*   iii] Lythraceae
     D] *Cassia angustifolia*   iv] Caesalpiniaceae
     E] *Ocimum sanctum*   v] Lamiaceae

112. Match the following :-
     A] Alternate bearing   i] Banana
     B] Heterodichogamy   ii] Pistachionut
     C] Frog eye leaf spot   iii] Pecan
     D] Indigenous fruit   iv] Bael
     E] Cigar end tip rot   v] Chilli

113. Match the following :-
     A] Anar butter fly   i] *Idioscopus atkinsoni*
     B] Mango hopper   ii] *Virachola isocrates*
     C] Lemon butter fly   iii] *Diphornia citri*
     D] Citrus psylla   iv] *Papilo demoleus*
     E] Banana aphid   v] *Pentalonia nigronervosa*

114. Match the following :-
     A] Andromonoecious   i] Perfect, Staminate & Pistillate flowers on same plant
     B] Gynomonoecious   ii] Perfect & Staminate flowers separately on same plant
     C] Trimonoecious   iii] Male & Female flowers separately on same plant

Hybridization
A process of forming a hybrid by cross pollination/ interbreeding of plants of different species, races or varieties.

Mowing strip
Edging placed between a lawn and a planting bed or patio, set just below ground level to provide a flat surface for one wheel so the lawnmower blades can cut to the edge of the lawn, saving the later step of snipping the edge grass with scissors or a weed-whacker. Mowing strips are commonly made of bricks, pavers, or poured cement.

Margin
the edge of a leaf blade

Lobe

a portion of a leaf that projects outward and divides the leaf into distinct parts, but not enough to make them separate leaflets; lobes may be rounded or pointed

Blade

The expanded part of a leaf.

Rib

A conspicuous vein of a leaf; or a prominent ridge.

Vernation

Arrangement of leaves within a leaf-bud.

Toothed

the condition of a margin broken into small projecting segments, either serrations, dentations, or crenations

D] Monoecious   iv] Male & Hermaphrodite flowers on separate plants

E] Androdioecious   v] Pistillate & Perfect flowers separately on same plant

115. Match the following :-

A] Anthurium   i] Pre-cooling temp. 3°C
B] Gerbera   ii] Pre-cooling temp. 2°C
C] Gladiolus   iii] Pre-cooling temp. 1-2°C
D] Rose   iv] Pre-cooling temp. 1°C
E] Carnation   v] Pre-cooling temp. 4-5°C

116. Match the following :-

A] *Antigonan leptopus*   i] Layering
B] *Allamanda cathartica*   ii] By seed
C] *Campsis chinensis*   iii] Suckers
D] *Petrea volubilis*   iv] Rhizome
E] *Gloriosa superba*   v] Cutting

True / False

117. Premature heading is problem of cauliflower.   T/F

118. Shoot gall psylla of mango is serious insect in the Tarai region of U.P.   T/F

119. Tea flowers arise from axils of leaves.   T/F

120. Umran is a variety of ber.   T/F

121. Verbena is a tall flowering annual.   T/F

122. Caladium is a shade loving bulbous annual.   T/F

123. Chrysanthemum is propagated by suckers.   T/F

124. Cocoa trees are highly cross-incompatible.   T/F

125. Colour development in tomato occurs due to accumulation of 'Fe'.   T/F

126. Grubs of mango stem borer feeds inside the stem from down to upwards.   T/F

127. In North India, phalsa is pruned twice a year.   T/F

128. Leaf curl is a serious disease of peach.   T/F

129. Niranjan variety of mango is an odd-season variety.   T/F

130. Orissa is ideally suitable for cashewnut cultivation. T/F
131. Strawberry fruits are born on the fleshy thalamus. T/F
132. Apple is a cryophilous plant. T/F
133. Bitterness in bitter gourd is due to Tetracyclic triterpenes. T/F
134. C.P.R.I. is located at Shimla. T/F
135. Cauliferous bearing is unique feature of cocoa. T/F
136. Colour in ripe tomato is due to xanthophyll. T/F
137. *Jasminum sambac* flowers during summer. T/F
138. Little leaf of brinjal is a nutritional disorder. T/F
139. Panama wilt is a disease of guava. T/F
140. Pineapple is commercial crop in Assam. T/F
141. Proper time for phalsa pruning in N. India is Dec.-Jan. T/F
142. 'Spongy tissue' in mango is caused due to 'Ca' deficiency. T/F
143. Bunchy top is a se;ious disease of banana. T/F
144. Fire blight is a limiting factor for pear growing. T/F
145. Flower colour of Kochia is red. T/F
146. In S. India, phalsa is pruned twice a year. T/F
147. Kokette and Frances are single flowered varieties of dahlia. T/F
148. Mango bears two types of flowers. T/F
149. Niranjan variety of mango flowers in Feb.-March. T/F
150. Onion is an example of tunicated bulb. T/F
151. Thimma is a variety of bougainvillea. T/F
152. Avenue trees are indoor plants. T/F
153. Cabbage is a rich source of Vit-A & Vit-C. T/F
154. Fire blight in pear is caused by fungus. T/F
155. Kufri Deva is a late variety of potato. T/F
156. Late blight is common disease of cauliflower. T/F
157. Maximum area under banana is in A.P. T/F
158. Potato is root tuber vegetable. T/F
159. *Quisqualis indica* is a shrub. T/F

Phytopathogenic
Term applicable to a microorganism that can incite disease in plants.

Stripling
A sapling grown in a nursery which is stripped of its leaves and branches and sometimes root-pruned before planting out.

Alkaline soil
A soil with a pH higher than 7.0 is an alkaline soil. (a soil pH lower than 7.0 is acidic) Basically, pH is a measure of the amount of lime (calcium) contained in your soil.

Pistillate
An imperfect flower with a pistil, or seed organ, but having no functional stamens (male pollen producing organs).

160. Robusta banana is susceptible to panama wilt. T/F
161. The portion of banana plant above the ground made of sheaths of leaves is known as true stem. T/F
162. 'Sanna Chenkadali' is native diploid banana. T/F
163. Banana can be grown successfully under severe winter. T/F
164. Breeding is very difficult and time taking in banana due to parthenocarpy & polyploidy. T/F
165. Edible portion of almond is kernel. T/F
166. Grape is propagated by herbaceous cutting. T/F

Fill in the Blanks

167. Jonathan is a variety of _____ .
168. Y.V.M. is serious disease of _____ .
169. Confidence is an export quality variety of _____ .
170. Chrysanthemum is _____ plant.
171. _____ is most suitable for saline soils.
172. Snowball is a variety of _____ .
173. IIVR is located at _____ .
174. Floating gardens are found in _____ .
175. Canna is propagated by _____ .
176. _____ is suitable for pineapple growing.
177. _____ is summer season annual.
178. *Phoenix* is genus of _____ .
179. Shoot & fruit borer is serious problem in _____ .
180. Pectin content is important constituent of _____ .
181. Asian Vegetable Research & Development Centre (AVRDC) is located at _____ .
182. _____ is the most tolerant to salt soils.
183. Careless Love is popular variety of _____ .
184. Little leaf of brinjal is caused by _____ .
185. Blanching is practiced in _____ .
186. _____ is intervarietal hybrid of mandarin.

GLOSSARY

187. Acid content is important consideration in _____ preparation .
188. Leaf curl is serious disease of _____.
189. _____ is commercially propagated by ring budding.
190. _____ is the most tolerant to salt.
191. Canker is serious disease of _____.
192. Double working is commonly followed in _____.
193. _____ is rich source of Vit-A.
194. _____ is commonly propagated by air-layering.
195. Red colour of jelly is due to _____.
196. Montezuma is popular variety of _____.
197. Arka Rajhans is variety of _____.
198. Kinnow is related to _____.
199. Which C/N ratio produces vegetative growth? _____.
200. Hexagonal system accommodates about _____ more plants.

Carrier
An individual heterozygous for a recessive and usually detrimental gene.

Kurumba
An immature coconut containing a refreshing, clear liquid.

Antocarp
A fruit obtained as a result of self-fertilization.

Autocarp
A fruit obtained as a result of self-fertilization.

Finisher
A machine composed of a rotating screw inside a close fitting cylindrical or conical housing perforated as a screen; used to squeeze the juice.

Radicle
The embryonic root of a seed.

The retiring president of a company after a standing farewell, gave two envelopes marked No. 1 and No. 2 to the incoming president, and said, "Whenever you run into a management crisis you cannot handle by yourself, open envelope No. 1. At the next crisis, open the second one."

A few years later, a major crisis came. The president went into the safe and pulled out the first envelope. It said, "Blame it on your predecessor." A few years later a second crisis came. The president went for the second envelope, and it said, "Prepare two envelopes for your successor."

Script your success story

Responsible **P**eople **A**ccept and **L**earn from their **M**istakes.

Step-Three

The Loser says, "That is not my job.";
The Winner says, "Let me to do it for you"

Multiple Choice

1. Amrapalli is a cross of-
    A] Dashehari x Neelam   B] Neelam x Dashehari
    C] Neelam x Alphonso   D] Neelam x Mulgoa

2. Which is a dioecious cucurbit-
    A] Cucumber   B] Bottle gourd
    C] Muskmelon   D] Pointed gourd

3. Bottleneck is a disorder of
    A] Cucumber   B] Bottle gourd
    C] Carrot   D] Rose

4. Which is a heavy climber-
    A] *Antigonon leptopus* B] Sweet Pea
    C] *Petrea vollubilis*   D] All of them

5. 'Ikebana' is a Japanese art of-
    A] Modern painting   B] Tree farming
    C] Free style wrestling D] Flower arrangement

6. Cymbidium is a kind of -
    A] Vegetable   B] Fig
    C] Orchid   D] Rose

7. Which is considered as a method of 'pruning for form'-
    A] Thinning out   B] Heading back
    C] Wintering   D] Modified leader

8. 'Ikebana' is a method of-
    A] Growing of roses   B] Growing of vegetables
    C] Planting of   D] Flower arrangement
       chrysanthemum

9. Which is a monoembryonic species of citrus-
    A] *C. limon*   B] *C. sinensis*
    C] *C. grandis*   D] *C. aurantium*

10. 'Sheetal' is a newly developed variety of-
    A] Tomato   B] Watermelon
    C] Cucumber   D] Winter squash

Aestivation
The induction of a state of dormancy by drought or the use of high temperature to promote flowering.

Hibernation
The induction of a state of dormancy by low temperature.

Axil
The location on a stem between the upper surface of a leaf or leafstalk and the stem from which it is growing.

Internode
the portion of a stem between two nodes

Ray flower
A flower with a strap-shaped corolla, usually found on the outer edge of a Compositae inflorescence.

11. Which element is not a part of chlorophyll but essential for its synthesis-
    A] Boron   B] Zinc
    C] Iron   D] Mn

12. 'Chanchal' is a popular variety of-
    A] Capsicum   B] Cauliflower
    C] Radish   D] Tomato

13. Gardenia is a popular
    A] Foliage plant   B] Succulent
    C] Flowering shrub D] Winter annual

14. Methionine is a precursor of-
    A] GA_3 B] I A A
    C] Ethylene   D] ABA

15. Adenine is a precursor of-
    A] NAA   B] ABA
    C] GA_3 D] Cytokinin

16. 'Pithiness' is a problem of-
    A] Carrot   B] Beet root
    C] Onion   D] Radish

17. Which is a rich source of fat-
    A] Avocado   B] Litchi
    C] Walnut   D] Cashew nut

18. Which is a rich source of riboflavin-
    A] Bael   B] Aonla
    C] Apple   D] Karonda

19. Guava is a rich source of
    A] CHOs   B] Vitamin-C
    C] Vitamin-A   D] b-Carotene

20. Carrot is a rich source
    A] Vitamin A   B] Vitamin B
    C] Vitamin C   D] Vitamin D

21. Which is a scented variety of gladiolus-
    A] Eurovision   B] Friendship
    C] Oscar   D] Lucky Star

22. The 'endosperm' of a seed is-
    A] Haploid (n)   B] Diploid (2n)
    C] Triploid (3n)   D] Tetraploid (4n)

23. Which is a self-unfruitful variety of peach-
    A] J. H. Hale   B] Alexander
    C] Elberta   D] Triumph

24. Which is a serious pest of peach-
    A] Case worm   B] Leaf curl aphid
    C] Black aphid   D] Wooly aphid

25. Bollworm is a serious pest of
    A] Wheat   B] Rice
    C] Cotton   D] Maize

26. An asexual progeny of a single homozygous plant is known as-
    A] Variety   B] Race
    C] Strain   D] Clone

27. 'Caveri' is a soft seeded variety of-
    A] Grape   B] Banana
    C] Jamun   D] Pomegranate

28. 'Clove', a spice is taken from which part of plant-
    A] Root   B] Leaf
    C] Stem   D] Flower

29. Which is a surface flowering aquatic plant-
    A] *Nelumbo nucifera*   B] *Nymphaea alba*
    C] *Vectoria regia*   D] All of them

30. 'Tapka' is a traditional maturity index of-
    A] Jamun   B] Mango
    C] Apple   D] Banana

31. N-53, a variety of *Allium cepa,* is suitable for-
    A] Kharif   B] Rabi
    C] Zaid   D] Saline soils

32. 'Ganesh' is a variety of pomegranate, while 'Ganesh Kirti' is a variety of -
    A] Ber   B] Phalsa
    C] Loquat   D] Anar

Base
the bottom of a structure, such as a leaf

Acid soil
A soil with a pH lower than 7.0 is an acid soil. (a soil pH higher than 7.0 is alkaline) Basically, pH is a measure of the amount of lime (calcium) contained in your soil.

Fulcrum
A supporting organ like a tendril or stipule.

Implant
A grafted portion of a tissue, infix or insert.

Protoxin
A molecule which yields a toxin on being processed in a certain specific manner.

Pith
The central part of a twig, usually lighter or darker than the surrounding wood.

Cuspidate
With an apex somewhat abruptly and concavely constricted into an elongated sharp-pointed tip.

Ethylene absorbants
The chemicals which absorb the emitted ethylene form around the fresh perishables in the package and thus delay the senescence process and prolong the shelf-life of the commodity *e.g.,* brominated activated carbon, celite with $KMnO_4$ purafil, etc.

Acanthaceous
Bearing thorns or prickles.

33. McIntosh is a variety of -
    A] Pear   B] Peach
    C] Plum   D] Apple

34. 'Snowball' is a variety of-
    A] Cabbage   B] Onion
    C] Cauliflower   D] Football lily

35. 'Independence' is a variety of-
    A] Gladiolus   B] Chrysanthemum
    C] Rose   D] Canna

36. 'Angoorlata' is a variety of-
    A] Grape   B] Lobia
    C] Tomato   D] Kidney bean

37. 'Lahar' is a variety of-
    A] Grape   B] Rose
    C] Hibiscus   D] Pointed Gourd

38. 'Mahara' is a variety of-
    A] Gulmohar   B] Marigold
    C] Bougainvillea   D] China rose

39. 'Mallika' is a variety of-
    A] Jasmine   B] Rose
    C] Mango   D] Loquat

40. 'Sathgudi' is a variety of-
    A] Lemon   B] Mandarin
    C] Grape fruit   D] Sweet Orange

41. 'Neelam' is a variety of-
    A] Litchi   B] Grape
    C] Guava   D] Aonla

42. 'Priya' is a variety of-
    A] Muskmelon   B] Watermelon
    C] Cucumber   D] Bitter gourd

43. 'Thompson' is a variety of-
    A] Pummelo   B] Grapefruit
    C] Lime   D] Grape

44. 'Y2K' is a variety of-
    A] Rose   B] Chrysanthemum
    C] Tuberose   D] Dahila

45. 'Friendship' is a variety of-
   A] Rose   B] Gladiolus
   C] Dahlia   D] Chrysanthemum

46. 'Sathgudi' is a variety of-
   A] Sweet orange   B] Mandarin
   C] Lime   D] Grape fruit

47. 'Chipsona' is a variety of-
   A] Tomato   B] Potato
   C] Chilli   D] Guava

48. 'Shakti' is a wilt resistant variety of-
   A] Tomato   B] Chilli
   C] Okra   D] Brinjal

49. Dendrobium is a
   A] Anthrium   B] Carnation
   C] Orchid   D] Rose

50. Oendrobium is a
   A] Succulent   B] Cactus
   C] Orchid   D] None

51. 'Umran is a/an _____ variety of_____.
   A] Early, Ber   B] Late, Ber
   C] Early, Peach   D] Dwarf, Banana

52. 'Palmolein' is a/an-
   A] Antitranspirant   B] Protein
   C] Edible oil   D] Essential oil

53. A 'fruit' is a/an-
   A] Ripened Ovary   B] Ripened Ovule
   C] Edible part of plant   D] Thalamus

54. Squash contains about _____ per cent sugar.
   A] 15-18%   B] 65-68%
   C] 25-33%   D] 10-15%

55. Indole Acetic Acid (IAA) was discovered by-
   A] Nelson (1930)   B] Mullar (1938)
   C] Frist Went (1934)   D] McClintoff (1941)

56. Which amino acid is precursor of ethylene
   A] Cystine   B] Purine
   C] Methionine   D] Proline

Acetic acid
Also called pyroligneous acid. A colourless organic liquid with pungent odour, caustic when concentrated, and used for destroying warts. Germicide. Chief active component of vinegar. Product of lactic acid fermentation, important constituent of flavour in many milk products.

Antimetabolite
Anti-vitamins that act by inhibiting the coenzyme function of a vitamin also known as antagonists.

Green pruning
Pruning of actively grown rose plants without benefit of a dormancy period.

Tropophyte
A plant that is adapted to a climate characterised by a very dry season followed by a wet one.

Codon
A group of three adjacent nucleotides of mRNA which code for one amino acid.

Air-lock
Bubble of air in the cut stem, which causes obstruction in the flow of water resulting in premature wilting of cutflower.

Twining
A stem winding around a support.

Vegetable Forcing
This is a type of gardening which is concerned with the production of vegetables out of their normal season.

57. Addition of 'citric acid' in vase solution is helpful in-
A] Causing stem closing
B] Increasing water uptake
C] Preventing water uptake
D] Reducing bending of bud

58. Now- a-days which is the major problem in coconut cultivation
A] Lack of good varieties
B] Stem bleeding
C] Eryophide mite
D] Red palm weevil

59. Gypsum is added to correct
A] Saline soil
B] Alkali soil
C] Sodic soil
D] Acidic soil

60. Which is a new addition under plantation crops-
A] Rubber
B] Red oil palm
C] Tea
D] Coconut

61. Ratio of additive variance to phenotypic variance is called as
A] Heterobaltosis
B] Heritability
C] Linkage relation
D] None

62. Pear is adopted to-
A] Temperate climate
B] Asiatic climate
C] European climate
D] Tropical climate

63. Which is an ' aesthetic branch' of horticulture-
A] Pomology
B] Olericulture
C] Floriculture
D] Hydroponics

64. Which plant hormone is alkaline in nature-
A] NAA
B] GA_3
C] Ethylene
D] Kinetin

65. 'Best of All' is a cultivar of-
A] Rose
B] Cauliflower
C] Radish
D] Tomato

66. Bitterness in almond kernel is governed by-
A] Single recessive gene
B] Single dominant gene
C] Multiple genes
D] Mutagens

67. Neelam x Alphonso, Cross produce which variety of mango
A] Ratna
B] Neelphonso
C] Sindhu
D] None

68. Glycolysis is also called as
   A] EMP pathwey   B] Kreb cycle
   C] Both   D] None

69. Canna is also known as
   A] Indian shot   B] African lily
   C] Spanish day   D] Century plant

70. ' Amrapali' mango was released during-
   A] 1975   B] 1978
   C] 1980   D] 1981

71. Topioca is a/ an _____ crop.
   A] Fruit crop   B] Pulse
   C] Tuber   D] Oilseed

72. Vitamin A requirement for an adult man is-
   A] 1000 IU/day   B] 2000IU/day
   C] 1500IU/day   D] 2500IU/day

73. Removal of cambium is an essential practice in-
   A] Wedge grafting   B] Budding
   C] Gootee   D] Hard wood cutting

74. Mulching is an important practice in orchard related to-
   A] Floor management B] Canopy management
   C] Pruning   D] Training

75. Bulbophyllum is an important
   A] House plant   B] Suculent
   C] Palm   D] Orchid

76. RRIM-1 is an improved variety of-
   A] Rubber   B] Coffee
   C] Date palm   D] Coconut

77. ' Anab-e-Shahi' grape was bred by ....through selection.
   A] G.S. Cheema   B] R.S. Pillai
   C] C.B.S. Rajput   D] None of these

78. 'Janam Picking' and 'Shear picking' are done in-
   A] Tobacco   B] Ocimum
   C] Mentha   D] Tea

Punch
An alcoholic or non -alcoholic beverage, variable in composition but usually based on wine or some sort of distilled liquor, fruit, soda, or fruit juices with sugar, lemon, and other ingredients added to form a flavoured combination.

Caffeine
A purine base stimulant alkaloid found in tea and coffee.

Dominance
Ability of an allele to express itself in heterozygous state.

Dominant
A member of an allelomorphic pair of characters with the quality of manifesting itself wholly or largely to the exclusion of the other member (recessive) in heterozygous state.

Conicine
An alkaloid causing acridness along with calcium oxalate, found in *Amorphophallus* sp.

Aeration
To increase the amount of air space in the soil by tilling or otherwise loosening the soil.

Pick
1. The total amount of crop harvested, or the yield of an individual tree. 2. To pull or pluck ripe fruit, as berries, apples, etc.

Kernrot
Dutch term for an advanced stage flower abortion of tulips.

Basipetal
Developing sequentially from an apical position towards base.

79. 'Sri Nandani' and 'Sri Vardhini' are varieties of-
    A] Sweet potato   B] Dioscorea
    C] Sugar beet   D] Cassava

80. 'Sri Pallavi' and 'Sri' Reshmi' are improved varieties of-
    A] Cassava   B] Radish
    C] Sweet potato   D] Arvi

81. 'Kinnow' 'Atemoya' and 'Strawberry' are considered as-
    A] Natural fruits   B] Man made fruits
    C] Plantation crops D] Salad crops

82. The origin and family of guava is
    A] India, solanaceae   B] America, Myrtaceae
    C] China, Vitaceae   D] None

83. 'Cauliferous' flowering and fruiting is a unique feature of-
    A] Coconut   B] Cocoa
    C] Coffee   D] Cauliflower

84. When *Euvitis* and *Muscadinia* grapes are crossed, the resultant progeny becomes-
    A] Fertile   B] Self-compatible
    C] Sterile   D] Hermaphrodite

85. Term 'Two leaf and one bud' is related with-
    A] Coffee   B] Tea
    C] Cocoa   D] Tobacco

86. When the main stock and scion are incompatible, another root stock is used which is known as-
    A] Inter stock   B] Stion
    C] Bud   D] Polarity

87. ' Andro-dioecious' form of sex is found in—
    A] Fig   B] Rambutan
    C] Papaya   D] None of these

88. Which flowering annual have Indian origin
    A] Zinnia   B] Balsam
    C] Portulaca   D] Carcopsis

89. Transplanting of annual plants is known as
    A] Bricking   B] Pricking
    C] Potting   D] None

90. Winter season 'foliage annual' is-
    A] Coleus   B] Kochia
    C] Croton   D] Poinsettia

91. Family of anthurium is
    A] Scrophulariaceae   B] Labiatae
    C] Vervanaceae   D] Araceae

92. Inflorescence of Anthuriums called as
    A] Hypanthodium   B] Spadix
    C] Sorosis   D] None of these

93. In non-recurrent apomixis, an embryo develops directly from haploid egg and so the resultant embryo will be-
    A] Haploid   B] Diploid
    C] Triploid   D] Tetraploid

94. Per capita arable land is highest in
    A] Australia   B] Japan
    C] Canada   D] India

Matching

95. Match the following :-
    A] Apple   i] Citric acid
    B] Grape   ii] Malic acid
    C] Blue berry   iii] Quinic acid
    D] Cran berry   iv] Tartaric acid
    E] Raspberry   v] Benzoic acid

96. Match the following :-
    A] Apple   i] Marmalade
    B] Guava   ii] Amchur
    C] Citrus fruits iii] Jam
    D] Mango   iv] Juice
    E] Pomegranate   v] Jelly

97. Match the following :-
    A] Apple   i] Mexico
    B] Plum   ii] Native of S-W Asia
    C] Mexican Aster iii] South Europe
    D] Larkspur   iv] Native of France
    E] Sweet William v] Native of N. America

Insulation
Technique of isolating an area usually from heat or cold.

Landscape architecture
It is an art of arranging land and landscape for human use convenience and enjoyment.

Trisomic
A diploid organism having an extra chromosome $(2n + 1)$.

Drupel
A small drupe. An individual component of aggregate fruit as of raspberry.

Strobilus
A cone which is an inflorescence made up of imbricated scales as in *Equisetum*, *Selaginella*, etc.

Abortion
Arrested development of an organ in plants, production of a few or no seeds, or dropping of fruit prematurely.

Lesion
Any structural change in an organ resulting from an injury or disease.

Photoperiodic response
Behavior of an organism to the length of day.

Locellus
A small compartment of an ovary.

Pedigree
A record of ancestry.

Grading
Classification of plants and flowers for market, based on size and quality.

98. Match the following :-
A] Arjun (Bougainvillea)   i]   Mutant of Flirt
B] Man Bhavan (Chrysanthemum)   ii]   Mutant of Montezuma
C] Lalita (Portulaca)   iii]   Mutant of Wildrose
D] Angara (Rose)   iv]   Mutant of Partha
E] Shobha (Gladiolus)   v]   Mutant of Vibhuti

99. Match the following :-
A] Arka Harit   i]   Selection of Bottle gourd
B] Pusa Sandesh   ii]   Hermaphrodite Ridge gourd
C] Arka Chandan   iii]   Selection of Bitter gourd
D] Arkajeet   iv]   Selection of Muskmelon
E] Priya   v]   Hybrid of Bottle gourd
F] Satputia   vi]   Selection of Pumpkin
G] Pusa Chikni   vii] F_1 Hybrid of Cucumber
H] Punjab Komal   viii]   Winter Squash
I] Arka Suryamukhi   ix]   Hybrid of Ridge gourd
J] Arka Sujat   x]   Selection of Sponge gourd

100. Match the following :-
A] Arka Krishna, Arka Neelmani   i]   Anab-e-Shahi x Thompson Seedless
B] Arka Soma; Arka Kanchan   ii]   Black Champa x Thompson Seedless
C] Arka Sweta   iii]   Bangalore Blue x Black Champa
D] Arka Hans   iv]   Anab-e-Shahi x Queen of Vine Yards
E] Arka Shyam   v]   Bangalore Blue x Anab-e-Shahi

101. Match the following :-
A] Arka Ravi, Arka Swarna   i]   Garlic
B] Sunrise Solo   ii]   Chrysanthemum
C] Ambika   iii]   Onion
D] Agrifound Parvati   iv]   Papaya
E] Agrifound Red   v]   Mango

102. Match the following :-
     A] Aroma in Cranberry   i] Benzaldehyde
     B] Aroma in Almond   ii] Isopentyl acetate
     C] Colour in Bael   iii] 2-methylbutyric acid
     D] Aroma in Banana   iv] 3-methyl propionate ester
     E] Aroma in Pineapple   v] Leucoanthocynin

103. Match the following :-
     A] Aroma in pears   i] Polyphenolase
     B] Browning of cut pears ii] Native of India
     C] *Pyrus communis*   iii] Ancestor of modern pear
     D] *Pyrus pashia*   iv] Central China origin
     E] *Pyrus pyrifolia*   v] Methyl & Ethyl esters

104. Match the following :-
     A] Aroma in Raspberry   i] Ethyl-hexanoate
     B] Aroma in Strawberry   ii] Methyl salicylate
     C] Aroma in Cranberry   iii] Geraniol
     D] Aroma in Cherry   iv] 2-Methyl butylate
     E] Aroma in Apple   v] 2-Benzyl benzoate

105. Match the following :-
     A] Asepsis   i] Y.V.M.
     B] Pusa Sawani   ii] Bolting
     C] Cauliflower   iii] T.P.S.
     D] Onion   iv] Temporary preservation
     E] Potato   v] Buttoning

106. Match the following :-
     A] Asia Minor   i] Papaya, Rubber
     B] Central America   ii] Guava
     C] China   iii] Apple, Fig
     D] Brazil   iv] Bael, Banana
     E] Peru   v] Litchi, Peach
     F] Central Asia   vi] Pomegranate
     G] India   vii] Pineapple, Cashewnut
     H] Abyssinia   viii] Gladiolus
     I] South Africa   ix] Grape, Pear
     J] Iran   x] Coffee

Pinching back
Utilizing the thumb and forefinger to nip back the very tip of a branch or stem. Pinching promotes branching, and a bushier, fuller plant

Land Scaping Gardening
Designing and laying out home gardens, public gardens, parks, road side plantation, avenues, etc. is called land scaping gardening.

Smog
A combination of smoke and natural fog causing air pollution.

Floriculture
Cultivation of flowers and ornamental plants for commercial purposes or merely for getting pleasure and as a hobby.

Active absorption
Absorption of water and other substances against the concentration gradient; an energy requiring process.

Netted venation
The veins reticulated and resembling a fish net; the interstices close.

Pruning
It is art and science of cutting away a portion of a plant to improve its shape, to influence growth, flowering and fruitfulness, to improve the quality of the product.

Fasciated
Abnormally much flattened, and seemingly several units fused together.

Corky
Soft and springy.

107. Match the following :-
A] Attributes of 'Table' grapes   i] Thin skin, igh juice, High Peel, Pulp ratio
B] Attributes of 'Raisin' grapes   ii] High sugar, High/low acid
C] Attributes of 'Canning' grapes   iii] Attractive colour, Compact bunch, High sugar
D] Attributes of 'Wine' grapes   iv] Seedless, Smooth Skin, Maleable, large
E] Attributes of 'Juice' grapes   v] Seediess, Thin skin, High Sugar: Acid ratio

108. Match the following :-
A] Axillary bearing   i] Mango, Papaya
B] Terminal bearing   ii] Ber, Datepalm
C] Mixed bearing   iii] Phalsa
D] Climacteric fruit   iv] Aonla, Apple
E] Non-climacteric fruit   v] Gladiolus
F] Salinity tolerant   vi] Loquat, Litchi
G] Tiliaceae   vii] Grape, Pineapple
H] Moraceae   viii] Cocoa
I] Irridaceae   ix] Pomegranate, Citrus
J] Cauliferous bearing   x] Jack fruit

109. Match the following :-
A] Bael   i] Rutaceae
B] Cashewnut   ii] Sapindaceae
C] Rambutan   iii] Anacardiaceae
D] Rose apple   iv] Nymphaeaceae
E] Gorgon nut   v] Myrtaceae

110. Match the following :-
A] Bangalore Blue   i] Introduction from California
B] Pusa Seedless   ii] Seedling selection of Pandhari Sahebi
C] Thompson Seedless (Sultania)   iii] A labrusca type grape
D] Cheema Sahebi   iv] Selection from Anab-e-Shahi
E] Dil Khush   v] A clonal selection of Thompson Seedless

111. Match the following :-
     A] Banganpalli x   i]   Arka Nilkiran
        Alphonso
     B] Alphonso x   ii]   Mallika
        Janardan Pasand
     C] Alphonso x Neelam   iii]   Arka Anmol
     D] Ratna x Alphonso   iv]   Sindhu
     E] Neelam x Dashehari   v]   Arka Aruna

112. Match the following :-
     A] Bark eating caterpillar i]   *Heliothis armigera*
     B] Diamond back moth   ii]   *Drosicha mangiferae*
     C] Red pumpkin beetle   iii]   *Leucinodes orbonalis*
     D] Fruit borer   iv]   *Inderbela tetraonis*
     E] Fruit fly   v]   *Virachola isocrates*
     F] Leaf hopper   vi]   *Plutella maculipensis*
     G] Mealy bug   vii]   *Bemisia tabaci*
     H] Anar butter fly   viii] *Meloidogyne incognita*
     I] Nematode   ix]   *Amarasca bigutula*
     J] Shoot and fruit borer   x]   *Raphidopalpa foveicollis*

113. Match the following :-
     A] Bathroom fruit   i]   Papaya
     B] Poor man's apple   ii]   Mango
     C] Ascorbic fruit   iii]   Phalsa
     D] Indigenous vegetable   iv]   Aonla
     E] Anti-pallegra fruit   v]   Avocado
     F] Delicious fruit   vi]   Guava
     G] Spices Bowl   vii]   Brinjal
     H] Butter fruit   viii] Calicut
     I] Dioecious fruit   ix] Bael
     J] Indigenous fruit   x]   Banana

114. Match the following :-
     A] Bela   i]   *Jasminum grandiflorum*
     B] Chameli   ii]   *Ryostonia regia*
     C] Juhee   iii]   *Jasminim sambac*
     D] Mallika   iv]   *Jacaranda mimosaefolia*
     E] Swarna Chameli v] *J. auriculatum*

Venation
Arrangement of the veins and the veinlets in the leaf-blade.

Topiary
A method of pruning and training certain plants into formal shapes such as animals.

Steppe
A vast usually level and treeless transitional zone between humid and desert climates; found especially in southeast Europe and Asia.

Blanching
Heating of fruits and vegetables in water or in live steam before canning.

Monocot
A plant of the angiosperm group having one cotyledon.

Dicot
A plant of the angiosperm group having two cotyledons.

Acute
With sides forming an angle of less than 90° ; shape pointed.

Obtuse
With sides forming an angle of more than 90° ; blunt.

Bedding plant
Plants (mainly annuals), nursery grown and suitable for growing in beds. Quick, colorful flowers.

Pickling
Preservation of fruits and vegetables in common salt or vinegar is called pickling.

Adpressed
lying close to another organ, but not fused to it

F] Flame of forest   vi] *Delonix regia*
G] Golden shower   vii] *J. angustifolium*
H] Peacock flower   viii] *Butea monosperma*
I] Blue gulmohar   ix] *Jasminum humile*
J] Royal palm   x] *Cassia fistula*

115. Match the following :-
A] Benzoic acid   i] Jam & Jelly
B] Potassium   ii] Vegetable canning
    metabisulphite
C] Salt   iii] Fruit canning
D] Sugar syrup   iv] Naturally coloured fruit products
E] Brine solution   v] Pickling of vegetables

116. Match the following :-
A] Beri-beri   i] Vit-K
B] Night blindness   ii] Vit-B$_1$
C] Scurvy   iii] Vit-C
D] Sterility   iv] Vit-A
E] Blood coagulation   v] Vit-E

True / False

117. High temperature promotes sugar content in fruits.   T/F
118. K. Badshah & K. Bahar are early varieties of potato for plains.   T/F
119. Pusa Makhmali is a variety of radish.   T/F
120. Sathgudi is a variety of grape.   T/F
121. Spongy tissue is common in Alphonso mango.   T/F
122. 'Nandran' banana belongs to AAB group.   T/F
123. Arka Hans is a hybrid variety of muskmelon.   T/F
124. Black heart of potato is caused by bacteria.   T/F
125. *Cassia fislula* and *Cassia nodosa*, both flower at same time.   T/F
126. Edible part of litchi is thalamus.   T/F
127. Flowers of cucumber are hermaphrodite.   T/F
128. Litchi is the oldest fruit plant.   T/F

129. Rough lemon is suitable as root stock for sandy soils. T/F

130. The best propagation material in banana is water sucker. T/F

131. Thrips affects onion & chilli. T/F

132. Annuals are commercially propagated by cutting. T/F

133. Black tip of mango affects proximal end of fruit. T/F

134. Botanically apple fruit is a false fruit. T/F

135. Citrus is the richest source of Vit-C. T/F

136. Gilas is a variety of mango. T/F

137. Leaf curl is a disease of carrot. T/F

138. Medjool and Shamran are cultivars of datepalm. T/F

139. Mohini var. of rose is an aneuploid $(2n = 21 + 1)$. T/F

140. *Rosa indica* is a common rose root stock in N. India. T/F

141. 'Safed Velchi' banana belongs to AAA group. T/F

142. Banana fruits are allowed to ripe on plant. T/F

143. Bombay Green mango is highly susceptible to malformation. T/F

144. Hazelnut is commonly grown in Tropical Coastal region. T/F

145. *Jacaranda mimosaefolia* is suitable for road side planting. T/F

146. Mango mealy bug is controlled by grease banding. T/F

147. Plant population of pineapple in H.D.O. is 63798/ha. T/F

148. Small tubers are tolerant to hollow heart of potato. T/F

149. Troyer/Citrange is resistant to cold. T/F

150. Up-to-Dlite and Phulwa are varieties of sweet potato. T/F

151. *Eupatorium canabinum* is suitable for edge making. T/F

152. Gladiolus is propagated by bulbs. T/F

153. Good jelly can be prepared from litchi. T/F

154. Incidence of' Scurf is less when potato is grown in acidic soil. T/F

155. Moringa is a quick growing plant. T/F

156. Ney Poovan is commercial diploid (AB) banana. T/F

157. Papaya is an andromonoecious plant. T/F

158. Sweet orange requires dry and semi-arid conditions. T/F

Epiphyte
A plant growing on another without being a parasite (orchid) or on a rock or tree trunk (moss, lichen). Contrasted usually with plants rooted in the soil; sometimes parasites.

Phytoalexin
A phenolic substance having antifungal principle, synthesised by plants in response to parasite invasion or infection by certain fungi, *e.g.*, pisatin, phaseolin, trifolirhizin, orchinol and isocumarin from pea and bean pods, red colour orchid tubers and carrot root, respectively.

Incomplete flower
One which lacks any one or more of these parts; calyx, corolla, stamens, and pistils.

Spray treatment
Removal of apical flower bud to stimulate development of lateral flowers.

Emasculation
Removal of immature anthers form a hermaphrodite flower.

Dentate
Having marginal teeth whose apices are perpendicular to the margin and do not point forward.

Lateral
Said of buds which appear along the sides of the twig; borne at or along the side.

159. Washington Naval orange produces parthenocarpic fruits.   T/F
160. Base of commercial banana cultivation in India is Dwarf Cavendish.   T/F
161. Botanically apple & pomegranate both fruits are pome.   T/F
162. Catkins in nut fruits are staminate inflorescence.   T/F
163. Jasmine is propagated by shield budding.   T/F
164. Puree is a product of tomato fruits.   T/F
165. Pusa Red & Pusa Jwala are mosaic resistant chillies.   T/F
166. Solanine formation in potato is beneficial to health.   T/F

Fill in the Blanks

167. Amrapalli is related to _____.
168. _____ is the most susceptible to cold injury.
169. _____ is the most susceptible to waterlogging.
170. Rose is prunned in the month of _____.
171. _____ is variety of ber.
172. Commonly banana is propagated by _____.
173. _____ is used as dwarfing rootstock.
174. _____ is widely grown in tropics of coastal region.
175. Cauliflower variety _____ can be successfully grown even during rainy season in hilly areas.
176. _____ is variety of brinjal.
177. _____ is classified as vegetable.
178. _____ is rich source of Vit-A.
179. Good jelly can be prepared from _____ .
180. A product known as *puree* is prepared from _____.
181. _____ is foliage ornamental vine.
182. _____ system of planting accommodates maximum number of plants.
183. Larkspur flowers during _____ season.
184. Marry Palmer is variety of _____ .

185. The growth substance that breaks the dormancy in seed:_____.
186. Arka Shyam is hybrid variety of _____.
187. _____is ideally suited for growing in scanty rainfall areas.
188. Beautiful flowering climber _____is annual climber .
189. Edible banana is seedless due to _____.
190. _____is ideal crop for growing in coastal region.
191. Knolkhol is _____ vegetable.
192. Pusa Delicious is variety of _____ .
193. CO-1 is variety of _____released at Coimbatore.
194. Florida Sun is variety of _____.
195. Carnation is _____season annual.
196. Baby Masquerede is a _____variety.
197. Spinach is rich source of _____.
198. Where climate remains drier for most of the year, ideal fruit crop is _____.
199. _____ is generally used for breaking bud dormancy in potato.
200. _____ is example of plant growth inhibitor.

Dysploidy
Abnormal ploidy as the appearance of triploid and tetraploid ones in a normally diploid population.

Reticulate
net-like appearance; for example, referring to a leaf having veins that branch many times in a dense pattern, often more prominently noticed when the veins are impressed

Awn
A bristle-like appendage.

Crinkle
A disorder of apples in which the surface of the fruit becomes roughened. Supposed to be a form of drought injury.

A young man asked Socrates the secret to success. Socrates told the young man to meet him near the river the next morning. They met. Socrates asked the young man to walk with him toward the river. When the water got up to their neck, Socrates took the young man by surprise and ducked him into the water. The boy struggled to get out but Socrates was strong and kept him there until the boy started turning blue. Socrates pulled his head out of the water and the first thing the young man did was to gasp and take a deep breath of air. Socrates asked, "What did you want the most when you were there?" The boy replied, "Air." Socrates said, "That is the secret to success. When you want success as badly as you wanted the air, then you will get it." There is no other secret.

A burning desire is the starting point of all accomplishment. Just like a small fire cannot give much heat, a weak desire cannot produce great results.

Script your success story

[**W**henever the mind of **M**an can **C**onceive and **B**eliefe, the **M**ind can **A**chieve.]

Step-Four

The Loser sees a problem for every answer;
The Winner sees an answer for every problem.

Multiple Choice

1. Per capita arable land is maximum is
   A] Australia   B] Canada
   C] India   D] USA

2. Which fruits are 'climacteric' in nature-
   A] Papaya, Jack fruit   B] Pineapple, Lemon
   C] Grape, Orange   D] Litchi, Pomegranate

3. Multistyled flowers are common in
   A] Brinjal   B] Okra
   C] Chilli   D] Lima bean

4. Gynoecious lines are common in
   A] Water melon   B] Musk melon
   C] Long melon   D] Cucumber

5. 'Drupe' fruits are developed for...........ovary.
   A] Superior   B] Inferior
   C] Epigynous   D] Hypogynous

6. Mango pieces are dipped before pickling in the solution of-
   A] 2% salt   B] 15% salt
   C] 2% sugar   D] 10% sugar

7. 'Fusiform' roots are found in-
   A] Carrot   B] Sugar beet
   C] Sweet potato   D] Radish

8. 'Exalbuminous seeds' are found in-
   A] Fern   B] Orchid
   C] Cacti   D] Xerophytes

9. Litchi fruits are harvested in bunches and with few leaves to avoid-
   A] Rupturing   B] Ripening
   C] Fruit drop   D] All the above

10. When scion and stock are incompatible which is practicable-
    A] Layering   B] Top working
    C] Cutting   D] Apomixis

Fragmented
Not continuous, as applied to bundle scars.

Broken
Not continuous, as applied to bundle-traces.

Indehiscent
Not opening, as applied to fruits.

Persistent
Not deciduous, as applied to leaves; not disappearing, as applied to pith, pubescence, epidermis, etc. Adhering to a position instead of falling, whether dead or alive.

Separate
1. To put apart, make sever, disunite, keep from union or contact.
2. Divide into constituent parts such as fruit, vegetable, flower etc. into sizes

Excavated
Hollowed out, as applied to pith, making the stem fistulous.

Enzymes
Chemical substances that are associated with the living cell. Its presence is necessary for working as a catalyst.

Potgarden
A garden where potherbs are grown.

Green House
Green house are man made devices treating condition favourable for the growth of plant.

Arecoline
An alkaloid derived from areca, used as anthelmintic and cathartic.

11. Dehydrated grapes are known as-
A] Raisin
B] Resin
C] Nut
D] Dried berry

12. When bulbs of onion are planted, then one hectare field requires-
A] 250-500 kg bulbs
B] 500-1000 kg bulbs
C] 1000-1200 kg bulbs
D] 1200-1500 kg bulbs

13. Cole crops are predominantly-
A] Self pollinated
B] Cross pollinated
C] Often cross pollinated
D] Dioecious

14. 'Resinous canals' are the characteristic features of the plants of family-
A] Moraceae
B] Sapindaceae
C] Anacardiaceae
D] Euphorbiaceae

15. Which crop is/ are the most suitable as 'Filler'-
A] Papaya
B] Banana
C] Pineapple
D] All of above

16. Total cultivated area in India is close to-
A] 180 mh
B] 60 mh
C] 400 mh
D] 500 mh

17. Net sown area in the country is about
A] 105 MH
B] 205 MH
C] 142 MH
D] 166 MH

18. How much area is required for raising nursery for planting one hectare of rice
A] 1/5th of total area
B] 1/10th of the total area
C] 1/3th of the total area
D] 1/5 th of the total area

19. How much area is shared by Kerala under cassava-
A] 60%
B] 40%
C] 75%
D] 90%

20. How much area is suitable for kitchen gardening for a family of five
A] 50 m²
B] 100 m²
C] 200 m²
D] 250 m²

21. The largest area under fig is found in-
    A] M.P.   B] Karnataka
    C] Bihar   D] Maharashtra

22. Total cultivated area under wheat in India is
    A] 25 mh   B] 28 mh
    C] 32 mh   D] 24 mh

23. Directorate of Arecanut & Spices Development was renamed as Directorate of Cocoa, Arecanut & Spices Development during-
    A] 1970   B] 1974
    C] 1978   D] 1982

24. National Research Centre on Arid Zone Horticulture (NRCAZH) is located at-
    A] Jodhpur   B] Abohar
    C] Bikaner   D] Hissar

25. The Central Arid Zone Research Institute is located at
    A] Jhansi   B] Jodhpur
    C] Jabalpur   D] Bikaner

26. ' Arka Chitra' is a variety of-
    A] Grape   B] Mango
    C] Pomegranate   D] Pineapple

27. ' Arka Garima' and 'Pusa Komal' are varieties of-
    A] Sem (Bean)   B] Cow Pea
    C] Radish   D] Grape

28. The parents of ' Arka Kanchan' grape are-
    A] Thompson Seedless x Queen of Vine Yards
    B] Anab-e-Shahi x Queen of Vine Yards
    C] Black Champa x Bangalore Blue
    D] Anab-e-Shahi x Black Champa

29. ' Arka Nishant' is a variety of-
    A] Grape   B] Radish
    C] Carrot   D] Beet leaf

30. ' Arka Surabhi' is a variety of-
    A] Marigold   B] Jasmine
    C] Gladiolus   D] Rose

Cormel
A small corm arising from the base of a parent corm.

Flavour
A composite term of aroma and taste of any food.

Ring
A cut or girdle around the trunk, branch or roots of a tree. 2. The annual growth of a tree seen as a ring in the cross-section of the stem, branch or root.

Herbarium
A collection and arrangements of plant specimens, according to taxonomy.

Offset
A side shoot used as a means of vegetative propagation.

Disk
An enlarge tip, as applied to tendrils.

Rooting hormone
Compound such as IBA, NAA, which stimulates rooting.

Oblique
Lop-sided, as one side of a leaf base being larger, wider or more rounded than the other.

Cyathium
Inflorescence of plants as Poinsettia. Relatively inconspicuous but bears pistils, stamens, and nectary glands.

Parachute
A special structure such as tuft of hairs, bristles, scales, etc. of seeds as aril, caruncle, pappus, wing, which assists in dispersal.

Impressed
having sunken veins as viewed from the upper leaf surface

31. ' Arka' Ajit' is a variety of-
A] Tomato   B] Pea
C] Bottle gourd   D] Chilli

32. Agricultural Research Services (ARS) was started in-
A] 1950   B] 1975
C] 1980   D] 1970

33. ' Artificial Pollination' is gel!erally practised in-
A] Chestnut   B] Macademia nut
C] Walnut   D] Hazelnut

34. Which region is known as 'spices bowl'-
A] Northern zone   B] Arid zone
C] Temperate zone   D] Southern zone

35. Food industry as a separate department of government of India was established in year
A] 1960   B] 1965
C] 1980   D] 1995

36. Sugar used as preservative have concentration of
A] 60%   B] 70%
C] 40%   D] 50%

37. The richest source of ascorbic acid is-
A] Aonla   B] Barbados cherry
C] Dahlia   D] Guava

38. Eminent scientists associated with citrus classification are-
A] Burns & Prayag   B] Cheema & Deshmukh
C] Tanaka & Swingle   D] H.B.Frost & Swingle

39. 'Proping' is associated with-
A] Banana   B] Grape
C] Coconut   D] Peach

40. Dried litchi is known as-
A] Nutmeg   B] Litchi nut
C] Drupe   D] Aril

41. The precursor of ' auxin' in plant tissues is-
A] Tryptophane   B] Mevalonic acid
C] Methionine   D] Purine

42. On an average a good variety of papaya may yield (t/hac)
    A] 60   B] 50
    C] 40   D] 20

43. Carambola (*Averrhoa carambola*) belongs to the family-
    A] Rutaceae   B] Bombacaceae
    C] Palmaceae   D] Oxalidaceae

44. Potato is a
    A] Allohexaploid   B] Autohexaploid
    C] Autotetraploid D] Allotetraploid

45. Calanchoe is a
    A] Cactus   B] Succulent
    C] Both   D] None

46. Gardenia is a
    A] Flowering shrub B] Tree
    C] Climber   D] None

47. Malathion is a
    A] Organophosphate B] Organochlorin
    C] Cabamate   D] None

48. Broccoli is a-
    A] Root crop   B] Legume
    C] Cole crop   D] Fruit crop

49. Underground stem banana is a-
    A] Runner   B] Corm
    C] Rhizome   D] Pseudostem

50. Chrysanthamum is a-
    A] Short day plant B] Long day plant
    C] Day neutral   D] None

51. Diffenbachia is a
    A] Shrub   B] Climber
    C] Foliage plant   D] Tree

52. Cyclamen is a-
    A] Temperate plant B] Tropical plant
    C] Annual   D] None

Cladode
Leaf like stem as with asparagus.

Bulb production phase
All aspects of bulb production, which lead to the sale of forcing, sized bulbs. The production phase may tame 1-3 years depending on the species.

ELISA
Enzyme linked immunosorbent assay, used for testing viral diseases.

Telocentric
A chromosome having centromere at one end.

Terminal buds
A bud at the end of a stem or branch

Yearlings
Lily planting stock at the end of the first growing season form a bulb let.

53. The full form of B.V.O is-
    A] Brown Virus of Orange
    B] Browning of Vegetable Oils
    C] Bacteria like Viral Organisms
    D] Brominated Vegetable Oil

54. 'Die- back' is a serious problem in-
    A] Mango   B] Litchi
    C] Guava   D] Pomegranate

55. Mango & banana are-
    A] Climacteric   B] Nonclimacteric
    C] Nature of ripening   D] None
       depend upon var

56. In India banana contributes following % of total fruit production-
    A] 10   B] 80
    C] 60   D] 31

57. 'Seedlessness' in banana is due to-
    A] Mutation   B] Absence of ovules
    C] Vegetative   D] Stimulative parthenocarpy
       parthenocarpy

58. Most of Banana varieties are
    A] Diploid   B] Triploid
    C] Tretraploid   D] None

59. Largest gene bank of coconut is located at
    A] CTCRI   B] CPCRI
    C] IIHR   D] IARI

60. 'Syrup' should be filled at-
    A] 60-65°C   B] 79-82 °C
    C] 90-100 °C   D] 110-116 °C

61. Heterosis can be fully exploited in the form of
    A] Hybrids   B] Composites
    C] Synthetics   D] Multilines

62. Almost all fruits can be processed satisfactorily at a temperature of-
    A] 80°C   B] 100°C
    C] -5°C   D] 121°C

63. Acidic fruits can easily be sterilized at-
    A] 88°C temp.   B] 100°C temp.
    C] 112°C temp.   D] 116°C temp.

64. Which group of roses bears a 'single flower bud' in a shoot-
    A] Hybrid perpetual roses B] H.T. roses
    C] Polyantha roses   D] Floribunda roses

65. Rice has been classified as
    A] Short day plant   B] Long day plant
    C] Day neutral plant   D] None

66. Lady bird beetle attack on family
    A] Cucurbitaceae   B] Solanaceae
    C] Lilliaceae   D] None

67. The coffee belong to family
    A] Rutaceae   B] Rubiaceae
    C] Euphorbiaceae   D] None

68. The fruit belonging to sapindaceae family
    A] Litchi   B] Avocado
    C] Apple   D] Peach

69. Which vegetable belongs to family 'Liliaceae'-
    A] Onion   B] Garlic
    C] Asparagus   D] None of these

70. Which fruit belongs to family Juglandaceae-
    A] Almond   B] Walnut
    C] Persimon   D] Bread fruit

71. Which of the following belongs to family Rutaceae-
    A] Mulberry   B] Rutabaga
    C] Wood apple   D] Lasoda

72. 'Jack fruit' belongs to family-
    A] Malvaceae   B] Euphorbiaceae
    C] Moraceae   D] Sapotaceae

73. *Ficus benghalensis* belongs to the family-
    A] Palmaceae   B] Moraceae
    C] Apocynaceae   D] Myrtaceae

74. Guava plants are ' bended' to facilitate-
    A] Better growth   B] Better fruiting
    C] Better India   D] Better sprouting

Distal
Toward the apex, away from the base.

Calycle
A group of co-axial leaves around the base of a calyx in some flowers, has the appearance of an outer calyx.

Phyllotaxis
Arrangement of leaves on axis or stem.

Balled and burlapped (B&B)
a field plant whose root has been balled and burlapped for transplanting

Keel
A ridge on the back of a leaf or bud scale.

Dorsal
Of or on the back or outer surface of a leaf, etc.

75. Vanilla grows best at a height of
    A] 1 m   B] 2 m
    C] 3 m   D] 0.5 m

76. Phalsa grows best in _____ climate.
    A] Temperate   B] Tropical
    C] Sub-tropical   D] Anywhere

77. Magnolia is best propagated through-
    A] Cutting   B] Seedling
    C] Air-layering   D] None

78. A cross between a F_1 hybrid with any of its parent is known as-
    A] Back cross   B] Hybridization
    C] Crossing over   D] Multiple cross

79. A cross between an inbred line and an open pollinated variety is called as
    A] Test cross   B] Back cross
    C] Poly cross   D] Top cross

80. ' Bhasinda' is an underground stem of-
    A] Chinese palm   B] Lotus
    C] Lily   D] Banana

81. ' Big Vein' is a seed bore viral disease of-
    A] Radish   B] Cauliflower
    C] Lettuce   D] Menthi

82. ' Bitter pit' in apple occurs due to-
    A] Deficiency of Ca   B] Deficiency of B
    C] Deficiency of Mo   D] Deficiency of Mn

83. ' Black Champa x Thompson Seedless' are the parents of-
    A] Arka Kanchan   B] Arka Vati
    C] Arka Neel   D] None of these

84. Why does blanching is practiced in vegetables
    A] Micro organism removal   B] In-activate enzyme
    C] For softening the vegetables   D] All

85. In tomato blossom end rot is due to
    A] Ca deficiency   B] High temperature
    C] Low temperature   D] High humidity

86. National Horticulture Board (NHB) was established during-
    A] 1976   B] 1980
    C] 1984   D] 1986

87. Coconut Development Board is located at
    A] Mumbai   B] Chennai
    C] Cochin   D] Bangalore

88. Coconut development board works under
    A] ICAR   B] Govt of India
    C] Independently D] None

89. In India, first book on horticulture was written by-
    A] W.B. Hayes   B] B.P. Paul
    C] J.E. Knott   D] M.S Randhawa

90. The Indian Botanical Garden, Kolkata is an example of-
    A] Mughal style   B] Italian style
    C] English style D] Japanese style

91. *Eriobotrya japonica* is the botanical name of -
    A] Sapota   B] Loquat
    C] Wood apple D] Lasora

92. *Areca catechu* is the botanical name of-
    A] Betel nut   B] Betel vine
    C] Oil palm   D] Chinese palm

93. *Althea rosea* is the botanical name of-
    A] China Aster B] Carnation
    C] Holly hock   D] China rose

94. *Vitis vinifera* is the botanical name of-
    A] Grape   B] Grapefruit
    C] Grape wine D] Wood apple

Acropetal
Developing sequentially from basal position to an apical one.

Stipellate
Having stipules at the base of the leaflets.

Stipular
Having stipules at the base of the leaves.

Truncate
with a flat base or apex, perpendicular along the margin to the midrib

Triploid
An organism with three basic sets of chromosome.

Samara
A dry indehiscent fruit bearing a wing (Maple, Ash).

Sours
A succulent stalk bearing an inflorescence as in aroid family.

Spadix
A succulent stalk bearing an inflorescence as in aroid family.

Caulescent
Having an evident leaf-bearing stem above ground.

Nascent
In the act of being formed.

Campanulate
A botanical term for bell-shaped flowers. The genus *Campanula* gets its name from the fact that majority of its members have campanulate flowers.

Ovate
Egg shaped, broadest below the middle.

Plastochron
The time interval between two successive similar occurrences as rhythmic initiation of leaves by the mreistem.

Matching

95. Match the following :-

A]	Best system of planting	i]	Costly land
B]	Hill region	ii]	Quincunx
C]	Hexagonal	iii]	Irrigation system
D]	Flller crop	iv]	Square
E]	Ring method	v]	Contour
F]	Sod culture	vi]	Plains
G]	Drip system	vii]	Training system
H]	Clean cultivation	viii]	Temperate region
I]	River bed cultivation	x]	Israel
J]	Bower method	x]	Cucurbits

96. Match the following :-

A]	Bhadauran	i]	Susceptible to malformation
B]	Alphonso	ii]	Tolerant to mango-malformation
C]	Bombay Green	iii]	Dwarfing rootstock
D]	Olour	iv]	Susceptible to spongy tissue
E]	Sindhu	v]	Seedless mango

97. Match the following :-

A]	Bitter pit in Apple	i]	'Mo' deficiency
B]	Exanthema in Citrus	ii]	O.Y.D. Virus
C]	Coluere in Grape	iii]	Delayed harvesting
D]	Yellow spot in Citrus	iv]	Waterlogging
E]	Greening in Citrus	v]	'Ca' deficiency
F]	Bitterness in Radish	vi]	MLO's
G]	Stocky roots in Carrot	vii]	'Cu' deficiency
H]	Yellow Dwarf in Onion	viii]	High temperature
I]	Bolting in Cauliflower	ix]	'B' deficiency
J]	Corkiness in Chilli	x]	'N$_2$' deficiency

98. Match the following :-

A]	Bitter principle in Olive	i]	Leucoanthocyanidin
B]	Tannin in Strawberry	ii]	Methyl salicylate
C]	Flavour in Strawberry	iii]	Ethyl butanoate

D] Colouring principle in   iv] Oleuropein
Cherry

E] Aroma in Cherries   v] Kerayanin chloraldehyde

99. Match the following :-
A] Bitterness in cucurbits   i] Capsanthin
B] Pungency in chillies   ii] Tetracyclic triterpenes
C] Red colour in chilli   iii] Solasodine
D] Colour in onion   iv] Capsicutin
E] Bitterness in brinjal   v] Quercetin

100. Match the following :-
A] Black pepper   i] $2n=22$
B] Turmeric   ii] $2n=20$
C] Ginger   iii] $2n=52$
D] Cardamom   iv] $2n=24$
E] Cacao   v] $2n=62$

101. Match the following :-
A] Blanching temperature   i] 82 to 87°C for
6-10 minutes
B] Syruping   ii] 100°C
C] Exhausting   iii] 100°C for 2-5 minutes
D] Processing of fruits   iv] 115-121°C
E] Processing of root   v] 79°C to 82°C
vegetables

102. Match the following :-
A] Blanching   i] Yellow flowers
B] *Cassia fistula*   ii] Grape
C] Dahlia   III] Rose
D] Muscat flavour   iv] Heat treatment
E] Shield budding   v] Tuber

103. Match the following :-
A] Blindness   i] Deficiency of Nicotinic acid
(Vit-B$_5$)
B] Beri-beri   ii] Deficiency of Vit-A
C] Cracked lips &   iii] Deficiency of Pyridoxine
mouth corner   (Vit-B$_6$)
D] Pellagra   iv] Deficiency of Thiamine
(Vit-B$_1$)
E] Anaemia   v] Deficiency of Riboflavin
(Vit-B$_2$)

Boulevard
A strip of land between two wide roads protected with low fence and beautified with ornamental plants.

Beer
A class of alcoholic beverage brewed from malt or malt substitute with the addition of hops to give a bitter taste.

Limolin
A glycoside responsible for bitter taste of citrus fruit juice.

Ephemerals
Plants that emerge and bloom during one season, then die back for the remainder of the year. Many spring ephemerals bloom in woodlands, including trillium and ladyslipper.

Spore
A minute reproductive body comprised of a single gametophytic (sexual) cell.

Colour
A technical term in Botany which means any colour except green ; here white is regarded as a colour and green is not.

Apogameon
A species that contains both apomictic and non-apomictic individuals.

Perfect flower
A flower having both functional stamens and pistils; a plant with both functioning male and female parts.

Monoecious
A plant having both male and female flowers on the same plant but at different locations.

104. Match the following :-
A] Blossom end rot in grape
B] 'Bud killing' in grape
C] 'Hen & Chicken' in grape
D] 'Barrenness' of vine
E] Interveinal chlorosis

i] Due to 'B' deficiency
ii] Due to excess of 'N$_2$'
iii] = Zn & Fe deficiency
iv] Due to defective 'Ca' nutrition
v] Due to imbalance in Cytokinin GA.

105. Match the following :-
A] Bombay Green, Swarna Rekha
B] Dashehari, Langra
C] Mulgoa, Neelam
D] Raspunia, Rajapuri
E] Olour, Kurukkan

i] Polyembryonic mango
ii] Mid season (July)
iii] Sucking type mango
iv] Late ripening (August)
v] Early ripening (June)

106. Match the following :-
A] Browning of cut fruits
B] Millard reaction
C] Caramelization
D] Spoilage of canned fruits
E] Spoilage of bread

i] Browning due to sugar-amines reaction
ii] Mucor
iii] Over heating of sugars
iv] Phenolase enzyme
v] *Byssochlamys* mould

107. Match the following :-
A] *C. aurantifolia* (Kagzi lime)
B] *C. medica* (Citron)
C] *C. reticulata* (Mandarin)
D] *C.paradisi* (Grape fruit)
E] *C.aurantium* (Sour orange)

i] Susceptible to frost & cold injury
ii] Most resistant to high temp. (upto 48 °C)
iii] Tolerant to damp heavy soil
iv] Tolerant to low temperature
v] Very susceptible to Canker & Tristeza

108. Match the following :-
- A] Calcium carbide
- B] Butter fly
- C] Spongy tissue
- D] Leaf blight
- E] Leaf curl
- F] Sanjose Scale
- G] Powdery mildew
- H] Late blight
- I] Bunchy top
- J] Fruit fly

- i] Pomegranate
- ii] Colocasia
- iii] Peach
- iv] Fruit ripening
- v] Grape
- vi] Mango
- vii] Banana
- viii] Guava
- ix] Apple
- x] Potato

109. Match the following :-
- A] Calcium deficiency
- B] Magnesium deficiency
- C] Sulphur deficiency
- D] Copper deficiency
- E] Boron deficiency

- i] Death of the shoot tips
- ii] Rigid/brittle cell wall
- iii] Interveinal chlorosis of leaves
- iv] Necrosis of tip of younger leaves
- v] Chlorosis of younger leaves

110. Match the following :-
- A] Cape lily
- B] Sword lily
- C] Belladona Lily
- D] Fox glove
- E] Blanket flower
- F] Chalk plant
- G] Day Lily
- H] Monkey flower
- I] Chinese lantern
- J] Devil's tree

- i] Amaryllis
- ii] Alstonia
- iii] Digitalis
- iv] Gypsophila
- v] Mimulus
- vi] Crinum
- vii] Gaillardia
- viii] Physalis
- ix] Gladiolus
- x] Hemerocallis

111. Match the following :-
- A] *Capsicum annuum*
- B] *Capsicum frutescens*
- C] Pendant fruit in chilli
- D] Erect fruit character in chilli
- E] Progenitor of *C. frutescens*

- i] Cluster bearing
- ii] Single recessive gene
- iii] Solitary bearing
- iv] *Capsicum chinensis*
- v] Single deminant gene

Hermaphrodite
A flower with both stamens and pistil *e.g.,* tomato, chilli, brinjal, etc.

Arm
1. The main branches or extensions of the trunk of a tree or a shrub. 2. A branch of a grapevine which is two years old or older.

Bracing
A method of strengthening branches that form weak crotches in which a small branch to be supported by bark or bridge grafting and sometimes small branches from the two limbs, are twisted together and allowed to unite naturally.

Deliquescent
Breaking up into fine branches.

Stem
The trunk and its branches; one of the three fundamental parts of a higher plants- root, stem and leaf.

Brittle
A condition of easy breaking or snapping.

Scarification
Any process of breaking, scratching, mechanically altering or softening the seed coverings to make them permeable to water and gases.

Chasmogamy
A built-in breeding mechanism where pollination follows only after the opening of flowers which favours self-pollination as seen in tomato, brinjal, chilli, etc.

112. Match the following :-

A]	Carbohydrate	i]	Apricot > Dates > Banana
B]	Protein	ii]	Vit B & C
C]	Fat	iii]	Cashew > Almond > Walnut
D]	Carotene	iv]	Litchi > Karonda > Wood apple
E]	Riboflavin	v]	Walnut > Almond > Cashewnut
F]	Ascorbic acid	vi]	Dry Karonda > Date > Walnut
G]	Calcium	vii]	Mango > Papaya > Persimon
H]	Iron	viii]	Vit -A, D, E & K
I]	Fat Soluble Vitamins	ix]	Barbados cherry > Aonla > Guava
J]	Water Soluble Vitamins	x]	Bael > Papaya > Cashew

113. Match the following :-

A]	Carbohydrate	i]	Carrot
B]	Protein	ii]	Sweet potato
C]	Vitamin-C	iii]	Chilli
D]	Vitamin-A	iv]	Dehydrated pea
E]	Calcium	v]	Colocasia

114. Match the following :-

A]	*Carica monoica*	i]	Hardiest species of papaya
B]	*Carica quercifolia*	ii]	Cultivated species of papaya
C]	*Carica pentagona*	iii]	Monoecious species of papaya
D]	*Carica cauliflora*	iv]	Frost resistant species of papaya
E]	*Carica papaya*	v]	Virus resistant species of papaya

115. Match the following :-

A]	*Carya illinoensis*	i]	Chestnut
B]	*Corylus maxima*	ii]	Black Currants
C]	*Castanea mollisima*	iii]	Pecan
D]	*Fragaria ananassa*	iv]	Filbert
E]	*Ribes nigrum*	v]	Cultivated strawberry

116. Match the following :-

A]	Cashew nut	i]	Myristicaceae
B]	Nutmeg	ii]	Rubiaceae
C]	Tea	iii]	Arecaceae
D]	Cocoa	iv]	Umbelliferae
E]	Coffee	v]	Papaveraceae
F]	Rubber	vi]	Theaceae
G]	Coconut	vii]	Moraceae
H]	Asafoetida	viii]	Sterculiaceae
I]	Khas-khas	ix]	Solanaceae
J]	Bell pepper	x]	Anacardiaceae

True / False

117. Sweet Pea is an ornamental annual.   T/F

118. The origin place of ' Mosambi' var. of sweet orange is said to be Mozambique.   T/F

119. All vegetables can be grown profitably by raising seedlings. T/F

120. Avocado is monocotyledonous fruit tree.   T/F

121. Banana leaf spot disease is more serious in humid tropics. T/F

122. Carrot & Beet leaf are rich source of Vit-A.   T/F

123. Maximum number of plants are accommodated in Hexagonal system.   T/F

124. Micro propagation is common in pineapple.   T/F

125. Mosaic is a serious disease of papaya.   T/F

126. Pineapple is a sub-tropical crop.   T/F

127. Pusa Anmol brinjal is a cross of P.P.L. x Hyderpur.   T/F

128. Sweet orange is commercially propagated by budding.   T/F

129. Datepalm and coconut, both are dioecious plants.   T/F

130. Early blight in tomato is caused by *Pythium sp.*   T/F

131. K.Chandramukhi is an early variety of potato.   T/F

132. Nut fruits are rich source of fat.   T/F

133. Pointed gourd is propagated by cuttings.   T/F

Domestication
The process to bring a wild species under human management.

Solasodine
A glycoalkaloid present in brinjal which causes bitter taste and off-flavour. It exhibits an antibiotic activity.

Elliptical
having a relatively broad middle and tapering at both ends

Oval
Twice as long as broad, widest at the middle, both ends rounded.

Terminal
Applied to the end bud beyond which no further growth takes place normally until the following season. At the tip or distal end.

Accessory bud
An additional axillary bud, a bud formed on a leaf.

Bullhead
Spherically shaped flower bud, usually resulting in a malformed flower.

Budding
Type of grafting in which a scion (vegetative bud) is placed in the stock/stock plant.

Gemmation
Act or manner of budding, arrangements of buds, reproduction by gemmae, formation of new individual by protrusion and separation of part of the parent.

Self-branching
Axillary buds may initiate growth without pinching.

134. *Pothos* is a foliage climber. T/F
135. Puse Jwala chilli is resistant to nematode. T/F
136. Rasthali and Poovan banana are susceptible to fruit cracking from sun. T/F
137. Sweet orange is highly monoembryonic T/F
138. Umran is a late variety of ber. T/F
139. Avocado is also known as 'butter fruit'. T/F
140. Chitra is a variety of H. T. rose. T/F
141. Citrus canker is a viral disease. T/F
142. Cracking in cabbage head occurs due to late harvesting. T/F
143. In sweet orange, pre-harvest fruit drop is a common problem. T/F
144. Larkspur flowers in winter. T/F
145. Like other cucurbits, pointed gourd is also monoecious. T/F
146. Odd season mango 'Romani' gives fruits in February. T/F
147. Poovan banana is the most suitable for problem soils. T/F
148. Seed variation is greatly observed in papaya. T/F
149. Chrysanthemum is a short day plant. T/F
150. CIMAP is situated at Lucknow. T/F
151. Jalgaon district of M.S. is famous for banana. T/F
152. Monthan (culinary var.) banana is tolerant to salts. T/F
153. Persimmon is a dioecious plant. T/F
154. *Phlox* is dwarf flowering annual. T/F
155. Sigatoka is a problem in banana. T/F
156. Spongy tissue is a disorder of banana. T/F
157. Sweet orange fruit is a non-climactic one. T/F
158. Whiptail is a common disorder in all cole crops. T/F
159. 'Oleocellosis' is a storage disorder of Kinnow. T/F
160. Astringency in persimmon flesh is due to tannin. T/F
161. *Gaillardia* can be grown round the year. T/F
162. Grape is considered as a fruit of temperate region. T/F

163. *Grewelia robusta* and *Cassia fistula* are yellow flowered trees.   T/F
164. I.I.V.R. is situated at Varanasi.   T/F
165. Karpuravalli is the sweetest cultivar of banana.   T/F
166. Marry Palmer is a variety of *Poinsettia*.   T/F

Fill in the Blanks

167. _____ is known as ripening hormone.
168. _____ seeds loose their viability if they are exposed to sun.
169. Sweet orange is commercially propagated by _____.
170. _____ is classified as shrub.
171. For lye peeling in orange, 0.5% _____ solution is used.
172. Arka Manik is improved variety of _____.
173. Seedlessness in Thompson Seedless grape is due to _____.
174. Whiptail in _____ occurs due to deficiency of Mo.
175. The most common growth regulator for rooting is _____.
176. _____ is plant growth regulator used for ripening of fruits.
177. _____ is polyembryonic root stock of mango.
178. _____ preservative is used for highly coloured juices.
179. _____ is improved variety of radish which does not develop pithiness.
180. _____ is *in situ* method of grafting.
181. Sioux is a hybrid variety of _____ developed in India.
182. _____ is ideal plant for topiary.
183. Mango variety _____ is suited for high density planting.
184. African violet is a _____ plant.
185. _____ breaks seed dormancy.

Limb up
Prune off lower branches of a woody plant to (further) expose its main trunk or trunks.

Puffer
A completely rotted bulb under an intact skin which, when crushed released a cloud of spores, caused by Penicullium.

Precooling
Dry storage of bulbs at temperatures between 2° and 9°C after floral initiation and development is completed but prior to planting.

Retarded bulb
Storage of bulbs in such a way, which prevents flower development, especially at high temperature.

186. _____ is seedless fruit.

187. Indian gooseberry is good source of Vit-_____.

188. _____ is avenue tree.

189. Cavendish banana is propagated by _____.

190. _____ is suitable for sub-tropical dry climate.

191. Apple belongs to family _____.

192. _____ is applied for uniform ripening in pineapple.

193. Alternate bearing is problem of _____.

194. Yellow Vein Mosaic is caused by _____.

195. Okra belongs to _____ family.

196. _____ headquarters is situated at New Delhi.

197. Indian Horticulture is published from _____.

198. Granulation is physiological disorder of _____.

199. Seeds of _____ are viviparous.

200. Panama wilt is a disease of _____.

Once there was a lark singing in the forest. A man came by with a box full of worms. The lark stopped him and asked, "What do you have in the box and where are you going?" The farmer replied that he had worms and that he was going to the market to trade them for some feathers. The lark said, "I have many feathers. I will pluck one and give it to you and that will save me looking for worms." The farmer gave the worms to the lark and the lark plucked a feather and gave it in return. The next day the same thing happened and the day after and on and on until a day came that the lark had no more feathers. Now it couldn't fly and hunt for worms. It started looking ugly and stopped singing and very soon it died.

The lark thought of easy way to get food turned out to be the tougher way after all.

[The Easier way may Actually be the Tougher way.]

Step-Five *The Loser says, "It may be possible but it is too difficult"; The Winner says, "It may be difficult but it is possible".*

Multiple Choice

1. *Syzygium cumuni* is the botanical name of-
   A] Monkey bread B] Custard apple
   C] Jamun D] Khirni

2. *Nerium odorum* is the botanical name of-
   A] Yellow Kaner B] Red Oleander
   C] Balsam D] Arjun

3. Fruit of brassica is called as
   A] Pome B] Berry
   C] Siliqua D] Drupe

4. In heterosis breeding one should capitalize on-
   A] Inbreeding depression B] Homozygosity
   C] Over dominance D] High seed cost

5. Two basic steps of breeding programme are-
   A] Hybridization and Selection B] Variation and Selection
   C] Mutation and Selection D] Pollination and Fertilization

6. The cornerstone of all breeding programmes is-
   A] Hybridization B] Selection
   C] Selfing D] Introduction

7. In grape breeding seedlessness is achieved by
   A] Triploidy B] Aneuploidy
   C] Doubling of haploids D] None

8. Origin of Brinjal is
   A] Africa B] America
   C] India D] China

9. ' Bromelin' alkaloid is found in-
   A] Custard apple B] Pineapple
   C] Ber D] Sapota

10. 'Brown heart/Heart rot' in garden beet develops due to deficiency of-
    A] Zinc   B] Boron
    C] Magnesium   D] Sulphur

11. 'Browning' in cole crops occurs due to deficiency of-
    A] Zinc   B] Molybdenum
    C] Nitrogen   D] Boron

12. T- budding is common method of propagation in
    A] Citrus   B] Peach
    C] Pomegranate   D] Apple

13. 'Budha Jayanti Park' is situated at-
    A] Varanasi   B] Mysore
    C] Chandigarh   D] New Delhi

14. Removal of undesirable auxillary buds at initial stage of flowerings is known as-
    A] Notching   B] Deblossoming
    C] Clipping   D] Topiary

15. Which onion variety produces bulbs in short-day conditions-
    A] N-53   B] Patna Red
    C] Nasik-Red   D] Patna White

16. ' Bunchy top' is a disease of-
    A] Grape   B] Mango
    C] Banana   D] Tomato

17. ' Bunchy Top' of banana was first recorded in-
    A] India (1920)   B] Fiji (1891)
    C] Malaysia (1895)   D] U.S.A. (1875)

18. ' Buttoning' in cauliflower occurs due to-
    A] N_2 deficiency   B] Late planting
    C] Over aged seedling   D] All the above

19. The rose variety developed by mutation breeding is-
    A] Avon   B] Sonia
    C] Super Star   D] Abhisarika

20. Polyploidy can be induced by-
    A] Ethyl methane sulphonate   B] Nitrous acid
    C] Colchicine   D] Methyl methane sulphonate

21. Brix degree is measured by-
    A] Lactometer   B] Refractometer
    C] Hydrometer   D] Jel-meter

22. Gladiolus is commercially propagated by-
    A] Tuber   B] Bulb
    C] Cutting   D] Corm

23. Dahlia is commercially propagated by-
    A] Tuber   B] Cutting
    C] Sucker   D] Budding

24. Vitamin- C is commonly known as
    A] Citric acid   B] Acetic acid
    C] Lactic acid   D] Ascorbic acid

25. For Vitamin C rich source is
    A] Banana   B] Mango
    C] Guava   D] All

26. Theobromine ($C_8N_4O_2$) is found in-
    A] Coffee   B] Tea
    C] Cocoa   D] All of these

27. Pollination in cacao is done by
    A] Wind   B] Insect
    C] Water   D] Bird

28. ' California Wonder' is a variety of-
    A] Tomato   B] Sweet pepper
    C] Cauliflower   D] Litchi

29. Which is called as a protective food
    A] Vegetables   B] Fruits
    C] Fruits & vegetables   D] Cereals

30. Vegetables are called as protective food because they are rich in -
    A] Fats   B] Vitamins & minerals
    C] Proteins   D] Sugar

31. Mango fruits can be best stored at-
    A] -5°C   B] 16°C
    C] 5°C   D] 8°C

Aflatoxin
A toxic substance formed by *Aspergillus flavus*, one of several myxotoxins produced by fungi.

Cleistogameon
A species that reproduces by cleistogamy.

Spreading grass
Grass that spreads by creating new plants from existing plants by running roots across or under the ground (*stolons* or *rhizomes*, respectively), creating a dense turf. Also called *sod-forming grass*. Contrast with *bunchgrass* (also called *clump-forming grass*), in which single plants get larger over time but do not create new plants.

Brandy

An alcoholic product obtained by distillation of almost any fermented fruit juice.

Arrack

A commercial product obtained by distillation of fermented toddy.

Sun scald

Injury caused by excessively high light intensity and radiant heat.

Hermetic

Air-tight closure by fusion.

Opium

A brownish solid harvested by incising the fully swollen capsule (fruit) of poppy (*Papaver somniferum*) which contain several alkaloids namely, morphine, codeine, thebaine, marcotine, narceine and papavarine.

32. 'Mango malformation' can be controlled by spray of-
    A] 200 ppm MBA   B] 200 ppm GA
    C] 200 ppm IBA   D] 200 ppm NAA

33. 'Mango malformation' can be controlled by the spray of-
    A] Karathane   B] GA_3
    C] N.A.A.   D] Cytokinin

34. Seed production of cabbage can be done successfully in-
    A] Kashmir Valley   B] Kullu Valley
    C] Punjab   D] Nilgiri Hills

35. Seed production of cabbage can be done successfully in-
    A] Phylloxera   B] Powdery mildew
    C] Chinch bug   D] Wooly aphid

36. Sulphur fungicide can be freely used on all vegetables except one
    A] Peas & beans   B] Root vegetables
    C] Cucurbits   D] Okra

37. Transgenic plants can be produced under
    A] Open field condition B] Controlled condition
    C] In a small plot   D] None

38. Ratio of protein and carbohydrate in diet should be-
    A] 1:1   B] 2:5
    C] 2:1   D] 1:6

39. The large cardamom is propagated by
    A] Seeds   B] Suckers
    C] Cutting   D] All

40. The quickest method of carrot peeling is-
    A] Flame peeling   B] Lye peeling
    C] Mechanical peeling D] Steam peeling

41. 'Carrot' is originated from-
    A] Europe   B] China
    C] Afghanistan   D] Mexico

42. In which category does apple come according to respiration data
    A] Low   B] Medium
    C] High   D] Very high

43. In which category does mango come according to rate of respiration
    A] Low   B] Medium
    C] High   D] Very high

44. 'Curd' of cauliflower is an example of
    A] Hypertrophy   B] Hypotrophy
    C] Hyperplasea   D] Hypoplasea

45. Clubrot of cauliflower is caused by
    A] *Xanthomonas*   B] *Plasmodiophora brassicae*
       compestris
    C] *Fusarium spp*   D] None

46. Whiptail of cauliflower is caused due to deficiency
    A] Mo   B] Cu
    C] Zn   D] B

47. Whiptail of cauliflower is due to deficiency of
    A] Mo   B] Zn
    C] Co   D] Cu

48. Browning in cauliflower is due to
    A] Cu   B] Bo
    C] Ca   D] Mo

49. Ricyness in cauliflower is due to
    A] Growing in rice field   B] Unfavourable temperature
    C] Mo deficiency   D] None

50. 'Whiptail' in cauliflower occurs due to deficiency of-
    A] Molybdenum   B] Boron
    C] Iron   D] Manganese

51. *Alternaria solani* is a causal organism of-
    A] Leaf blight   B] Leaf curl
    C] Early blight   D] Late blight

52. 'Pellagra' is caused due to deficiency of-
    A] Thiamine   B] Riboflavin
    C] Niacin   D] Thyroxin

53. 'Scurvey' is caused due to deficiency of-
    A] Vit-A   B] Vit-B
    C] Vit-C   D] Vit-D

Basic seed
Seed produced by mass selection (with progeny test) in a pure line variety or clone.

Honeydew
The sticky secretion produced by sucking insects such as aphids.

Double digging
Preparing the soil by systematically digging an area to the depth of two shovels.

Certified seed
Seed produced by the breeder or the institute, which develop the variety.

Aggregate fruit
A fruit formed by the coherence or the connation of pistils that were distinct in the flower.

Submerged bud
A bud hidden by the petiole or embedded in the callus of the leaf scar.

Auxin
A substance synthesized by the plant influencing growth at some point other than the point of its synthesis, $C_{18}H_{32}O_5$.

Hardened
A plant condition created by various factors enabling it to withstand environmental stresses; contrast with succulent growth which is very vulnerable to environmental stress and damage.

Blight
Any plant disease characterized by withering and shrivelling of plant parts without rotting.

54. 'Cavity spot' in carrot develops due to deficiency of-
A] Ca
B] Na
C] K
D] Mg

55. Sugar per cent in jelly should be-
A] 20-30%
B] 10%
C] 60-65%
D] 50-55%

56. National Research Centre for Spices (NRCS-1986) is reorganised as IISR in-
A] 1988
B] 1990
C] 1992
D] 1995

57. Most important cereal crop in India is
A] Rice
B] Wheat
C] Sorghum
D] Barley

58. The isolation distance for certified seed production in radish should be-
A] 2000 m
B] 1600 m
C] 1000 m
D] 400 m

59. 'Chakaiya' is a variety of-
A] Grape fruit
B] Pummelo
C] Lime
D] Sour Orange

60. The botanical name of Chameli is-
A] *Jasminum sambac*
B] *Jasminum auriculatum*
C] *Jasminum grandiflorum*
D] *Jasminm pubescens*

61. The 'Champaine' is a type of
A] Juice
B] Wine
C] Beverage
D] Pod

62. 'Chaubatia Princes' and 'Chaubatia Anupam' are hybrids of-
A] Peach
B] Litchi
C] Pear
D] Apple

63. Botanical family of Barbados cherry is-
A] Juglandaceae
B] Malpighiaceae
C] Lauraceae
D] Mytaceae

64. GMS in chilli is maintained through-
A] Heterozygous pollinator
B] Homozygous pollinator
C] Mutation
D] Polyploidy

65. Variety of chilli resistant to leaf curl virus
    A] Bhaskar　　　　　　B] Pusa Deepti
    C] Jwalamukhi　　　　 D] Pusa Jawala

66. For seed production in chilli, isolation distance should be-
    A] 50-200 m　　　　　 B] 250-400 m
    C] 800-1000 m　　　　 D] 1200-1600 m

67. The somatic chromosome numbers in pea are-
    A] 7　　　　　　　　　 B] 14
    C] 28　　　　　　　　　 D] 48

68. The hereditary material in chromosomes is-
    A] RNA　　　　　　　　 B] DNA
    C] ATP　　　　　　　　 D] ADP

69. In annual chrysanthemum, which 'flower type' is found-
    A] Single　　　　　　　 B] Quilled
    C] Spoon　　　　　　　 D] Incurved

70. Harvesting of cinchona is called as
    A] Tapping　　　　　　 B] Coppicing
    C] Beheading　　　　　 D] Lancing

71. The shape of *Xanthomonos citri* is-
    A] Spherical　　　　　　 B] Oval
    C] Rod Shaped　　　　　 D] Cylindrical

72. ' Citrus Canker' is caused by-
    A] MLO's　　　　　　　 B] Bacteria
    C] Fungi　　　　　　　 D] Nematode

73. 'Gummosis' in citrus is caused by-
    A] Virus　　　　　　　 B] Fungi
    C] M.L.O.s　　　　　　 D] Nematode

74. Exanthema in citrus is due to deficiency of
    A] Mg　　　　　　　　 B] Cu
    C] Zn　　　　　　　　 D] Cl

75. 'Exanthema' in *Citrus* spp. occurs due to lack of-
    A] Zinc　　　　　　　 B] Potash
    C] Copper　　　　　　 D] Molybdenum

76. The genetic classification of banana was given by-
    A] Swingle & Swingle　 B] Simmonds & Shepherd
    　　(1917)　　　　　　　 (1955)
    C] Habelandt (1948)　 D] Cheema & Mukherjee
    　　　　　　　　　　　　 (1907)

Inulin
A storage polysaccharide $(C_6H_{10}O_5)$ found in the roots of Jerusalem artichoke, dahlia and a few other plants.

Hakuran
An artificial amphidiploid of cabbage, as Chinese cabbage produced through embryo culture technique, which is a good leafy vegetable.

Trimonoecious
A sex form also called gynoandromonoecious, where staminate, pistillate and hermaphrodite flowers are produced separately in the same plant in varying proportions; found in papaya.

Stripping

Chlorotic stripes that can be caused by pesticides phytotoxicity or nutrient deficiencies.

Serophyte

A plant that can endure extreme drought where the evaporation stress is high and the water supply low.

Zygomorphic

An irregular flower capable of being divided in half, but in only one plane.

Spore

A unit of protoplasm capable of developing asexully into a new individual. In higher plants it is the haploid product of meiosis that gives rise to male and female gametes.

77. The citrus classification was given by
A] Bruns
B] Tanaka & Swingle
C] Simmonds
D] None

78. Which is classified as 'preserve'-
A] Preserve
B] Candy
C] Glazed fruit
D] All of above

79. Size of clay particle is
A] .2- .02 mm
B] .2- .02 mm
C] < .002 mm
D] .02 -.002 mm

80. The sparkling clear liquid free from all suspended solid is called as
A] Concentrate
B] Cordial
C] Squash
D] Beverage

81. Non- climacteric fruit is
A] Which ripe after the maturity
B] Which ripe before the maturity
C] Which ripe at the time of maturity
D] None

82. Which is a ' Climacteric' type fruit-
A] Litchi
B] Papaya
C] Grape
D] Orange

83. Which requires the warmest climate among following temperate fruits-
A] Apple
B] Plum
C] Peach
D] Pear

84. The best climber for cool dry places is-
A] *Quisqualis indica*
B] *Wisteria chinensis*
C] *Hiptage madablata*
D] Morning glory

85. Pusa purple cluster var of brinjal is resistant to-
A] Purple blotch
B] Bacterial wilt
C] Drought
D] Mite

86. Coffee, Cocoa, Rubber and Cardamom are grouped under-
A] Commercial Crops
B] Ratoon Crops
C] Plantation Crops
D] Spices & Condiments

87. ' Coconut Development Board' came into existence in-
A] 1960 B] 1969
C] 1981 D] 1986

88. Inflorescence of coconut is called as
A] Raceme B] Spike
C] Spadix D] Catkin

89. ' Coconut root wilt' was first observed in-
A] 1891 B] 1882
C] 1855 D] 1890

90. Vinegar from coconut today is prepared by
A] Yeast B] Bacteria
C] Fermentation D] Fungus

91. ' Coffee Act' was formed during-
A] 1960 B] 1975
C] 1965 D] 1942

92. A major problem of coffee cultivation is-
A] White stem borer B] Aphid
C] Nematodes D] Powdery mildew

93. Aroma in coffee is due to
A] Fermentation B] Drying
C] Roasting D] None

94. Acridity in colocasia is due to
A] Calcium oxalate B] Potassium
C] Sulphur D] None

Matching

95. Match the following :-
A] *Cassia fistula* i] Scarlet Red
B] *Delonix regia* ii] Edging
C] *Cassia nodoso* iii] Yellow
D] *Lagerstroemia speciosa* iv] Blue
E] *Bauhinia variegata* v] Hedge
F] *Erythrina indica* vi] Mauve-Purple
G] *Jacaranda* vii] Scarlet Orange

Bulb maturity
Measure of capacity of a healthy daughter stem axis to sprout delay and to respond to flower-inducing treatments.

Sedges
Plants of the genus *Carex*, which includes over 1,500 species; sedges look similar to grasses.

Passive absorption
Absorption of water by the roots as a result of forces originating in the leaves due to transpiration pull.

Botulism
A food poisoning caused by the toxin produced by bacterium (*Clostridium botulinum*) in preserved foods.

Salb-side
A carnation flower where the petals elongate and expand first on one side of the calyx to look a lopsided flower.

Bonsai
The art of growing carefully trained, dwarf plants in containers.

Leafyness
A physiological disorder of cauliflower in which green bracts grow out of the curd due to exposure of the curd at temperature higher than optimum for its development.

Physiological disorders
Undesirable affects caused by a non-pathogenic agent or factor.

mimosaefolia

H] *Eranthemum* viii] Floatlng Garden
I] *Clerodendron inerme* ix] Pink
J] *Typha angustifolia* x] White-Pink

96. Match the following :-
A] Cauliflower i] Tiliaceae
B] Sunflower ii] Cruciferae
C] Grape fruit iii] Rutaceae
D] Spinach iv] Compositae
E] Phalsa v] Chenopodiaceae

97. Match the following :-
A] Chaubatia Alankar i] Turkey x Charmagz
B] Chaubatia Madhu ii] Alphonso x Neelam
C] Chaubatia Kesri iii] St. Ambroise x Charmagz
C] Arka Puneet iv] Kaisha x Chamagz
E] Arka Neelkiran v] Banganpalli x Alphonso

98. Match the following :-
A] Chaubatia Kesri i] Peach hybrid
B] Saharanpur Prabhat ii] Pollinizer of apple
C] Early Rivers iii] Cracking resistant sweet cherry
D] Crispin iv] Apricot hybrid
E] Golden Delicious v] Triploid apple

99. Match the following :-
A] Chilli seed rate i] 400-500 g/ha
B] Brinjal seed rate ii] 1000-1200 g/ha
C] Tomato seed rate iii] Raported in 1939
D] Seedless watermelon iv] 500-750g/ha
E] 'Little leaf' in brinjal v] Dr. Kihara

100. Match the following :-
A] Chilli i] Arka Navneet, Pant Samrat
B] Sweet pepper ii] Arka Basant, Arka Lohit
C] Round Brinjal iii] Arka Vikas, Arka Meghali
D] Long Brinjal iv] Arka Gaurav, Arka Mohini
E] Tomato v] Arka Shirish, Arka Sheel

101. Match the following :-
     A] Chilli   i] Brinjal
     B] Capsicum   ii] Paprika
     C] Brinjal   iii] Often cross pollination
     D] Heterostyly   iv] Self-pollination
     E] Arka Abhir   v] Cross-pollination

102. Match the following :-
     A] Citrange   i] Trispecific hybrid
     B] Citrangequat   ii] Intergeneric hybrid
     C] Citrangedin   iii] Trigeneric hybrid
     D] Tangor & Tangelo   iv] Intrageneric (Interspecific)
                                  hybrid
     E] Kinnow   v] Intervarietal hybrid

103. Match the following :-
     A] Citrus Canker   i] Virus
     B] Tristeza & Psorosis   ii] Bacterium (*X. citri*)
     C] Greening in citrus   iii] Physiological disorder
     D] Foot rot & Pink   iv] MLOs/BLOs
        disease
     E] Granulation in citrus v] Fungi

104. Match the following :-
     A] Cleistogamy   i] Jack fruit, Banana
     B] Homogamy   ii] Sapota, Walnut
     C] Dicliny   iii] Grape, Papaya
     D] Protandry   iv] Chestnut
     E] Protogyny   v] Loquat, Apple
     F] Heterodichogamy   vi] Citrus, Peach
     G] Duodichogamy   vii] Mango, Aonla
     H] Sporophytic Self-   viii] Peach, Pear
        incompatibility
     I] Gametophytic Self-   ix] Pistachionut
        incompatibility
     J] Self-sterility   x] Plum, Pomegranate

105. Match the following :-
     A] Coconut   i] 2n = 36
     B] Cahewnut   ii] 2n = 62

Loculus
A small chamber or cavity; cavity of an ovary or of an anther.

Apiin
A glycoside present in celery.

Meristem
Growing point where cell divisions occur. The undifferentiated plant tissue for whcih new cells arise by division.

Amphisarca
A superior, many - celled and many seeded fruit with woody pericarp; the placenta and the inner layer of the pericarp forms the pulpy edible part *e.g.* wood-apple, Katbel (*Feronia limonia*).

Pepo
An inferior, one - celled, many-seeded pulpy fruit, as a cucurbit.

Glossary

Pod
A superior one-celled, one or many-seeded fruit of two valves which split open or a legume like a husk.

Meristem
Areas of actively dividing cells and plant growth; capable of developing into specialized tissues.

Brine
1-2 per cent solution of common salt which is generally used as covering liquid in canning vegetables.

Disc florets
Florets in center of chrysanthemum inflorescence, conspocuous in daisy or anemone flowered types.

Axis
The main stem or central support of a plant.

C] Rubber   iii] $2n = 42$
D] Tea   iv] $2n = 24$
E] Coffee   v] $2n = 44$
F] Blackpepper   vi] $2n = 32$
G] Cardamom   vii] $2n = 24$
H] Turmeric   viii] $2n = 52$
I] Ginger   ix] $2n = 30$
J] Cinnamon   x] $2n = 22$

106. Match the following :-
A] Coconut   i] Seed potato technique
B] Shimla   ii] Pectin
C] Onion   iii] Sprouting
D] Guava   iv] April & Oct spruning
E] Grape   v] Coastal areas

107. Match the following :-
A] Coffee leaf rust   i] *Corticium salmonicolor*
B] Black pod in Cocoa   ii] *Phytophthora areceae*
C] Thanjavur wilt of Coconut   iii] *Hamelia vestatrix*
D] Mahali disease in Arecanut   iv] *Ganoderma luciderm*
E] Pink disease in Rubber   v] *Phytophthora palmivora*

108. Match the following :-
A] Coffee rust   i] *Elsinoe ampelina*
B] Guava wilt   ii] *Hemeleia vestarix*
C] Ring rust of aonla   iii] *Gliocladium roeum*
D] Panama wilt of banana   iv] *Fusarium oxysporum*
E] Grape anthracnose   v] *Ravenelia emblicae*

109. Match the following :-
A] Coffee   i] Native to Ethiopia
B] Cocoa   ii] Native to South America
C] Cashewnut   iii] Native to Brazil
D] Arecanut   iv] Native to West Africa
E] Oil palm   v] Native to Malayan Archipeiago

110. Match the following :-
     A] Coleus   i]   Hardwood cutting
     B] Bigonia   ii]   Softwood cutting
     C] Grape   iii] Ring budding
     D] Croton   iv]   Leaf cutting
     E] Ber   v]   Air layering

111. Match the following :-
     A] Colour in pumpkin   i]   Beta-xanthins
     B] Colour in brinjal   ii]   Carotenoids
     C] Colour in sugarbeet iii] Mutant of All Green
     D] Pusa Jyoti   iv]   Anthocyanin
     E] Indian spinach   v]   Pusa Bharati

112. Match the following :-
     A] Component of protein   i]   Calcium
     B] Component of ATP   ii]   Boron
     C] Opening of stomata   iii]   Magnesium
     D] Constituent of cell wall iv] Nitrogen
     E] Photosynthesis   v]   Manganese
     F] Constituent of amino   vi]   Phosphorous
        acid
     G] Transportation of   vii]   Iron
        carbohydrate
     H] Biosynthesis of auxin viii] Potassium
     I] Chlorophyll synthesis ix]   Zinc
     J] Respiration   x]   Sulphur

113. Match the following :-
     A] Cordial   i]   25 % juice & 30% T.S.S.
     B] Squash   ii]   25% juice & 65% T.S.S.
     C] Crush   iii] 25% juice & 45% T.S.S.
     D] Syrup   iv]   25% juice, 30% TSS &
                             0.25% starch
     E] Barley water   v]   25% juice & 55% T.S.S.

114. Match the following :-
     A] Crazy top & Mule tail i]   Disorders of Peach
     B] Split pit & Gumming   ii]   Disorders of Pecan
     C] Sharka/Plum pox   iii] Fungal disease of Apricot

Heterozygote
An organism whose chromosomes carry unlike members of any given pair of alleles or series of allelomorphs, consequently producing unlike gametes.

Enterotoxin
A toxin produced by certain strains of *Staphylococcus aureus* during their growth which causes gastroenteritis or inflammation of the lining of the intestinal tract. It causes commonly occuring food poisoning.

Mutation
A sudden heritable changes in an organism. The term is used to include point mutation involving a single gene change as well as a chromosome change.

D]	Apoplexy	iv]	Genetic disorders of Almond
E]	Mouroe ear & Rosette	v]	Viral disease of Plum

115. Match the following :-

A]	Croton	i]	Budding
B]	Jasminum	ii]	Tissue culture
C]	Rose	iii]	Air layering
D]	Carnation	iv]	Seed
E]	Banana	v]	Cutting

116. Match the following :-

A]	Crystalization of jam	i]	Sugar <30%
B]	Gummy jam	ii]	Low T.S.S.
C]	Premature setting of jam	iii]	>90% R.H.
D]	Surface shrinkage	iv]	High T.S.S.
E]	Microbial spoilage	v]	High evaporation

Fill in the Blanks

117. Sweet potato is propagated by tuber. T/F

118. China is the origin place of Japanese persimmon. T/F

119. *Clerodendron inerme* is best for evergreen hedge. T/F

120. Gibberellin is used for dormancy breaking. T/F

121. I.A.A. is a rooting hormone T/F

122. Kagzi lime is propagated by seed. T/F

123. *Saraca indica* is a tall bush. T/F

124. Scarification is synonym of stratification. T/F

125. Seeded grapes are most suitable for raisin making. T/F

126. Sweet lime is hardy than Kagzi lime. T/F

127. The use of CaC_2 for banana ripening is banned. T/F

128. 'Meetha Chikna' is a popular variety of sweet lime. T/F

129. Arka Shyam is a hybrid variety of brinjal. T/F

130. Bolting in cauliflower is due to late planting of early varieties. T/F

131. Confidence is a variety of rose. T/F

132. Ethrel is used for banana ripening.	T/F
133. Fig is an evergreen subtropical bush.	T/F
134. In S. India, grape is pruned once a year.	T/F
135. Nasik district of M.S. is known as 'Vine Yard of India'.	T/F
136. *Poinsettia* produces colourful bracts.	T/F
137. Strawberry is a simple fruit.	T/F
138. Acroclinum is suitable for dry flowers.	T/F
139. Bael is rich source of Riboflavin.	T/F
140. Black Ischia and Excell are varieties of fig.	T/F
141. Curing is useful in onion.	T/F.
142. Date palm is a dicotyledonous evergreen tree.	T/F
143. Kinnow is a variety of sweet orange.	T/F
144. Large tubers of potato are susceptible to black heart.	T/F
145. Leaf curl disease in chilli is transmitted through aphid.	T/F
146. Mango and citrus produces parthenocarpic seedlings.	T/F
147. Triploid variety of apple are self-fruitful.	T/F
148. Arka Sahan is a variety of custard apple.	T/F
149. Cavendish banana is most suitable for Chips.	T/F
150. Common fig is also known as Adriatic fig.	T/F
151. Garlic is propagated by planting bulbs.	T/F
152. Pairy is a variety of mango.	T/F
153. Papaya is the most suitable 'Filler' crop.	T/F
154. Papery membrane covering seeds in pomegranate is mesocarp.	T/F
155. Soft seeded var. of pomegranate, 'Ganesh' is developed at IARI.	T/F
156. Streams & Waterfalls are main features of Mughal gardens.	T/F
157. Sweet lime, when used as root stock, reduces acidity of fruits.	T/F
158. 'Jhumka' is a disease of grape.	T/F
159. *Adenocalymma allicea* is suitable for pergola.	T/F

GLOSSARY

EMS
A chemical mutagen, chemically ethylmethane sulphonate.

Design
Determination f a character of an object to serve certain purpose known in advance.

Legume
A pod; the characteristic fruit of the pea family.

Allelopathic
Plant that produces chemicals affecting other nearby plants' growth. Usually used to indicate a negative effect, inhibiting the growth of nearby plants. Black Walnut (*Juglans nigra*) trees inhibit the growth of many other plants.

Glossary

Chelate
A formulation in chemicals, which gradually releases elements.

Chlormequat-(2-chlorethyl)Trimethyl ammonium chloride
The active ingredient in the commercial product Cycoel.

Photomorphs
All living organisms containing chlorophyll which have the ability to transform the kinetic energy of light into the potential chemical energy in foods and other manufactured compounds.

Quadruplochromosome
A chromosome with eight chromatids, produced by endoreduplication.

160. Adriatic fig produces parthenocarpic fruits.   T/F
161. Banana is a succulent berry.   T/F
162. Cucumber is suitable for glass house cultivation.   T/F
163. Emasculation is practiced in tomato.   T/F
164. Lemon is a deciduous tree.   T/F
165. Onion & garlic are also exported in powdered form.   T/F
166. Papaya is propagated by sexual method.   T/F

Fill in the Blanks

167. Mango and Citrus produces _____ seedlings.
168. _____ produces seedless fruits.
169. Cucumber is _____ in sex form.
170. Shield budding is practised in _____.
171. 'Greening' is a common disease of _____.
172. Muscat is a variety of _____.
173. _____ is regular bearing variety of mango.
174. Malformation is a problem of _____.
175. _____ induces rooting in stem cuttings.
176. _____ induces artificial ripening in mango.
177. _____ requires heavy manuring due to shallow root system.
178. Unlike most of fruit crops _____ is herbaceous.
179. Alphonso mango thrives best in _____ climate.
180. _____ requires severe pruning to get optimum yield.
181. _____ is the fourth most important fruit of India in area & production.
182. Bunchy top is a disease of _____.
183. Pineapple can be propagated by _____.
184. Premature heading is a problem of _____.
185. Golden Delicious is a variety of _____.
186. Application of _____ increases growth and yield of cabbage.

187. Serious disease of tomato is _____.
188. Underground stem of gladiolus is _____.
189. CO-2 variety of papaya is _____.
190. Umran is a _____ variety of ber.
191. Ganga Bondam is a _____ variety of coconut.
192. Powdery mildew is a major problem in _____.
193. Citrus canker is a _____ disease.
194. In acidic soil, _____ toxicity is common.
195. Sigatoka is major problem in _____.
196. Deficiency of Vit-B$_1$ causes _____ disease.
197. Red rust of mango is a _____ disease.
198. Green plants are _____ producers.
199. _____ is the best planting system near a city.
200. Attached method of grafting in mango is _____.

Genome
A complete set of chromosome inherited as a unit from one parent.

Chromosome aberration
A change in chromosome number from the diploid state, or a change in chromosome structure from the normal chromosome complement.

Phylloclade
A leaf like or chlorophyll containing stem that performs the photosynthetic function like ordinary leaves e.g., *Opuntia, Euphoribia tirucalli,* etc.

Taxonomy
The science of classification of plants and animals.

Three people were laying bricks and a passerby asked them what they were doing. The first one replied, "Don't you see I am making a living?" The second one siad, "Don't you see I am laying bricks?" The third one said, "I am building a beautiful monument." Three people doing the same thing gave totally different replies. The question is: did they have different attitudes? And would their attitude affect their performance? The answer is a clear yes.

Excellence comes when the performer takes pride in doing his best. Every job is a self-portrait of the person who does it, regardless of what the job is, whether washing cars, sweeping the floor or painting a house. Do it right the first time, every time. The best insurance for tomorrow is a job well done today.

[**H**alf-hearted effort does not produce **H**alf results; it **P**roduces no results.]

Step-Six

When a Loser makes a mistake, he says, "It wasn't my fault.";
When a Winner makes a mistake, he says, "I was wrong".

Multiple Choice

1. For quick colour development in a garden, which is the most suitable-
    A] Annuals   B] Shrubs
    C] Trees   D] Climbers

2. In a garden which colour dominates round the year-
    A] Green   B] Red
    C] Yellow   D] Blue

3. The best combination for better fruiting in loquat is-
    A] Pale Yellow & Golden Yellow   B] Tanaka & Golden Yellow
    C] California Advance & Tanaka   D] Thames Pride & Pale Yellow

4. The only commercial fruit crop originated in Australia is-
    A] Persimmon   B] Macademia nut
    C] Durian   D] Passion fruit

5. 'Heterosis' is commercially exploited in-
    A] Cross pollinated crops   B] Self pollinated crops
    C] Vegetatively propagated crops   D] Sexually propagated crops

6. Litchi is commercially multiplied by-
    A] Seed   B] Ground Layering
    C] Marcottage   D] Grafting

7. Tea is commercially propagated by
    A] Grafting   B] Budding
    C] Cutting   D] None

8. Mango is commercially propagated by
    A] Inarchi g   B] Veneer, grafting
    C] Side grafting   D] Stone grafting

9. Canna is commercially propagated by-
    A] Seed   B] Cutting
    C] Rhizome   D] Layering

Syrup
This is a clear sugar syrup, which has been artificially flavoured.

Cleistogamous
Describes a small, closed self-fertilized flower, usually near the ground.

Species
A category of taxonomic classification lower than a genus or subgenus and above that of a subspecies of variety.

Nocturnal
Opening at night and closing during the day.

Involucrate
With an involucre or cluster of bracts.

Fascicle
A close or compact cluster of leaves, blossoms, or fruits.

Coffelite
A plastic manufactured from coffee beans.

Whiptail
A nutritional disorder of cole crops due to deficiency of molybdenum resulting in severe reduction of the lamina leaving the large bare mid-rib in acute condition.

Dingy
Neither white nor brighly colored, as applied to pubescence.

Quercetin
A pigment which imparts colouration to the outer skin on onion bulb.

Double nose bulb
A commercial bulb with two daughter bulbs and axial meristems with a common basal plat.

10. Grape is commercially propagated by
A] Seed   B] Hard wood cutting
C] Soft wood cutting   D] Grafting

11. 'Litchi' is commercially propagated by-
A] Simple layering   B] Serpentine layering
C] Mound layering   D] Air-layering

12. Tea is commercially propagated by
A] Single node cutting   B] Seed
C] Layering   D] None

13. Guava is commercially propagated by
A] Stooling   B] Inarching
C] Air layering   D] None

14. Rose is commercially propagated by
A] T budding   B] seed
C] grafting   D] layering

15. Fruit splitting is a common disorder of-
A] Mango   B] Pear
C] Apple   D] Sweet orange

16. Incompatibility is common in family
A] Malvaceae   B] Rosaceae
C] Liliaceae   D] Poaceae

17. Iron deficiency is more common in-
A] Flower crops   B] Vegetables crops
C] Fruit crops   D] Pulses

18. Disbudding is common practice for obtaining better bloom in
A] Rose   B] Chsysanthemum
C] Antirrhinum   D] Jasmine

19. If chromosome compliment of 2 diploid spp. is combined in one the resultant spp. would be called-
A] Monogenic   B] Polygenic
C] Amphidiploid   D] Haploid

20. To determine the salt concentration in a brine solution useful instrument is-
A] Brix hydrometer   B] Jelmeter
C] Thermostat   D] Salometer

21. Which is considered as 'neutral colour'-
    A] Red　　　　　　　B] White
    C] Yellow　　　　　　D] Blue

22. Which Is considered as a cool/soft colour-
    A] Red　　　　　　　B] Yellow
    C] Orange　　　　　　D] Blue

23. Which is considered as a warm colour-
    A] Violet　　　　　　B] Green
    C] Blue　　　　　　　D] Orange

24. Maximum pesticide consumption in India is in which crop
    A] Barley　　　　　　B] Maize
    C] Cotton　　　　　　D] Sugarcane

25. Jam should contain atleast % of sugar
    A] 60%　　　　　　　B] 70%
    C] 50%　　　　　　　D] 80%

26. Cashew nut contains following % of protein
    A] 21%　　　　　　　B] 22.7%
    C] 27%　　　　　　　D] 47%

27. Which fertilizer contains maximum % of nitrogen
    A] Urea　　　　　　　B] Ammonium sulphate
    C] Ammonium chloride　D] CAN

28. Maximum oil contents are in
    A] Oil palm　　　　　B] Soyabean
    C] Coconut　　　　　D] None

29. Which is useful to control broad leaf weeds-
    A] Thiodone　　　　　B] 2, 4-D
    C] Agroson　　　　　D] Pyrethrum

30. For the control of gram pod borer, which chemical is used
    A] Aldrin　　　　　　B] BHC
    C] Malathion　　　　D] Seed treatment

31. Pre- cooling is related to
    A] Fruit preservation　B] Tissue culture
    C] Cut flower　　　　D] Flower arrangement

32. Which is copper fungicide-
    A] Captan　　　　　　B] Dithane M 45
    C] Blitox　　　　　　D] Streptomycin

Sanctuary
An area constituted by competent authority in which killing and capturing of any form of wildlife is prohibited and the boundaries and character of which are sacrosanct.

Complete flower
One that has corolla, calyx, stamens and one or more pistils.

Neutrotoxin
An endotoxin produced by *Clostridium botulinium*, which affects the peripheral nervous system by adversely disturbing the production of acetylcholin necessary for conduction of nurve impulses.

Proteins
A class of foods composed of carbon, hydrogen, oxygen and nitrogen and often sulphur and phosphorus, made of repeating amino acid units linked together by peptide bonds.

Pedate
A palmately divided or compound leaf whose two lateral lobes are again cleft or divided.

Pinna
A leaflet of a compound leaf; when applied to ferns, the primary division attached to the main rachis; feather-like.

Depressed
Flattened, as if compressed somewhat.

33. Which is/are correctly matched-
    A] Pusa Sawani-Okra   B] Pusa Jyoti-Watermelon
    C] Arka Hans-Grape   D] All the above

34. Which is not matched correctly-
    A] Solanine-Potato   B] Black pepper-Spice
    C] Scurf disease-Acidic   D] ICMR, New Delhi
    soil

35. In temperate countries sugar is obtained from which of following
    A] Sugarcane   B] Maple
    C] Wheat   D] Sugar beet

36. 'Creeper' is a dwarf variety of-
    A] Strawberry   B] Pear
    C] Guava   D] Mango

37. Which vegetable crop are rich source of niacin
    A] Spinach   B] Cowpea
    C] Radish   D] Sweet potato

38. Which fruit crop can be grown in high pH soil
    A] Banana   B] Amla
    C] Papaya   D] Banana

39. Which fruit crop can be grown under extreme draught condition
    A] Date palm   B] Banana
    C] Mango   D] Guava

40. Which fruit crop does not belong to Anacardiaceae family
    A] Mango   B] Cashew
    C] Pistachonut   D] Pecanut

41. The most popular cash crop in Kerala is-
    A] Rubber   B] Tea
    C] Tapioca   D] Strawberry

42. Which vegetable crop is day neutral
    A] Tomato   B] Potato
    C] Sweet potato   D] Spinach

43. Which vegetable crop is propagated by stem cutting
    A] Raddish   B] Carrot
    C] Cucumber   D] Sweet potato

44. Which fruit crop is richest source of Vitamin C
   A] Guava
   B] Banana
   C] Citrus
   D] Aonla

45. Cental Tuber Crop Research Institute (CTCRI) was established during-
   A] 1975
   B] 1987
   C] 1956
   D] 1963

46. In which crop the adventitious root is present
   A] Sweet potato
   B] Turnip
   C] Raddish
   D] Carrot

47. In which crop tissue culture is commer cially used
   A] Rose
   B] Carnation
   C] Orchid
   D] Jasmine

48. In which crop, flower borne terminally
   A] Mango
   B] Jackfruit
   C] Coffee
   D] Cocoa

49. In citrus crop, granulation can be reduced by spray of-
   A] Copper Sulphate
   B] Urea
   C] Lead Arsenate
   D] Copper Oxide

50. Kinnow is cross between
   A] Willoleaf x King
   B] King x Willowleaf
   C] Citrus reticulata
   D] *Citrus sinensis*

51. Genus of croton is
   A] Crossandra
   B] *Erythrina*
   C] Cosdium
   D] None

52. Anther mutant cultivar of tomato is
   A] Pusa Divya
   B] Pusa Gaurav
   C] Pusa Ruby
   D] None

53. Furete is cultivar of
   A] Avocado
   B] Persimmon
   C] Loquat
   D] Phalsa

54. Rabi pulse cultivated on large scale is
   A] Chick pea
   B] Pigeon pea
   C] Cow pea
   D] None

Bark-bound
A condition in which trees or shrubs, growing in dearth of moisture and nutrient in the soil and with root injury etc. fails to expand normally and almost stop further growth of trunk and branches. In nature, trees and shrubs effect their own cure by splitting their bark longitudinally; the best remedy for this, is to slit the bark right up the affected trunk or branch at the onset of new growth in spring; subsequent watering, manuring and wound dressings are important.

Tracheid
Any of the water - conducting, tube-like cells in the xylem of a plant also act as a support.

Aggregated
Joined together, confluent, as applied to bundle traces.

Cormels
A specialized underground consisting of an enlarged stem axis with distant nodes and internodes and enclosed by dry, scale like leaves.

Polyembryony
Some-seeds contains more than one embryo is known as polyembryony.

Order
A taxonomic group which contains one or more families.

Food
An organic substance which contributes materially to growth and repair of tissues.

55. The commercial cultivation of rubber trees in India was commenced in-
A] 1895   B] 1902
C] 1920   D] 1927

56. 'Green house cultivation' is mostly applied for growing of-
A] Vegetables   B] Cut flowers
C] Vine crops   D] Shade loving trees

57. An intermediate phase 'curdling' is found in-
A] Cabbage   B] Turnip
C] Cauliflower   D] Kohlrabi

58. 'Browning' in cut potato occurs due to-
A] Enzymatic browning   B] Millard reaction
C] Caramelization   D] Heat

59. The most popular 'cut-flower' sold in Indian market is-
A] Dahlia   B] Marigold
C] Croton   D] Poinsettia

60. The main effect of cytokinin in plants is-
A] Abscission   B] Cell enlargement
C] Cell elongation   D] Cell division

61. 'Cytokinin' induces-
A] Rooting   B] Abscission
C] Cell division   D] Ripening

62. 'Cytokinin' is generally synthesized in-
A] Leaves   B] Flower
C] Young fruit   D] Stem

63. First time cytoplasmic genetic male sterility is used in
A] Onion   B] Pea
C] Ground nut   D] Carrot

64. 'Cytoplasmic male sterility' is found in-
A] Carrot   B] Onion
C] Sugarbeet   D] Tomato

65. The residual moisture in dehydrated fruits should be-
A] < 10-20%   B] < 20-25%
C] > 25-28%   D] > 30-40%

66. Inbreeding is deleterious in
    A] Tomato   B] Maize
    C] Onion   D] Cucumber

67. Which region has been developed as 'Mango-belt'-
    A] Bangalore region   B] Lucknow region
    C] Pune region   D] Kutch region

68. The root developed at above sugarcane set from ground is called as
    A] Primary roots   B] Secondary roots
    C] Set roots   D] None

69. An improved guava variety developed by IIHR is-
    A] Arka Puneet   B] Arka Anmol
    C] Arka Mridula   D] Arka Vati

70. ' Development of Floriculture' was recommended by which committee-
    A] K.L. Chadha   B] P.P. Trivedi Committee
       Committee
    C] S.P.S. Raghava   D] M.S. Swaminathan
       Committee   Committee

71. For better growth and development of indoor plants, ideal illumination combination is-
    A] Red & Green light B] Red & Blue light
    C] Blue & Green light D] Red & Yellow light

72. In our diet staple vegetable is
    A] brinjal   B] potato
    C] tomato   D] chilli

73. Select the dioecious cultivar of papaya from the following
    A] Pusa Giant   B] Pusa Delicious
    C] Pusa majesty   D] None

74. Which is dioecious vegetable is propagated by stem cutting
    A] Yams   B] Spinach
    C] Pointed gourd   D] All

75. In a diploid spp. generally following no. of chromosomes are involved in pollen mitosis-
    A] 2n   B] 3n
    C] n   D] 4n

Year-round flowering
Control of day length and temperature to produce flowering plants throughout the year.

Bronzing
A metallic bronze or coppery color, especially of foliage after a winter.

Lenticel
a dot or corky spot on stems or branches, usually of different color than the stem

Shuck
1. The outer covering of certain nuts or a pod. 2. The calyx of the flower of stone fruit (as in peach) which for a time encloses the fruit but gradually dries, splits, and is pushed off by the growing fruit.

76. First Deputy Director General (Horticulture) in ICAR was-
   A] G.S. Randhawa   B] K.L. Chadha
   C] B.P. Pal   D] G.L. Kaul

77. 'Leaf curl' disease in chilli is transmitted by-
   A] Aphid   B] Whitefly
   C] Thrips   D] Housefly

78. 'Leaf curl' disease in tomato occurs severily during-
   A] Autumn season   B] Summer season
   C] Winter season   D] Rainy season

79. The viral disease of banana is
   A] Bunchy top   B] Panama wilt
   C] Sigatoka leaf spot   D] All

80. Seed borne disease of cauliflower is managed by
   A] Formaldehyde   B] Hot water treatment
   C] Methyl bromide   D] Soil sterilization

81. Most important disease of gladiolus is caused by
   A] Fusarium   B] Verticillium
   C] Virus   D] None

82. Most serious disease of grape is
   A] Mildew   B] Wilt
   C] Anthracnose   D] Leaf spot

83. Most serious disease of vegetable seedlings
   A] Root rot   B] Damping off
   C] Wilt   D] None

84. Appropriate method for transfering disease resistance from donor to an estalished variety is-
   A] Pure line method   B] Pedigree method
   C] Back cross method   D] Bulk method

85. Apical bud distorted, leaves small, dark green are symptom of deficiency of which nutrient
   A] N   B] P
   C] K   D] Ca

86. In which distribution mean, median and mode are same
   A] Normal   B] Poisson
   C] Binomial   D] None

87. 'Allyl propyle di-sulphide' is a volatile oil found in-
    A] Ginger   B] Onion
    C] Garlic   D] All of these

88. Best combination of ' diversified horticulture' is-
    A] Floriculture + Food   B] Vegetable growing + Agro
       processing   foresty
    C] Floriculture +   D] Fruit growing + Dairying
       Apiculture

89. Where does DNA synthesis occur
    A] Interphase   B] Prophase
    C] S phase   D] G_1 phase

90. Which colour dominates in winter in a garden-
    A] Yellow   B] Wihte
    C] Red   D] Blue

91. Cryopreservation is done by liquid nitrogen at temperature of-
    A] 200°C   B] 0°C
    C] -196°C   D] 4°C

92. Waxing is done to reduce
    A] Transpiration   B] Respiration
    C] Both   D] None

93. Randomization is done to remove-
    A] Bios   B] Degree of freedom
    C] LSD   D] Probability

94. Wintering is done to
    A] Expose plants to   B] It is alternative to root
       cold temperature   pruning in rose
    C] It protects plant   D] All
       against cold

Matching

95. Match the following :-
    A] Cytoplasmic male   i]   French bean, Cowpea
       sterility
    B] Staminal male sterility ii] Cauliflower, Knolkhol
    C] Sporophytic self-   iii] Sweet potato, Carambola
       incompatibility

Thermocouple
An instrument constructed by connecting two unlike metal wires (usually copper-constantan or nichrome and constantan wires) generally used to determine adequate processing for canned foods i.e., estimation of cold point in processing industry.

Conical
Cone shaped, as he young form of many spruces.

Culm
Stem of grasses and sedges.

Response group
Classification of cultivars in to various categories based on response to environment as in chrysanthemum.

Florist
A person who sells cut flowers and plants, often growing them himself.

Waxy
having a thick cuticle

Stele
A bulky strand or cylinder of vascular tissue contained in stem and root of plants, developed from plerome, the centre cylinder of the stem and roots in vascular plants, that part of stem which includes the vascular system in woody plants.

Critical day length
The day length above, which a plant will flower, depending on whether the plant is a short -or-long- day plant.

D] Gametophytic self-incompitibility   iv] Corn

E] Protoandry   v] Brinjal, Lathyrum

F] Protogyny   vi] Pointed gourd, Kakrol

G] Tristyly   vii] Gynoecious cucumber

H] Distyly   viii] Strawberry, Marigold

I] Dioecious cucurbits   ix] Radish, Turnip

J] Gynomonoecious   x] Carrot, Onion

96. Match the following :-

A] Cytoplasmic male sterility   i] Onion, Sugarbeet

B] Genetic male sterility   ii] Formaldehyde

C] Cytogenetic male sterility   iii] Carrot, Capsicum

D] Self-sterility   iv] Maleic hydrazide

E] Gametocide   v] Captan

F] Soil sterilant   vi] Radish, Tomato

G] Weedicide   vii] Filbert

H] Fruit ripening   viii] Sodium benzoate

I] Preservative   ix] 2,4-D

J] Fungicide   x] Ethylene

97. Match the following :-

A] D.G. (I C A R)   i] Dr. G. Kalloo

B] D. D. G. Horti. (I C A R)   ii] Dr. Mathura Rai

C] Director (C I S H)   iii] Dr. B. S. Dhillon

D] Director (IIVR)   iv] Dr. Mangala Rai

E] Director (NBPGR)   v] Dr. R. K. Pathak

98. Match the following :-

A] Date palm   i] Cotyledons

B] Rambutan   ii] Pericarp

C] Cashewnut   iii] Fruit seeds

E] Apple   iv] Fleshy aril

E] Cardamom   v] Fleshy thalamus

99. Match the following :-
A] Date   i] Pod
B] Coffee   ii] Pome
C] Tea   iii] One seeded berry
D] Pear   iv] Lomentum
E] Imli   v] Drupe

100. Match the following :-
A] Decline   i] Coconut
B] Sigatoka   ii] Citrus
C] Root wilt   iii] Mango
D] Yellow leaf disease   iv] Banana
E] Malformation   v] Arecanut

101. Match the following :-
A] Dioecious grape sp.   i] *Vitis champin, V. idaoniana*
B] Phylloxera resistant   ii] *Euvitis x Euvitis grapes*
   sp.
C] Nematode resistant   iii] *Vitis rotundifolia*
   sp.
D] Fertile progeny   iv] *Euvitis x Muscadinia grapes*
E] Sterile progeny   v] *Vitis rupestris, V. riparia*

102. Match the following :-
A] Dioecious   i] *Fragaria anallassa*
                               (Strawberry)
B] Androdioecious   ii] *Nephelium lappaceum*
                               (Rambutan)
C] Monoecious   iii] *Cucumis melo* (Musk melon)
D] Andromonoecious   iv] *Phoenix dactylifera* (Date
                               palm)
E] Gynomonoecious   v] *Cocos nucifera* (Coconut)

103. Match the following :-
A] *Duranta plumeri*   i] Hedge plant
B] *Clerodendron inerme*   ii] Flowering tree
C] *Ravenella*   iii] Flowering bush
   medagascarensis
D] *Poinciana regia*   iv] Edging plant
E] *Cassia alata*   v] Specimen plant

Growing season
The number of days between the average date of the last killing frost in spring and the first killing frost in fall. Vegetables and certain plants require a minimum number of days to reach maturity, so be sure *your* growing season is long enough.

Arborescent
Tree-like; defined arbitrarily as pertaining to a woody plant at least 20 ft. high as maturity with a single stem and more or less definite crown.

Shrub
Shrub may be defined as a perennial pant having many woody branches arising from the base of the plant.

104. Match the following :-

    A] Dwarfing R.S. of Mango   o] *Poncirus trifoliata*

    B] Vigorous R.S. of Mango   ii] Oppenheim

    C] Dwarfing R.S. of Citrus   iii] Vellaicollumban

    D] Vigorous R.S. of Citrus   iv] Sweet orange

    E] Dwarfing R.S. of Cherry   v] Chausa

105. Match the following :-

    A] Early Cabbage   i] Jersey, Wakefield

    B] Mid season Cabbage   ii] Improved Japanese, Pusa Synthetic

    C] Late season Cabbage   iii] Tolerant to 'Black-leg'

    D] Pointed Head Cabbage   iv] Snowball-16, Dania

    E] Early Cauliflower   v] Golden Acre, Pride of India

    F] Mid season Cauliflower   vi] Tolerant to 'curd' drying

    G] Late season Cauliflower   vii] Pusa Drum Head, Drum Head Savoy

    H] Punjab Kunwari (Cauliflower)   viii] Pusa Katki, Early Kunwari

    I] Synthetic-III (Cauliflower)   ix] Sensitive to Riceyness

    J] Pusa Drum Head (Cabbage)   x] Badger Shipper, Globe

106. Match the following :-

    A] Early variety of Apple   i] Red Gold, McIntosh, Golden Delicious

    B] Mid season Apple   ii] Rymer, Buckingham, Golden Delicious

    C] Late season Apple   iii] Beauty of Bath, Benoni, Fanny

    D] Culinary Apple   iv] Shamrock, Early Victoria, Grenadier

    E] Pollinizer varieties   v] King of Pippin, McIntosh, Cart Land

107. Match the following :-
    A] EDTA   i] Buffer
    B] Sodium benzoate   ii] Chelating agent
    C] Monosodium   iii] Stabilizer
       phosphate
    D] Pectin   iv] Emulsifier
    E] Citric acid   v] Preservative

108. Match the following :-
    A] Edward Rose   i] *R. chinensis* x *R. gallica*
    B] Hybrid China   ii] Tea Rose x Hb Perpetual
    C] Tea Rose   iii] Damask Rose x Hb China
    D] H. T. Rose   iv] Polyantha x H. T. Rose
    E] Polyantha Rose   v] *R. wordifolia*
    F] Modern Floribunda   vi] Noisette x *R. gigantia*
       Rose
    G] Miniature Rose   vii] Floribunda x H. T. Rose
    H] Garandiflora Rose   viii] *R. chinensis* x *R. multiflora*
    I] Rambler Rose   ix] *R. lawrentiana* x *R. rouletti*
    J] Thornless Rose   x] *R. multiflora* x *R. wituriana*

109. Match the following :-
    A] Export quality cashew   i] Allepey
    B] Arecanut variety   ii] Priyanka
    C] Soft seeded Anar   iii] Caveri
    D] Capsicum variety   iv] Sumrudhi
    E] Cardamom variety   v] Arka Mohini

110. Match the following :-
    A] Fig   i] Scaley bulb
    B] Bryophyllum   ii] Carprification
    C] Garlic   iii] Root cutting
    D] Pomegranate   iv] Leaf bud cutting
    E] Sweet Potato   v] Fruit splitting

111. Match the following :-
    A] First in mango   i] China
       production
    B] First in vegetable   ii] Mexico
       production

Training
Means developing a desired shape of the tree with particular objectives by controlling the habit of growth.

Saccharimeter
Any device used for determining the concentration of sugar in solution, particularly some form of polarimeter.

Phanerogam
A plant whose flowers develop pistils and stamens, a flowering plant.

Fruit-setting
A development of ovary and adjacent tissues following the blossoming period.

Flower differentiation
Complete morphological development of the floral organs following initiation.

Sorosis
A multiple fruit developing from a spike or spadix where the flower fuse together by their succulent sepals and at the same time the axis bearing them grows and becomes fleshy or woody. And as a result the whole inflorescence forms a compact mass e.g., mulberry, pineapple, jackfruit, etc.

Calorimeter
A simple instrument for determining quantities of heat evolved, absorbed or transferred.

Frond
The term used to describe the branch and leaf structure of a fern or members of the palm family.

C] First in fruit production   iii] Kerala
D] Second in mango   iv] India
    production
E] First in spices   v] Brazil
    production

112. Match the following :-
A] Floral bud   i] July to October
    differentiation in litchi
B] Chicken tongued   ii] January-February
    seeds in litchi
C] Flowering in litchi   iii] April-May
D] Fruiting in litchi in   iv] Early Seedless; Late
    S. India   Seedless
E] Air-layering in litchi   v] December-January

113. Match the following :-
A] Flower bud   i] 79ºC
    differentiation
B] Flowering in mango   ii] October-December
C] Viability in mango   iii] 4-5 weeks
D] Flowering duration in   iv] February-March
    mango
E] Storage temp. for mango   v] 2-3 weeks

114. Match the following :-
A] Flower spices   i] Cinnamon, Tejpat
B] Bud spices   ii] Mentha, Coriander
C] Rhizomes   iii] Nutmeg, Clove
D] Bark spices   iv] Garlic, Onion
E] Leafy spices   v] Clove
F] Seed spices   vi] Chilli, Dill
G] Tree spices   vii] Fenugreek, Cumin
H] Vine spices   viii] Ginger, Thumeric
I] Bulbs   ix] Black pepper, Vanilla
J] Fruit spices   x] Saffron

115. Match the following :-
A] Flowering tree   i] *Cocos nucifera*
B] Shade tree   ii] *Ficus benghalensis*
C] Foliage annual   iii] *Raphanus sativus*

D] Plantation crop   iv] *Poinciana regia*

E] Root vegetable   v] *Kochia scoparia*

116. Match the following :-

A] Flowering Trees   i] Dr. B.P. Pal

B] The Rose in India   ii] Dr. L.B. Singh

C] Garden Flowers   iii] Dr. B.S. Chundawat

D] The Mango   iv] Dr. M.S. Randhawa

E] Arid Fruit Culture   v] Dr. Vishnu Swarup

True / False

117. Seedlessness in grape is due to ovule abortion.   T/F

118. 'Tropical Beauty' apple can be grown in subtropical areas.   T/F

119. Butter cup & Apricot are yellow flowered varieties of canna.   T/F

120. Colour of ripe apple is due to lycopene   T/F

121. Custard apple is 'etaerio of berries'.   T/F

122. Ikebana is an art of bonsai making.   T/F

123. Lemon can tolerate frost better than Kagzi lime.   T/F

124. Neelam is a variety of aonla.   T/F

125. Niranjan is an off season cultivar of mango.   T/F

126. Smyrna fig is cross-pollinated crop.   T/F

127. Uttar Pradesh is a major onion producing state.   T/F

128. *Aceria ficus* is vector of fig mosaic   T/F

129. Cabbage head is a modification of flower stalk.   T/F

130. Chlorosis is yellow colour development due to lack of light.   T/F

131. Cocoa is suitable intercrop for arecanut plantations.   T/F

132. Edible part of apple is thalamus.   T/F

133. Generally, colour of candy tuft flower is red.   T/F

134. Gibberellins stimulate cell elongation.   T/F

135. Lemon is mainly propagated by seeds.   T/F

136. Mango malformation is caused by regular bearing.   T/F

Parameters

A numerical quantity which describes a particular characteristics of a population.

Potting soil

A soil mixture designed for use in container gardens and potted plants. Potting mixes should be loose, light, and sterile.

Callus

Wounded tissues which develops from cambium of other exposed meristem.

Runner

A specialised stem which develops from the axil of a leaf at the crown of a plant, grows horizontally along the ground and forms a new plant at one of nodes as seen in spider plants (*Chlorphytum comosum*)

Devernalization
Negation of vernalizing stimulus by temperatures above a critical level.

Off-type
Any notable deviation from the normal or standard in plants.

Allopolyploid
A polyploid containing genetically different sets of chromosomes derived from two or more species.

Gibberellic
acid $C_{19}H_{22}O_6$. Tetracyclic dihydroxy lactonic acid, one of the gibberellins. A plant growth regulator, effects elongation of shoots of even a genetic dwarf plant.

137.	Smyma fig receives pollen grain from Adriatic fig.	T/F
138.	'Niranjan' mango ripens in October.	T/F
139.	Almost all fruit crops are perennial in nature.	T/F
140.	Blastophaga wasp is chief pollinator of Smyrna fig.	T/F
141.	Carrot belongs to family Cruciferae.	T/F
142.	Early December is a variety of pea.	T/F
143.	Guava fruit is botanically berry.	T/F
144.	India is a major producer of spices in the world.	T/F
145.	Jatti Khatti & Karna Khatta are common root stock for lemon.	T/F
146.	Pineapple produces seedless fruits.	T/F
147.	Seed production of cabbage & cauliflower is done in Gangetic Plains.	T/F
148.	'Greening' disease in citrus is caused by virus.	T/F
149.	Aonla contains more Vit-C than mango.	T/F
150.	Bending is practised in erect type variety of guava.	T/F
151.	Chilli/Sweet pepper is main spice exported from India.	T/F
152.	*Clerodendron splendens* is an important hedge plant.	T/F
153.	Colourful parts of bougainvillea flowers are bracts.	T/F
154.	First main crop of fig is produced on current growth.	T/F
155.	Grape fruit and pummelo are same things.	T/F
156.	Phalsa is a self-pollinated fruit crop.	T/F
157.	Rataul is a famous variety of mango.	T/F
158.	Allahabad Safeda is a spreading type guava variety.	T/F
159.	Balausta is a modified form of berry.	T/F
160.	*Citrus limonia* is the botanical name of rough lemon.	T/F
161.	Forking in carrot develops due to application of fresh cow dung.	T/F
162.	Grape fruit bears fruits in bunches.	T/F
163.	Holland & Germany are main importers of Indian cut flowers.	T/F
164.	Litchi fruits are harvested individually.	T/F

165. Sapota tree propagated by layering has shallow roots than its seedling.　　　　　T/F
166. Terminal buds of grape are always vegetative.　　　T/F

Fill in the Blanks

167. Mostly cultivated banana varieties are _____.
168. Neelam x Alphonso are parents of _____ mango.
169. Grape is _____ under sub-tropical conditions.
170. _____ fruits have chilling requirement.
171. _____ thrives well on heavy soils.
172. _____ is tolerant to waterlogging.
173. _____ varieties are usually self-fruitful.
174. Winter Banana is a variety of _____.
175. Bartlett variety of pear is _____.
176. Leaf curl is a serious problem in the cultivation of _____.
177. _____ is a tropical nut crop.
178. Almost all varieties of almond are _____.
179. Pusa Majesty is a variety of _____.
180. Rose Scented is a _____ season variety of litchi.
181. Pineapple has its origin from _____.
182. _____ is a variety of loquat.
183. Grape is propagated from _____ cuttings.
184. _____ is the richest source of vitamin -C.
185. Ambri is a cultivar of _____.
186. _____ is the most suitable crop for sodic soils.
187. _____ is a typical example of dioecism.
188. _____ is the important content in jelly.
189. Arkas Shyam is _____ variety of grape.
190. Cross protection is related to _____.
191. Wooly Aphid is serious pest of _____.
192. Amrapalli is cross between _____.

Androdioecious
A sex form in dioecious species where only staminate flowers are borne on one plant and bisexual flowers on another plant of the same variety or species.

Frenching
A non-parasitical disease of crops, characterized by loss of colour, thickening, narrowing, and curling of leaves between veins. Considered due to some nutritional deficiency.

Exudate
A liquid discharge from diseased tissues of plants.

Bull nose
A physiological disorder of narcissus characterized by failure of the flower bud, often occurring on azaleas.

Soluble fertilizers
Fertilizers that dissolve easily in water and are immediately available for plant use.

Bunchgrass
Grass that grows as distinct plants that get larger over time. Also called *clump-forming grass*. Contrast with *spreading grass* (also called *sod-forming grass*), which expands using running roots that create new plants from existing ones.

Acotyledon
A plant with no distinct seed-lobes.

Tree
Perennial plant having distinct trunk and crown the top.

193. Loquat is a _____ fruit.

194. Papaya is a _____ plant.

195. J.H.Hale is a _____ variety of peach.

196. Jackfruit is a _____ plant .

197. Table banana cultivars are _____.

198. For nursery sowing of onion, seed rate is _____ per ha.

199. Seed rate of early cauliflower is _____.

200. _____ is the optimum time for rose pruning.

An executive called a company to check on a potential candidate. He asked the candidate's supervisor, "How long has he worked for you?" The man replied, "Three days." The executive said, "But he told me he was with you for three years." The man replied, "That is right, but he worked three days."

Excellence is not luck; it is the result of a lot of hard work and practice. Hard work and practice make a person better at whatever he is doing.

Success is the Result of Believing in asking how much Work and not how Little work, how many Hours not how few Hours.

Step-Seven

A Loser makes promises;
A Winner makes commitments.

Multiple Choice

1. ' Double sigmoid' growth curve is found in-
   A] Apple   B] Ber
   C] Pear   D] Fig

2. Which plant drop its flowers during sun-rise (morning)-
   A] Bougainvillia B] Balsam
   C] Harsingar   D] Portulaca

3. 'Heroine' drug is prepared from which medicinal plant
   A] Datura   B] Sarpagandha
   C] Opium   D] Safed musali

4. Central Research Institute for Dryland Agriculture (CRIDA) is situated at-
   A] Jodhpur   B] Bikaner
   C] Nagpur   D] Hyderabad

5. Aroma in fruits is due to-
   A] Acids   B] Hydroxy compounds
   C] Esters   D] Xanthophyll

6. Bitterness in chili is due to-
   A] Capsanthin   B] Capsinoid
   C] Capsaicin   D] Carotenoid

7. Which has the shortest duration to harvest-
   A] Tomato   B] Chilli
   C] Colocasia   D] Potato

8. Crossing over during meiosis results in-
   A] Promoting linkage   B] Breaking linkage
   C] Help in mutation   D] None

9. In aonla flowering occurs during-
   A] Jan-Feb.   B] April-May
   C] July-August   D] Oct-Nov.

10. In apple dwarfing rootstocks are good for -
    A] High density planting   B] Scab resistance
    C] Low chilling   D] Fruit colour

Divaricate
Spreading very wide apart.

Chambered
Said of pith when divided into small, empty compartments separated by transverse partitions. With cavities separated by plates or disks, as applied to pith; discoid.

Septum
A partition or membrane dividing two cavities or masses of tissue as in fruits, vegetables, etc.

Homotypic
As applied to cell division, involving the usual process of karyokinesis.

Perennials
Those plants which do not furnish their life cycle in one or two years, but lives for many years.

Evergreen
Evergreen trees do not have definite resting season and they do not shed their leaves during any particular season. eg. mango, lime, sapota, lemon, orange, etc.

Plasmid
A ring-shaped DNA molecule other than the chromosome of the bacteria (*e.g., Agrobacterium*) capable of independent replication and transmission. It is used as vector in genetic engineering.

Cultivar
A cultivated variety; does not occur naturally.

Punctate
With translucent or covered dots, depressions or pits.

11. 'Karathane' is effective for the control of-
A] Anthracnose   B] Downey Mildew
C] Powdery Mildew   D] Root Knot

12. The most effective gametocide to induce sterility in okra is-
A] MCPA   B] 2 4-D
C] MH   D] $NaNO_3$

13. The most effective method for the transfer of oligogenic-character is
A] Pedigree breeding   B] Bulk breeding
C] Back cross breeding D] None

14. Wind break effectiveness is of .... its height
A] 2 times   B] 4 times
C] 6 times   D] 8 times

15. 'Blossom end rot' of tomato is due to deficiency of
A] Mg   B] Bo
C] Mo   D] Ca

16. Non- endospermic seeds are found in-
A] Apple   B] Guava
C] Pear   D] Ber

17. Inactivation of enzyme is done by
A] Canning   B] Freezing
C] Blanching   D] Dehydration

18. The term epistasis is given by
A] Mendel   B] Halman
C] Bateson   D] Hull

19. For 'Scurvey' disease eradication, which fruit is the most suitable-
A] Guava   B] Aonla
C] Chilli   D] Barbados Cherry

20. The standard error of a sample mean is given by
A] s/n   B] sÖn
C] s^2/n   D] None of above

21. Which is the most essential for litchi production-
A] Frost free climate   B] High humidity
C] High soil moisture   D] All of above

22. Stratification is essentially a-
    A] Chilling treatment　　B] Storing
    C] Heating　　D] Ageing

23. Heterobeltosis is estimated over the
    A] Wild parent　　B] Better parent
    C] Popular variety　　D] Popular hybrid

24. Increase in ethylene production with ripening is characteristic of-
    A] Climacteric fruit　　B] Non climacteric
    C] Citrus　　D] Grape

25. ' Eurovision' is a -
    A] ICAR Agenda 2002　B] Gladiolus variety
    C] Agenda of European　D] None of these
    　　Countries

26. A typical example of 'Formal style' of gardening is-
    A] Rose garden　　B] Roof garden (Agra)
    　　(Chandigarh)
    C] Taj garden (Agra)　　D] Vrindavan garden (Mysore)

27. A typical example of 'tropical' vegetable crop is-
    A] Radish　　B] Lobia
    C] Cluster bean　　D] Dolichos bean

28. Ripining fruits exit which type gas
    A] Ammonia　　B] Ethylene
    C] Methane　　D] All

29. In pepper export market, India shares about-
    A] 25%　　B] 36%
    C] 50%　　D] 60%

30. Which mango variety is exported freshly-
    A] Neelam　　B] Amrapali
    C] Alphonso　　D] Dashehari

31. Fruit crop exported on large scale from India
    A] Grape　　B] Banana
    C] Mango　　D] Citrus

32. The largest exporter of black pepper is-
    A] India　　B] Brazil
    C] Indonesia　　D] Malaysia

Fimbriata
Term used for double tuberous begonia hybrids with frilled tepals.

Bi-
Twice or doubly.

Bhaj
An operation of lowering down the betelvine to the ground level and covering the stem with soil to increase the production of leaves.

Reflexed
Bent abruptly backward or downward.

Dropper
In tulips, a bulb formed at the distal end of a hollow stolon.

Scale
A small and usually dry bract or vestigial leaf or a structure resembling such.

Flipper
A mild positive pressure due to hydrogen swelling inside the can resulting swelling of the lid which can be brought down to its original position with finger pressure and lid remains in that place even after removal of finger pressure.

Oleocellosis
A disorder of citrus due to rough handling or harvesting where turgid fruits show necrosis of epdermis adjacent to oil glands.

Cyclic lighting
Intermittent illumination during the dark period, to simulate a long day and prevent flowering of short-day plants.

33. A chemical extracted from datura which withdraws the reaction of morphine
    A] Ajmachine   B] Hyocasmine
    C] Reserpine   D] Nicotine

34. 'Curcumin' is extracted from-
    A] Cucumber   B] Capsicum
    C] Turmeric   D] Cashew apple

35. 'Papain' is extracted from-
    A] Sapota   B] Papaya
    C] Pear   D] Persimmon

36. In bulk method, F_2 and subsequent generations are harvested-
    A] Individually   B] In mass
    C] Any of them   D] Randomly

37. Temperature is a limiting factor for the cultivation of-
    A] Banana   B] Pineapple
    C] Peach   D] Mango

38. Sweet pea falls under category of-
    A] Winter annual   B] Summer annual
    C] Hedge   D] Creeper

39. Loquat belongs to the family-
    A] Sapindaceae   B] Sapotaceae
    C] Anacardiaceae   D] Rosaceae

40. If a farmer is able to give only one irrigation to wheat crop, he should do it at
    A] Maximum tillering   B] CRI
    C] Dough stage   D] All

41. The richest source of fat is-
    A] Banana   B] Cashewnut
    C] Coconut   D] Almond

42. Grasses heavy feed upon
    A] N   B] P
    C] K   D] Ca

43. 'Femaleness' in plants can be induced by-
    A] GA_3 B] IBA
    C] I A A   D] ABA

44. The invert sugar in finished jam should be-
   A] 30-40%   B] 45-55%
   C] 65-68%   D] 10-20%

45. Thailand rank first in production of the fruit
   A] Banana   B] Mango
   C] Pineapple   D] Orange

46. Who noticed first time the 'internal fruit necrosis' in mango-
   A] R.N. Singh (1905) B] Sant Ram (1972)
   C] M.S. Randhawa   D] None of these
      (1925)

47. In which crops, flower buds borne terminally-
   A] Mango & Apple   B] Fig & Grape
   C] Papaya & Banana D] All the above

48. The precocious development of flower buds in the curd of cauliflower is known as-
   A] Ricyess   B] Blindness
   C] Fuzziness   D] Whiptail

49. For better flowering and fruiting in fruit trees, C: N ratio should be-
   A] 2:3   B] 1:1
   C] 1:2   D] 3:2

50. According the flowering habit data, palm is
   A] Monoecious   B] Dioecious
   C] hermaphrodite   D] Andromonoecious

51. To increase flowering in grapes which practice is followed
   A] Training   B] Pruning
   C] IAA   D] Interculture operation

52. Apple crop flowers during-
   A] Dec.-Jan.   B] Sept.-Oct.
   C] June-July   D] March-April

53. Colour of *Ervatomia divericata* flowers is-
   A] Red   B] Yellow
   C] White   D] Blue

54. During transportation 'cut -flowers' affected by-
   A] Chemotropism   B] Phototropism
   C] Geotropism   D] Hydrotropism

Confluent
Blending together, not easily distinguishable apart, as applied to bundle-traces.

Merogony
An individual with the egg cytoplasm from one parent and the egg nucleus from the other parent.

Amateur
A person who either personally or with assistance maintains gardens or grows plant, flower for pleasure and enjoyment and not his livelihood or for financial gains.

Pasterrization
Process used to eliminate harmful pathogens. Temperatue usually does not exceed 82°C.

Broad-elliptic
Wider than elliptic.

Apex
The tip or terminal end.

Springer
A can with both ends bulged but one or both ends will stay concave if pushed in or if a swollen end is pushed in the opposite end will pop out. This bulging is caused by the presence of hydrogen gas inside the can, poor exhausting, over-filling, denting of the can, changes in temperature, etc. Food in such cans generally remains fit for consumption.

Bromelin
A protein digestive enzyme present in the mature pineapple fruit.

55. In loquat, which flush sets maximum fruits-
A] August-September   B] October-November
C] December-January   D] Mrig bahar

56. 'Anar butter fly' can be controlled by-
A] Light traps   B] Fencing
C] Waxing   D] Bagging

57. Which of follgwing fruit type is selected for making jelly
A] Ripe   B] Over ripe
C] Unripe   D] Firm ripe

58. Potato contains following amount of vitamin-C (mg/100g of edible portion)
A] 16   B] 17
C] 100   D] 25

59. Which of following bear white colour flower
A] Bottle gourd   B] Bitter gourd
C] Cucumber   D] All

60. Which of following crop produce maximum quantity of oil
A] Til   B] Mustard
C] Groundnut   D] Soybean

61. Which of following fruit contain maximum fat
A] Avacado   B] Walnut
C] Apricot   D] Pecanut

62. Which of following hormone is used as a herbicide
A] 2, 4-D   B] NAA
C] ABA   D] IAA

63. Which of following hormone is used in cucurbits for inducing flowering
A] IBA   B] NAA
C] Gibbrilin   D] Auxin

64. Which of following is associated with vegetable research
A] Swaminathan   B] G. Kallo
C] T K Bose   D] Randhawa

65. What among following is called as y linked character
A] Holoandric   B] Sex linked
C] Both   D] None

66. Which of following is fire hazardous fertilizer
    A] Ammonium sulphate   B] Ammonium nitrate
    C] Calcium ammonium   D] Urea
       nitrate

67. Which of following is hybrid var of black pepper
    A] Penyur-3   B] Penyur-2
    C] Penyur-4   D] None

68. Which of following is monoembryonic spp
    A] Sweet orange   B] Grape fruit
    C] Pummelo   D] None

69. Which of following is resistant to 'Mn' deficiency-
    A] Strawberry   B] Pear
    C] Apple   D] All of above

70. Which of following is responsible for cracking in tomato
    A] B deficiency   B] Ca deficiency
    C] S deficiency   D] Zn deficiency

71. Which of following is responsible for sourness of radish
    A] Isothiosinate   B] Anthocynins
    C] Auxins   D] All

72. Which of following is responsible for spot disease in
    chrysanthemum
    A] Fungus   B] Bacteria
    C] Virus   D] Mycoplasma

73. Which of following is richest soil
    A] Black   B] Alluvial
    C] Laterite   D] None

74. Which of following is seedless variety of ango
    A] Langra   B] Alphanso
    C] Amrapalli   D] Sindhu

75. Which of following is sympodial orchid
    A] Venda   B] Archis
    C] Dendrobium   D] None

76. Which of following is the canning variety of pineapple
    A] Kew   B] Mauritius
    C] Jaldhup   D] All

Cuticle
The outermost layer of epidermal cell walls.

Trichome
An outgrowth of the epidermis which plays an important role in the water economy of the plant as seen in lower epidermis of pineapple leaf.

Bar
A unit of pressure equal to one million dynes per square centimeter.

Photodormancy
A type of physiological dormancy of seed where germination of seed is sensitive to light *i.e.*, seeds of some plants require light to germinate whereas others require darkness for germination. *e.g.*, lettuce seed require light and *Nigella, Allium, Amaranthus*, etc.

Horizontal resistance
Partial resistance equally effective against all races of a pathogen.

Fastigiate
A botanical term meaning erect and upright in habit chiefly applied to trees and shrubs; these trees and shrubs in their normal forms have spreading branches but in some particular varieties have fastigiate branches.

Ornamental Garden
An area established with valuable and pleasurable plants adjacent to a house or other building.

Barbed
Bristles, awns, etc. provided with terminal or lateral spine-like hooks that are bent sharply backward.

77. Which of following is the major flower grown in India
    A] Orchid   B] Chrysanthemum
    C] Jasmine   D] Rose

78. Which of following is used as a cut flower
    A] Marigold   B] Jasmine
    C] Gladiolus   D] All

79. Which of following is used as a surface sterilent
    A] $HgCl_2$ B] KNO_3
    C] KCl   D] none

80. Which of following is used as antioxidant-
    A] Fat   B] Protein
    C] Carbohydrates   D] Vitamin C

81. DNA contain following no. of nitrogenous base repeated in various sequences-
    A] 10   B] 2
    C] 4   D] I

82. Which of following pigment is responsible for red colour in rose
    A] Globuline   B] Anthocynin
    C] Xanthophylls   D] All

83. Which of following require regular pruning for fruiting
    A] Mango   B] Peach
    C] Pomegranate   D] None

84. Which of following river has maximum area under irrigation
    A] Ganga   B] Kaveri
    C] Brahmaputra   D] Narbada

85. Which of following variety of aster is developed by IIHR
    A] Red gold   B] Phule ganesh
    C] Poonam   D] All

86. Sulphur containing food is stored in
    A] Arcans   B] SR cans
    C] Both   D] None

87. The most suitable fruits for 'cordial' preparation are-
    A] Bael & Grape   B] Tomato & Mandarin
    C] Lime & Lemon   D] Grape fruit & Orange

88. Suitable variety of mango for 'kitchen garden' is-
    A] Mallika
    B] Sindhu
    C] Dashehari
    D] Amrapali

89. Chilling requirement for apple is-
    A] 1000-1600 hrs at 7.2°C
    B] 1000-1200 hrs at 9.3°C
    C] 600-800 hrs at 10°C
    D] 800-1100 hrs at 2°C

90. Best time for application of inorganic manure (Fertilizers) in most of the fruit trees is
    A] Feb-March
    B] May-June
    C] Aug-Sept
    D] Nov-Dec

91. Plants used for avenue are generally
    A] Shrubs
    B] Climbers
    C] Trees
    D] Annuals

92. Which is most suitable for avoiding 'scurvey disease'-
    A] Lemon
    B] Guava
    C] Aonla
    D] Apple

93. Standard spacing for ber planting is-
    A] 10x10m
    B] 12x12m
    C] 6x6m
    D] 8x8m

94. Suitable time for ber propagation is-
    A] June-Sept.
    B] Oct-Nov.
    C] Jan.-Feb.
    D] April-May

95. Match the following :-
    A] Formaldehyde   i] Disinfectant
    B] T.T.C.   ii] Fumigant
    C] Calcium hypochlorite iii] Fruit drop
    D] 2,4-D   iv] Dormancybreaking
    E] Calcium carbide   v] Viability test
    F] Captan   vi] Tissue culture
    G] Thiourea   vii] Fruit ripening
    H] H.Q.C.-8   viii] Preservative
    I] Kinetin   ix] Vase-life
    J] K.M.S.   x] Seed treatment

Euploid
An organisms having an exact multiple of the haploid (x) chromosome number is so euploid series are haploid (x), diploid (2x), triploid (3x) etc.

Exfoliate
peeling bark from a branch or trunk

Vestige
The remains of an exhausted or dead plant structure.

Stratify
Arrange in strata; expose seed to stratification.

Bark
1. The tissues external to the vascular cambium, collectively phloem, cortex and periderm. 2. A protective layer of dead corky cells on the outside of stems of woody plants.

96. Match the following :-

A]	*Fragaria viridis*	i]	Tetraploid strawberry(4x=28)
B]	*Fragaria orientalis*	ii]	Octaploid strawberry (8x=56)
C]	*F. moschata*	iii]	Diploid strawberry(2x=14)
D]	*F. chiloensis*	iv]	Interspecific hybrid
E]	*F. ananassa*	v]	Hexaploid strawberry

97. Match the following :-

A]	Fruit fly (*Bemisia tabaci*)	i]	Vector of 'Little leaf'
B]	Leafhopper (*Amrasca* spp.)	ii]	Tomato
C]	Thrips	iii]	Brinjal
D]	Buck eye rot	iv]	Vector of 'Leaf curl' virus
E]	Lace wing bug	v]	Vector of 'Spotted wilt' virus

98. Match the following :-

A]	Fruit necrosis	i]	Mango
B]	Hen & Chicken	ii]	Avocado
C]	Fruit cracking	iii]	Grape
D]	Vegetative malformation	iv]	Apple
E]	Spongy tissue	v]	Pineapple
F]	Decline	vi]	Pomegranate
G]	Sun scald	vii]	Citrus
H]	Pithiness	viii]	Aonla
I]	Bitter pit	ix]	Radish
J]	Dry neck	x]	Alphonso

99. Match the following :-

A]	Fruit plant for hedge	i]	Kochia
B]	Foliage annual for summer	ii]	Bougainvillia
C]	Foliage annual for winter	iii]	Coleus

D] Evergreen hedge plant   iv] Karonda
E] Flowering hedge plant   v] Clerodendron

100. Match the following :-
A] Fruits   i] Vinegar bacteria
B] Vegetables   ii] Sugar converted into alcohol
C] Acetobacter   iii] Acidic nature
D] Lactobacillus   iv] Lactic acid bacteria
E] Yeasts   v] Alkaline nature

101. Match the following :-
A] Ganesh   i] Rose
B] Arka Amulya   ii] Aonla
C] Arka Anmol   iii] Chrysanthemum
D] Neelam   iv] Lime
E] Super Star   v] Pomegranate
F] Chandrama   vi] Guava
G] N-53   vii] Gladiolus
H] Pusa Komal   viii] Mango
I] Eurovision   ix] Onion
J] Vikram   x] Cowpea

102. Match the following :-
A] Gaseous hormone   i] N.A.A.
B] Acetylene   ii] Banana
C] Bahar treatment   iii] Paclobutrazol
D] Floral malformation   iv] Calcium carbide
E] Vellaicolumban   v] Apple root stock
F] M-27   vi] GA_3
G] Sigatoka leaf spot   vii] Pomegranate
H] Quality improvement in grape viii] H.Q.C.-8
I] For regular bearing in mango   ix] Ethylene
J] Enhancing vase-life   x] Mango root stock

103. Match the following :-
A] Golden Shower   i] *Hiptage madablota*
B] Silver Lace Vine   ii] *Bignonia venusta*
C] Calico Flower   iii] *Porana paniculata*
D] Madhavi Lata   iv] *Polygonum auberti*
E] Bridal Bouquet   v] *Aristolochia elegans*

Turgid
Swollen to firmness.

Conelet
A young, immature first season cone, in the pines.

Corymb
A more or less flat-topped indeterminate inflorescence whose outer flowers open first. Ex. Viburnum

Flat
A shallow box or tray used to start cuttings or seedlings.

Pseudobulb
An elongated above ground fleshy plant stem in which food and moisture are stored *e.g.*, orchid species, etc.

Succulent
Thickened, juicy, fleshy tissues that are more or less soft in texture.

Shattering
Abscission of snapdragon florets, caused by ethylene or pollination.

Flower blasting
Phase of flower bud abortion occurring after flower differentiation is completed when blasting has occurred visible signs of the floral organs are evident.

Neutral flower
A sterile flower consisting of perianth without any essential organs.

Turkscap
A type of lily flower in which the perianth segments roll back in a turban-like manner.

Receptacle
Portion of the flower stalk or axis that bears the floral organs.

104. Match the following :-

A] Good female parent cvs.	i]	Beauty Seedless, Perlette, Pusa Seedless
B] Good male parents A	ii]	*Vitis vinifera*
C] Male sterile hybrid	iii]	'Hurt, Angoor Kalan, Banqui Abyad
D] Hyperploidy	iv]	*Vitis assami*
E] Poly/Tetraploidy	v]	Arka Trishna

105. Match the following :-

A] Grape pruning in N. India	i]	Oct.-Nov. (Maharashtra), Nov-Dec. (T.N.)
B] Grape pruning in S. India	ii]	Twice a year (Summer & Winter)
C] Foundation/Back/ Growth pruning	iii]	Once a year (Dec.-Jan.)
D] Fruiting/Winter pruning	iv]	Feb.-March (N. India), Oct.-Nov. (S. India)
E] Floral bud burst in Grape	v]	March-April (M.S., A.P.), June (T.N.)

106. Match the following :-

A] Greensil, Bruno, Monty	i]	Protandrous var. of Pecan
B] Matua, Tomuri	ii]	Hazelnut cultivars
C] Barcelona, Non-Pareil	iii]	Staminate var. of Kiwi fruit
D] Western, Desirable	iv]	Protogynous Pecan
E] Moreland, Stuart	v]	Pistillate cvs. of Kiwi fruit

107. Match the following :-

A] Guava	i]	Pericarp & Placentae
B] Grape	ii]	Mesocarp
C] Jack fruit	iii]	Thalamus & Pericarp
D] Ber	iv]	Meso & Epicarp
E] Papaya	v]	Bracts, Perianth & Seed

108. Match the following :-

| A] Guava, Crab Apple | i] | Low pectin & high acid |
| B] Grape fruit, Sour cherry | ii] | Low pectin & low acid |

C] Pineapple, Strawberry iii] High pectin & low acid

D] Pomegranate, Ripe   iv] High pectin & low T.S.S.
    peach

E] Premature setting of   v] High pectin & high acid
    jam

109. Match the following :-

A] Gumming   i] Citrus
B] Fruit cracking   ii] Mango
C] Leaf scorching   iii] Persimon
D] Blossom end rot   iv] Peach
E] Core breakdown   v] Apple
F] Exanthema   vi] Pear
G] Jonathan spot   vii] Guava
H] Calyx cavity   viii] Litchi
I] Fasciation   ix] Pineapple
J] Leaf bronzing   x] Grape

110. Match the following :-

A] H. T. rose varieties   i] Chattilon, Eco, Ideal
B] Grandiflora roses   ii] Lamarque, Morechel Neil, Cocktail
C] Polyantha roses   iii] Mutant of 'Kiss of Fire'
D] Miniature rose   iv] Suspense, Bazzazo, West Minister
E] Climbing rose   v] C!nderella, Twinkle, Anny
F] Floribunda varities   vi] Pink Perfait, Sea Pearl, Queen Alizabeth
G] Abhisarika   vii] Mutant of "Gulzar"
H] Pusa Christiana   viii] Lahar, Suryodaya, Banjaran
I] Madhosh   ix] Happiness, Super Star, Avon
J] Bicolour rose varieties x] Mutant of 'Christian Dior'

111. Match the following :-

A] Hand peeling   i] Potato & Peach
B] Steam peeling   ii] Carrot & Turnip
C] Mechanical peeling iii] Sweet potato & Mandarin
D] Flame peeling   iv] Mango & Papaya
E] Lye peeling   v] Onion & Garlic

Inflorescence
a grouping of flowers attached to a specialized stem (peduncle)

Dioecious
Male and female flowers on different plants.

Latex
Milky juice/ fluid found in stems, foliage and bracts or poinsettia.

Fluted
Grooved.

Chlorosis
The yellowing of foliage due to loss or breakdown of chlorophyll.

Fertilizer
Organic or inorganic plant foods which may be either liquid or granular used to amend the soil in order to improve the quality or quantity of plant growth.

112. Match the following :-

A]	Hard wood cutting	i]	Coleus, Eupatorium
B]	Herbaceous cutting	ii]	Begonia, Crassula
C]	Root cutting	iii]	Guava, Bael
D]	Leaf cutting	iv]	Aonla, Rose
E]	Apomixis	v]	Pomegranate, Rose
F]	Shield budding	vi]	Jack fruit, Karonda
G]	Ring budding	vii]	Black berry, Raspberry
H]	Air layering	viii]	Jamun, Citrus
I]	Seed	ix]	Ber, Peach
J]	Tip layering	x]	Litchi, Croton

113. Match the following :-

A]	Head to seed method	i]	Emergence of seed stalk without head formation
B]	Seed to seed method	ii]	Edible portion of cabbage
C]	Premature bolting (cabbage)	iii]	Kashmir valley
D]	Self-incompatibility in cole-crops	iv]	Edible part of cauliflower
E]	Cabbage seed production	v]	September-October
F]	Kohlrabi seed sowing in plains	vi]	Breeder seed production in cabbage
G]	Rossette of leaves	vii]	Edible part of Knol-khol
H]	Tuberous stem	viii]	Certified seed production in cabbage
I]	Fleshy apical meristem (curd)	ix]	Edible portion of kale
J]	Compact leafy 'head'	x]	'Sporophytic' type

114. Match the following :-

A]	Hen and Chicken	i]	Banana
B]	Powdery	ii]	Potato
C]	Panama wilt	iii]	Brinjal
D]	Fruit necrosis	iv]	Chilli
E]	Black heart	v]	Grape
F]	Fruit borer	vi]	Citrus
G]	Black tip	vii]	Aonla

H] Wooly aphid   viii] Apple
I] Greening   ix] Mango
J] Anthracnose   x] Pea

115. Match the following :-
   A] Herbaceous fruit crop   i] Jack fruit
   B] Temperate fruit crop   ii] Aonla
   C] Tropical fruit crop   iii] Banana
   D] Sub-tropical fruit crop   iv] Asparagus
   E] Bromeliaceae   v] Hazel nut
   F] Vitaceae   vi] Mango
   G] Euphorbiaceae   vii] Phalsa
   H] Bearing on old growth   viii] Grape
   I] High respiration   ix] Braccoli
   J] Edible flower   x] Pineapple

116. Match the following :-
   A] High salt tolerant   i] Broccoli, Pea
   B] Sensitive to salinity   ii] Cucumber, Tomato
   C] High acid tolerant   iii] Loquat, Phalsa
   D] High respiration   iv] Walnut, Cashew nut
   E] Warm forcing   v] Date palm, Aonla
   F] High ethylene   vi] Tamarind, Plum
      evolution
   G] Bearing on new growth   vii] Watermelon, Potato
   H] Bearing on old growth   viii] Pomegranate, Litchi
   I] Edible cotyledons   ix] Papaya, Sapota
   J] Aril   x] Peach, Avocado

True / False

117. Whiptail in cauliflower occurs due to 'B' deficiency.   T/F
118. Banana is grown in high humidity.   T/F
119. Bhim and Ganga H. T. rose are released from N.B.R.I.   T/F
120. *Citrus megaloxycarpon* is the acidiest citrus species. T/F
121. India ranks first in productivity of fruits.   T/F
122. Mature shoots of grape are called as Spurs.   T/F
123. *Opuntia* is suitable for rock garden.   T/F

Bin
An enclosed structure used for storage of seeds.

Capsanthin
A carotenoid pigment responsible for the characteristic orange-red colouration of ripe chilli.

Forest
An area set aside for the production of timber and other forest produce or maintained under woody vegetation for certain indirect benfits.

Mower
A machine mainly used for trimming grasses in the lawn.

Afforestation
Artificial establishment of a forest on an area from which forest vegetation has always or long been absent.

Axilary buds
Buds that form in leaf axils.

Capitulum
An inflorescence in the form of disc where disc florets are in the centre and the ray florets in the outer whorl ; characteristics of family Composite.

Pupation
Infreesia, the formation of a new corm form an old corm during storage.

Vermiculite
Mica platelets, formed by heating to about 738°C, and used as an ingredient growing media.

Enzyme
A complex chemical substance formed by living cells, accelerates specific changes of other substances without undergoing any change in itself.

124. Phalsa is very sensitive to Iron (Fe) deficiency. T/F

125. Potassic fertilizers in muskmelon increases sweetness. T/F

126. Pummelo (*C. grandis*) and Grape fruit (*C. paradisi*) were identified separately by James McFadyen (1830). T/F

127. Wintering is practised in rose apple. T/F

128. 'Foster' variety of grape fruit was discovered by R.B. Foster in 1906-07. T/F

129. Bahar treatment is applied in pomegranate. T/F

130. IIHR is situated at Pusa, New Delhi. T/F

131. Japanese follow as asymmetrical style of gardening. T/F

132. Phalsa is commercially propagated by cutting. T/F

133. Primary mandate of IIHR is to care of tropical & sub-tropical fruits. T/F

134. Sardar guava is a spreading type cultivar. T/F

135. Turnip belongs to family Umbelliferae. T/F

136. *Vitis vinifera* varieties are susceptible to Phylloxera. T/F

137. Arkel is an early variety of pea. T/F

138. Fruit juice concentrates are the base material for beverages. T/F

139. India ranks first in banana production in the world. T/F

140. Kagzi lime is used as indicator for *tristeza* virus. T/F

141. Litchi is cross-compatible. T/F

142. Protein content is important in jelly preparation. T/F

143. Red rot is a disorder of litchi. T/F

144. Tenderometer is used to measure texture in peas. T/F

145. The superior quality of guava fruits are obtained in Mrig bahar. T/F

146. *Vitis labrlusca* is phylloxera resistant root stock of grape. T/F

147. 'Metaxenia' is found in datepalm. T/F

148. Flower colour of *Ixora coccinea* is blue. T/F

149. Fruit drop in chillies can be controlled by Karathane spray. T/F

150. Island Gem is a cultivar of custard apple. T/F
151. Lime is the most susceptible citrus to cold injury. T/F
152. Little leaf of grape is caused due to virus. T/F
153. Pea is a self-pollinated crop. T/F
154. Sardar guava is synonym of Arka Amulya. T/F
155. The richest source of ascorbic acid is Barbados Cherry. T/F
156. Thinning of apple fruits encourages proper development of remaining fruits, their colour & taste. T/F
157. 'Cider' is a product of grape. T/F
158. *Carica cauliflora* is used as disease resistant parent of papaya. T/F
159. Chhuhara and Raisins, both are prepared from Datepalm. T/F
160. *Cordia myxa* (Lasora) is a temperate fruit. T/F
161. Jackfruit is the biggest fruit in the world. T/F
162. Keeping the mango in water increases sweetness. T/F
163. Phylloxera affects mainly roots of grape. T/F
164. Poona fig is a hybrid of *F. carica* x *F. glomerata*. T/F
165. Sardar Bahadur Lal Singh is the father of grape development in North India. T/F
166. Thompson Seedless grape is very suitable for head system of training. T/F

‖‖‖‖‖ Fill in the Blanks ‖‖‖‖‖

167. Jasmine should be pruned during _____.
168. Potato is a _____ crop grown maximum world wide.
169. Angoorlata is a famous variety of _____.
170. Tomato variety _____ is comparatively rich in Vit-A.
171. Bottle gourd belongs to _____ family.
172. Most of vegetables are _____ in action.
173. Germany is the biggest importer of _____.
174. _____ can be poisonous if eaten raw.
175. Ascorbic acid is mainly found in _____ parts of plants.

Graft-hybrid
An individual formed from graft and stock and showing characteristics of both, graft chimera.

Sinigrin
A glucoside containing sulphur formed in large quantities in cabbage and cauliflower making the produce pungent when raised under warm conditions.

Deshooting
Removal of newly formed shoots to prevent them from becoming branches. Also known as summer prunning.

Crown bud
Chrysanthemum inflorescence formed under adverse conditions, such as improper day length. The bud may abort.

Lateral bud
A bud forming along the side of a stem or branch rather than at the end.

Heterophyllous
Bearing leaves of different forms on the same plant.

Xylem
Water conducting vessels found throughout the plant. Xylem vessels transport water and minerals from the roots upward through the plant.

Tetraploid
An organism which contains four sets of chromosome or genome (4x).

Capsule
A dry fruit derived from a compound pistil, and opening in one of a number of ways.

176. Phomopsis disease damages to _____ of brinjal.
177. Damping off is caused by _____ .
178. Leaf curl in chilli is transmitted through _____.
179. _____ is common root stock of rose in North India.
180. Pithiness in radish occurs due to _____ deficiency.
181. Cider is fermented product of _____.
182. Citrus fruit is a modified form of _____.
183. Cashew apple is a modification of _____.
184. Burrowing nematode is serious problem of _____.
185. Cashew contains more _____.
186. Atleast _____ per cent pollinizers are recommended in apple orchard.
187. Apple exhibits _____ type apomixis.
188. Aonla contains _____ mg Vit-C/100g of edible part.
189. 'Hen and Chicken' in grape occurs due to _____ deficiency.
190. H.T. rose is commonly propagated by _____.
191. Beauty Seedless is a _____ pruning variety of grape.
192. Basrai Dwarf is a variety of _____.
193. Banana can be stored at _____ temperature at 95% R.H.
194. Nucellar Seedlings arise from _____.
195. _____ is day neutral variety of strawberry. .
196. _____ is male cultivar of kiwifruit.
197. _____ is an ornamental variety of mango.
198. _____ is dual purpose variety of banana.
199. Botanically, pineapple is _____.
200. _____ has the highest somatic chromosome number.

Once when Fritz Kreisler, the great violinist, finished a concert, someone came up to the stage and said, "I'd give my life to play the way you do." Kreisler replied, "I did!"

A horse that pulls cannot kick; a horse that kicks cannot pull. Let's pull and stop kicking.

Script your success story

**[There is
no **M**agic for
Success. In the
real **W**orld,
Success comes
to **D**oers, not
Observers.]**

Step-Eight

Losers have schemes;
Winners have dreams.

Multiple Choice

1. Potato requires acidic soils for better yield because these soils devoid it from-
    A] Late blight   B] Black heart
    C] Early blight   D] Scab disease

2. The strongest chemical agent for breaking dormancy is-
    A] Potassium nitrate   B] Gibberellin
    C] Ethylene   D] Cumerin

3. A 'limiting factor' for cashew nut cultivation is-
    A] Temperature   B] Water
    C] Quality Seedlings   D] Harvesting

4. The most suitable intercrop for citrus orchard is-
    A] Guava   B] Papaya
    C] Karonda   D] Banana

5. To breed for disease resistance which method is used
    A] Back cross   B] Mass selection
    C] Pure line selection   D] All

6. 'One tree for each child' Programme is related with-
    A] Agro forestry   B] Social forestry C]
    Horticulture   D] Silvi pasture

7. Suitable annual for fragrant flowers-
    A] Carnation   B] Hollyhock
    C] Petunia   D] Candytuft

8. The most common method for improvement of self pollinated crops is-
    A] Mass selection   B] Pure line selection
    C] Pedigree method   D] Bulk method

9. Seed rate of carrot for one hectare area is-
    A] 3-4 kg   B] 6-8 kg
    C] 10-12 kg   D] 15-16 kg

Radial
Arranged around and spreading from a common center.

Dammar
A resinous substance obtained from certain trees like members of Coniferae.

Diploid
apogamy
Development of embryos from diploid synergids or antipodal cells.

Cider
An alcoholic beverage made from fermented juice of apple.

Transperation
Evaporation of water from plant tissue to the atmosphere. Transpiration occurs mainly through the stomates in the leaves.

Branch
A subsidiary stem arising from the main stem or another branch.

Pollination
The transfer of pollen from the stamen *(male part of the flower)* to the pistil *(female part of the flower)*, which results in the formation of a seed. Hybrids are created when the pollen from one kind of plant is used to pollinate and entirely different variety, resulting in a new plant altogether.

Keiki
A small plant arising from the stem or pseudobulb of a mature plant.

Cut-over
An area from which some or all the tree have been removed.

10. Seed rate of papaya for planting in one hectare is-
    A] 50-100 g   B] 100-200 g
    C] 250-300 g   D] 400-500 g

11. The most common rootstock for rose budding in North India is-
    A] *Rosa multiflora*   B] *Rosa indica*
    C] *Rosa chinensis*   D] *Rosa foetida*

12. Which fruit is recommended for sugar patient-
    A] Aonla   B] Jamun
    C] Bael   D] Banana

13. The most suitable soil for tea cultivation is-
    A] Latosols   B] Vertisols
    C] Alluvials   D] Black soils

14. Ideal method for top working in mango is-
    A] Inarching   B] Veneer grafting
    C] Stone grafting   D] Side grafting

15. Indicator plant for tristiza in citrus is
    A] Sweet orange   B] Mandarin
    C] Acid lime   D] Grape fruit

16. Which soil is best for vegetable production-
    A] Sandy   B] Sandy loam
    C] Clay loam   D] Clay

17. First commodity for which grading & marketing rules were framed was-
    A] Mango   B] Citrus
    C] Apple   D] Grape

18. First commnodity for which grading and marketing rules were framed is-
    A] Mango   B] Tomato
    C] Grape   D] Potato

19. ' Forking' and 'Bitterness' are disorders of-
    A] Beet root   B] Carrot
    C] Radish   D] Sweet potato

20. 'Spongy tissue formation' in Alphonso mango is caused by-
    A] Bacteria   B] Algae
    C] Fungus   D] Physiological disorder

21. The correct formula for standard error is
    A] s^{-2}
    B] s/n
    C] s/n
    D] $2\ s$

22. Standard error formula is
    A] s/n
    B] s^2/n
    C] s
    D] None

23. 'Heterosis' is found in all stages of-
    A] Watermelon
    B] Bitter gourd
    C] Bottle gourd
    D] Sponge gourd

24. 'Zygodormancy' is found in-
    A] Apple
    B] Aonla
    C] Jack fruit
    D] Jamum

25. 'Caprification' is found in-
    A] Gular
    B] Fig
    C] Sapota
    D] Phalsa

26. The smallest seeds are found in-
    A] Onion
    B] Candytuft
    C] Orchid
    D] Mustard

27. Cytoplasmic male sterility is found in-
    A] Radish
    B] Limabean
    C] Carrot
    D] Tomato

28. The largest leaves are found in-
    A] *Typha angustifolia*
    B] *Nelumbo nucifera*
    C] *Victoria regia*
    D] *Vallisnaria*

29. The quick freezing is prefered over slow freezing, due to
    A] Formation of large crystal
    B] Formation of small crystal
    C] To save time
    D] None

30. A corm is differentiated from a bulb, because it is-
    A] Hollow
    B] Semi-hollow
    C] Solid
    D] Root

31. Which crop is free from red pumpkin beetle-
    A] Cucumber
    B] Pumpkin
    C] Watermelon
    D] Bitter gourd

32. Essential oil is extracted from-
    A] Citronella
    B] Khus
    C] Jasmine
    D] All of them

Husk
Outer covering of a fruit or seed.

Multiple fruit
A fruit formed from several flowers included in a single structure having a common axis. Ex. Pineapple, mulberry

Moss
A dried bog material from *Sphagnum* sp. or *Funaria* sp. usually used as rooting medium or to enrich the soil mixture.

Colchicine
A poisonous alkaloid obtained from seeds and bulbs of the autumn crocus (*Colchicum autumnale*), having the property to cause chromosome doubling.

Cracking

A disorder where fruit surface cracks mainly due to heavy irrigation or rain after long dry spell. This may occur due to varietal characters and micro nutrient deficiencies, generally found in tomato, litchi, cherry, apple, pomegranate, etc.

Orchard

A group of fruit trees grown in a specified area.

Arcure

A system of training fruit trees to promote fruitfulness. Branches are bent in the form of horizontal bows so that the upward flow of sap is checked.

Mummy

A dried, shriveled fruit.

33. KISHAN BHARTI is released from-
    A] CSAUAT   B] SVBPUAT
    C] GBPUAT   D] IARI

34. Foundation seed is produced from-
    A] Registered seeds   B] Certified seeds
    C] Quality seeds   D] Breeder seeds

35. In general fruit & vegetables are not rich source of-
    A] Vitamin A   B] Vitamin B
    C] Vitamin C   D] Mineral

36. Ripe papaya fruit contain
    A] Vitamin A   B] Vitamin B
    C] Vitamin C   D] Vitamin D

37. In hot summer which fruit is used in form of *sharbet* (juice)-
    A] Jamum   B] Phalsa
    C] Watermelon   D] Mango

38. Non climacteric fruit is
    A] Water melon   B] Musk melon
    C] Cucumber   D] None

39. Which is the largest fruit of the world-
    A] Bottle gourd   B] Watermelon
    C] Jack fruit   D] Snake gourd

40. In aonla fruit setting occurs during—
    A] February   B] March
    C] May   D] July

41. Which is spur bearing fruit tree-
    A] Citrus   B] Mango
    C] Loquat   D] Apple

42. Most of fruit vars are heterozygous in nature therefore for commercial production their propagation is recommended by-
    A] True seeds   B] Mutation
    C] Asexual propagation   D] All of above

43. For better fruiting and yield, which fruit crop requires pollinizers-
    A] Mango   B] Guava
    C] Loquat   D] Bael

44. In mango fruiting takes place at
A] Older shoots   B] Current season growth
C] Both   D] None

45. Treatment of fruits and vegetable by slight temperature for small time is known as
A] Sterilization   B] Blanching
C] Pasteurization   D] None

46. Immature coconut fruits are harvested after
A] 5-6 months   B] 8-9 months
C] 10-12 months   D] 18-29 months

47. Guava bears fruits mostly on-
A] Current season growth B] One year old
C] Two years old shoot   D] Spurs

48. Which is a seedless fruit-
A] Pineapple   B] Apple
C] Sapota   D] Strawberry

49. PK- ft-16 is the var of
A] Chilli   B] Tomato
C] Brinjal   D] Cucurbits

50. The main function of cytokinin is
A] Cell enlargement   B] Cell division
C] Cell elongation   D] All

51. ' Functional male sterility' is controlled by-
A] Dominant gene   B] Recessive gene
C] Cytoplasm   D] Jumping gene

52. Necterine is fuzzless variety of-
A] Peach   B] Plum
C] Apricot   D] Almond

53. In one gallon, how much water will accommodate
A] 2.55 litre   B] 3.55 litre
C] 4.55 litre   D] 5.55 litre

54. The most effective ' gametocide' to induced sterility in okra is-
A] 2,4-D   B] M.H.
C] IAA   D] CCC

Scape
A leafless peduncle arising from the basal rosette of a few or not basal leaves; sometimes a few scale-like leaves or bracts may be borne on it; a scape may be one or many-flowered.

Acinus
A bunch of fleshy fruits as of currants, grapes or berries.

Etaerio
An aggregate of simple fruits borne by single flower having apocarpous pistil (free carpels). The ovaries of respective carpel mature into fruitlets clustered together on the pedicel *e.g.* etaerio of berries (*Annona squamosa*), of drupes (*Rubus idaeus*) etc.

Tricarpous
Bearing three carpels or fruits.

Spores
Reproductive units of fungi and ferns, equivalent to seeds of higher plants.

Fissure
bark that is furrowed and ridged, or splitting lengthwise

Fertilization
The act of fusion of pollen nucleus and the ovule following pollination to form seed.

Graft
An artificially induced vegetative fusion or union of parts from different individuals, the rooted part being called stock and the part or parts inserted or otherwise vegetatively fused to it the scion.

55. 'Ganesh' is an improved variety of-
A] Cashewnut   B] Bael
C] Custard apple   D] Pomegranate

56. 'Ganesh' variety of pomegranate was developed by-
A] G.S. Cheema (1936)   B] G.S. Randhawa (1935)
C] Deshmukh (1932)   D] W.B. Hayes (1939)

57. The commercial method of gape propagation is-
A] Hard wood cutting   B] Seed
C] Layering   D] Herbaceous cutting

58. Tropical Botanical Garden and Research Institute (TBGRI) is located at-
A] Kolkata (W.B.)   B] Palode (Kerala)
C] Shillong (Meghalaya)   D] Lucknow (U.P.)

59. Biggest Rose Garden in the world is situated at
A] New Delhi   B] France
C] New York   D] Frankfurt

60. The Rastraphati Garden is example of which type of garden
A] Mughal   B] Italian
C] English   D] Japanese

61. In Persian gardens, *Juniperus chinensis* (Cypress) was used as a symbolic of-
A] Death & Eternity   B] Idea of Heaven
C] Persian paradise   D] Life & Youth

62. Pungency of garlic is due to
A] Allyl propyl disulphide   B] Diallyl disulphide
C] Isothiduinate   D] None

63. The characteristic odour in garlic is due to-
A] Gallic acid   B] Allicin
C] Capsicin   D] Sulphuric acid

64. Linkage between gene affect-
A] Independtdent assortment   B] Fertilization
C] Vernalization   D] Anaphase

65. First Director General (DG) of ICAR was-
A] Dr. B.P. Pal   B] Dr. G.S. Randhawa
C] Dr. K.L. Chadha   D] Dr. G.L.Kaul

66. Present director general of ICAR is
    A] Dr. R. S. Arora   B] Dr. M. S. Swaminathan
    C] Dr. B. P. Pal   D] Dr. K. L. Chadda

67. First Director General of ICAR was
    A] Dr. M.S Swaminathan   B] Dr. R.S. Paroda
    C] Dr. B.P. Pal   D] None

68. First Director General of IIHR was-
    A] G.S. Randhawa   B] I.S. Yadav
    C] K.L. Chadha   D] R.M. Pandey

69. Pinching is generally done in quality production of-
    A] Aster   B] Dahalia
    C] Gladiolus   D] Carnation

70. Araucaia is generally suited for-
    A] Cold desert   B] Sub tropical
    C] Marshy-land   D] None

71. Cactus is generally suited for-
    A] Shade   B] Sunny-sites
    C] Marshygapland   D] None

72. The chemical generally used as germicide in Vase solution is
    A] Sucrose   B] HQC
    C] $Ca(NO_2)_2$ D] MH

73. ' GF' series of root stocks for peach was developed by-
    A] Bemhard (1965)   B] Tanaka (1955)
    C] H.B. Frost (1925)   D] Simmonds (1939)

74. Which crop gives highest per hectare average production in India
    A] Wheat   B] Paddy
    C] Maize   D] Bajara

75. Chillis are good source of
    A] Protein   B] Fat
    C] Carbohydrates   D] Vitamin

76. Which variety of ridge gourd bears hermaphrodite form of sex-
    A] Pusa Nasdar   B] Arka Sujat
    C] Satputia   D] Arka Sumeet

Informal Gardening
Land Scaping Gardening It is defined as the application of garden method and materials for the improvement of any area, on which it is possible and desirable to developed new face.

Ethylene
Colorless, odorless gas that causes diseases of succulent tissue, often called soft rot.

Barbecue
A drying ground for gathered crops, as for coffee, cocoa, copra etc., close to the store or factory.

Tuebrous roots
Enlarged roots generally having primordia at proximal end and root primordia at the distal ends.

Resins

A distinct plant product generally secreted in plant tissue from special cavities called resin ducts. Resins are insoluble in water but soluble in alcohol and they burn with a smoky flame. Resins are obtained from *Shorea robusta, Pinus longifolia, Cedrus deodara,* etc.

Pome

A type of fleshy fruit represented by the apple, pear or related genera. A fruit with a papery or bony core at the center and with sepals or scars from which the sepals have fallen at the blossom end. Many "berry-like" fruits are really small pomes (cotoneaster).

77. 'Soft wood grafting' is standardized for propagation of-
   A] Jack fruit   B] Banana
   C] Cocoa   D] Grape

78. Food- grain production in India (1997-1998) in million tonnes was
   A] 170   B] 180
   C] 190   D] 200

79. ' Granulation' is a serious disorder of-
   A] Aonla   B] Guava
   C] Litchi   D] Citrus

80. Pruning in grape in North India is done in the month of
   A] May   B] Dec.-Jan.
   C] September   D] November

81. Seedlessness in grape is due to-
   A] Apomixis   B] Stenospermocarpy
   C] Vegetative   D] Ovule abortion
      parthenocarpy

82. Which training system of grape results the highest cost : benefit ratio-
   A] Head system   B] Kniffin system
   C] Pergola system   D] Telephone system

83. Cuttings of grape should be planted during-
   A] January   B] March
   C] July   D] September

84. International Working Group of Soilless Culture (IWGSC) is established at-
   A] Rome   B] New Delhi
   C] Mexico   D] Washington

85. Aristolochia represent group of
   A] Herbaceous plants   B] Perennial shrub
   C] Trees   D] Climbers

86. Spices and Condiments are grouped under-
   A] Industrial crops   B] Commercial crops
   C] Food adjuncts   D] Field crops

87. Indoor plants grow well at temperature of
    A] 15-21°C   B] 10-14°C
    C] 22-25°C   D] 25-30°C

88. Sequence of growing crop on a given piece of land is referred to as
    A] Crop management   B] Crop production
    C] Crop science   D] Crop rotation

89. The main objective of growing intercrops in an orchard is to-
    A] Better growth of fruit   B] Improve soil fertility
       crop
    C] Check soil erosion   D] Get quality fruits

90. Leading cardamom growing state in India is
    A] TN   B] AP
    C] Kerala   D] Karnataka

91. Which can be successfully grown in Arid Zones-
    A] Jack fruit   B] Papaya
    C] Sweet orange D] Ber

92. Commercially used growth retardent is
    A] Paclobutrazol   B] BHA
    C] Cytokinin   D] GA$_3$

93. The 'bending' in guava is a common practice in-
    A] South India   B] N-E States
    C] Western India   D] North India

94. 'Bending' in guava is practiced in-
    A] Allahabad   B] Punjab
    C] Mumbai   D] Haryana

Matching

95. Match the following :-
    A] Homogamy   i]   Cole crops, Sapota
    B] Dichogamy   ii] Tomato, Japanese radish
    C] Herkogamy   iii] Brinjal, Pomegranate
    D] Self-incompatibility   iv] Papaya, Mango
    E] Functional male sterility v]   Onion, Carrot
    F] Polygamy   vi] Lima bean
    G] Genic male-sterility   vii] Coconut

Operon
A set of structural genes whose transcription is regulated by a set of regulator, promoter and operator genes; their activity is manifested in the controlled synthesis (transcription) of mRNA.

Polygenes
A series of multiple genes with small and cumulative effect which produce continuous variation in the phenotypes; quantitative characters are governed by these genes.

Allotetraploid
An organism with four genomes (copies each of two or four different genomes) derived from hybridization of different species.

Hexaploid
An individual having six genomes of one (autohexaploid) or more (allohexaploid) kind.

Organic gardening
The method of gardening utilizing only materials derived from living things. *(i.e. composts and manures)*

Viability
Ability of seeds to germinate.

Transpiration
A physiological process of giving off water vapour from the internal tissues of living plants through the aerial parts under the influence of sunlight which is regulated to some extent by the protoplasm.

H] Plasmic-genic sterility   viii] Turnip, Sweet potato
I] Hermaphrodite flowers   ix] Cauliflower, Cabbage
J] *Androgynous flower*   x] Grape, Apple

96. Match the following :-
A] House fly   i] Pollinator of Kiwi fruit
B] Wind   ii] Pollinizer of Apple
C] Honey bee   iii] Pollination in Pecan
D] *Bombus vegans*   iv] Pollinator of Cranberry
E] McIntosh   v] Pollinator of Mango

97. Match the following :-
A] Hydrogen swell   i] Due to overfilling & under exhausting
B] Flipper   ii] Susceptible to 'Swell'
C] Plain cans   iii] 0.6 to 0.9 cm
D] Lacquered can   iv] Due to production of hydrogen gas
E] Proper head space   v] Safe for hydrogen Swell

98. Match the following :-
A] IIHR   i] Varanasi
B] CISH   ii] Jodhpur
C] IIVR   iii] Bangalore
D] CAZRI   iv] Mysore
E] CFTRI   v] Lucknow

99. Match the following :-
A] IISR   i] Lucknow (U.P.)
B] CIMAP   ii] Pusa (New Delhi)
C] CTRI   iii] Calicut (Kerala)
E] CPRI   iv] Shimla (H.P.)
E] CSIR   v] Trivandrum (Kerala)

100. Match the following :-
A] Indehiscent fruits   i] Bael, Woodapple
B] Lomentum   ii] Strawberry
C] Fleshy drupe   iii] Pomegranate
D] Amphisarca   iv] Raspberry, Blackberry
E] Etaerio of achenes   v] Date, Papaya

F] Etaerio of berries
G] Etaerio of drupes
I] Pome
I] Berry
J] Balausta

vi] Cashewnut, Hazelnut
vii] Quince, Loquat
viii] Tamarind
ix] Custard apple, Atemoya
x] Mango, Coconut

101. Match the following :-

A] Indian Horticulture
B] Scientific Horticulture
C] Indian Spices
D] Indian Jour. of Horticulture
E] Haryana Jour. of Horticulture

i] G.A.U. (Junagarh)
ii] Hissar (Haryana)
iii] Cochin (Kerala)
iv] ICAR (New Delhi)
v] IARI (New Delhi)

102. Match the following :-

A] Indicator for 'Tristeza' virus
B] Indicator of 'Greening' disease
C] Indicator for 'Exocortis' virus
D] Indicator for 'Psorosis' virus
E] Indicator for 'Xyloporosis' virus

i] Sweet orange (Pineapple, Valencia), Tangelo
ii] Rangpur lime, Etrog Citron
iii] Kagzi lime, West Indian lime
iv] Mexican lime, Sour orange
v] Sweet lime, Orlando tangelo

103. Match the following :-

A] Indigenous cucurbit
B] Fruit fly tolerant cucurbit
C] Perennial cucurbit
D] Powdery mildew resistant cucurbit
E] Free from Red Pumpkin Beetle
F] Susceptible to Anthranose

i] Pointed gourd
ii] Arka Manik
iii] Sweet gourd
iv] Watermelon
v] Bottle gourd
vi] Ivy gourd

Glandular-bristly
With stiff gland-tipped hairs.

Lustrous
Having a slight metallic gloss, less reflective than glossy.

Hardening off
The process of gradually acclimatizing greenhouse or indoor grown plants to outdoor growing conditions.

Germination
Beginning of growth, budding, sprouting, development, pertains to seed or embryo.

Uva
A soft, indehiscent grape-like fruit, a raisin or grape.

Sod Culture
Cultivation of grass in orchard is called sod culture.

Cool-season grass
A grass that greens up and grows during the spring, sets seed in early summer, then goes dormant until fall, when it begins growing again. A second dormancy may occur during cold winter weather.

Sod-forming grass
A grass that spreads by creating new plants from existing plants by running roots across or under the ground (*stolons* or *rhizomes*, respectively), creating a dense turf. Also called *spreading grass*. Contrast with *clump-forming grass* (also called *bunchgrass*), in which single plants get larger over time but do not create new plants.

G] Powdery mildew resistant water   vii] Pumpkin melon

H] Parthenocarpic fruits   viii] Cho-Cho

I] 'Single seeded' cucurbit   ix] Bitter gourd

J] 'Tuberous rooted' cucurbit   x] Cucumber

104. Match the following :-

A] Internal black spot   i] Turnip
B] Steckling roots   ii] Potato
C] Cavity Spot   iii] Cucumber
D] Forking   iv] Garden beet
E] 'Solanine' synthesis   v] Lettuce
F] 'Pox and Scurf,   vi] Carrot
G] 'Tip burn'   vii] Watermelon
H] Fibrous flesh   viii] Radish
I] 'Goose neck fruits'   ix] Sweet Potato
J] Riceyness   x] Cauliflower

105. Match the following :-

A] 1st Book of Horticulture in India   i] Dr. K.L. Chadha
B] 1st D.G. (ICAR)   ii] Dr. R.N. Singh
C] 1st D.D.G. (Hort.) ICAR   iii] H.B. Frost
D] 1st D.G.(IIHR)   iv] Tanaka & Swingle
E] Cage technique in Mango   v] H.P. Olmo
F] Spongy tissue disorder   vi] Dr. W.B. Hayes
G] Zero energy cool chamber   vii] Dr. G.S. Randhawa
H] 'Kinnow' mandarin   viii] Dr. G.S. Cheema
I] Grape breeding   ix] Dr. B.P. Pal
J] Citrus classification   x] Dr.S.K. Roy

106. Match the following :-

A] 1st.H.T. rose var.   i] Mohini
B] 1st Floribunda var.in India   ii] *R. damascena*

C] Brown/Chocolate colour rose   iii] Cri-Cri

D] Miniature rose variety   v] *R. bourboniana*

E] Rose water   v] La France

F] Rose root stock   vi] Weeping Standard

G] I[st] rose var. developed by Dr. B.P. Pal   vii] Delhi Princess

H] Budding at 1 m height   viii] Half standard

I] Budding at 50 cm height   ix] Rose Sherbet

J] Budding at 1.5 -2m height   x] Full Standard

107. Match the following :-

A] Ivy gourd   i] Indigenous to India

B] Cucumber   ii] Native to Tropical Africa

C] Watermelon   iii] Monotype genus

D] Sechium/Chayote   iv] Perennial cucurbit

E] Wax gourd   v] One seeded cucurbit

108. Match the following :-

A] Jam   i] Mango

B] Jelly   ii] Pea

C] Preserve   iii] Citrus

D] Pickle   iv] Guava

E] Sauce   v] Bael

F] Marmalade   vi] Apple

G] Raisin   vii] Tomato

H] Dehydration   viii] Aonla

I] Squash   ix] Ashgourd

J] Candy   x] Grape

109. Match the following :-

A] *Jasminum grandiflorum*   i] $2n = 14$

B] *Bougainvillea glabra*   ii] $2n = 26$

C] *Rosa chinensis*   iii] $2p = 52$

D] *Gladiolus tristis*   iv] $2n = 34$

E] *Piper nigrum*   v] $2n = 30$

Spikelet
A secondary spike of grasses often provided with a small pedicel and bearing one or few florets.

Stoniness
An abnormal development of grit cells as in the pear.

Procumbent
Lying flat on the ground but the stem not rooting at nodes or tip.

Haulm
Above ground portion of potato.

Root zone
The area of ground under which a given tree's (or other plant's) roots spread. Often the root zone covers the area encircled by the *drip line*; that is, the roots often spread underground to the same distance that the tree's branches extend above.

Aerial roots
Roots produced above ground, often used for climbing.

Cacti
Cacti are a group of plants with peculiar shape and size and mostly adapted for desert life.

Dwarf varieties
Varieties bred to grow smaller than their parent plants. Dwarf plants may lose the ability of the parents to set fruit (Bailey's dwarf highbush cranberry, *Viburnum trilobum 'Bailey's Compact'*). They may not resemble miniatures of their parents (dwarf Alberta spruce, *Picea glauca var. albertiana 'Conica'*).

110. Match the following :-
    A] Jelly failure to set   i] Due to over cooking
    B] Cloudy jelly   ii] Due to excess of acid
    C] Crystalization of jelly   iii] Due to excess sugar
    D] Syneresis of jelly   iv] Due to moulds
    E] Browning of marmalade   v] Due to lack of pectin

111. Match the following :-
    A] Jelmeter reading   i] 0.750 kg sugar/kg of extract
    B] Jelmeter reading 1   ii] 0.500 kg sugar/kg of extract
    C] Jelmeter reading   iii] 1 kg sugar/kg of extract
    D] Jelmeter reading   iv] <40% invert sugar
    E] Apple jam   v] 1.250 kg sugar/kg of extract

112. Match the following :-
    A] Juicy placental hairs   i] Litchi
    B] Mesocarp & Endocarp   ii] Coconut
    C] Fleshy Thalamus   iii] Jamun
    D] Aril   iv] Citrus
    E] Cotyledon   v] Fig
    F] Endosperm   vi] Mango
    G] Fleshy Receptacle   vii] Banana
    H] Seed   viii] Cashew nut
    I] Mesocarp   ix] Apple
    J] Pericarp & Thalamus   x] Almond

113. Match the following :-
    A] Khasi Mandarin in N-E India   i] Harvesting in Jan. to Feb.
    B] Darjeeling Mandarin in W.B.   ii] Harvesting time Sept. to Nov.
    C] Kinnow Mandarin in Punjab   iii] Harvesting time Oct. to Jan.
    D] Nagpur Mandarin in M.S.   iv] Harvesting time Nov. to Dec.
    E] Grape fruit in Central India   v] Ist crop harvesting April to June

114. Match the following :-
    A] Kochia, Portulaca   i] Winter season
    B] Balsam, Cock's Comb   ii] Fragrant flower

C] Tuberose, Carnation   iii] Colourful bracts
D] Petunia, Verbena   iv] Rainy season
E] Gladiolus, Gerbera   v]   Bulbous plants
F] Acroclinum,   vi] Cut-flower
    Helichrysum
G] Candytuft, Iresine   vii] Edging of beds
H] Pothos, Croton   viii] Dry-flower
I]   Dahlia, Tulip   ix] Summer season
J] Poinsettia, Bougainvillea   x]   Shade-loving plants

115. Match the following :-
A] *Lagerstoelmia indica*   i]   Blue flower
B] *Jatropha multifida*   ii]   Ornamental foliage
C] *Duranta plumieri*   iii] White & Pink flower
D] *Hibiscus rosa-sinensis* iv]   Red-Pink bracts
E] *Hamelia patens*   v]   Scarlet flower
F] *Cestrum nocturnum*   vi]   White corolla, Orange-tube
G] *Poinsettia pulcherima*   vii] Various colours
H] *Nyctanthes*   viii] Pale white, Scented
    arbor-tristis
I]   *Tecoma stans*   ix]   Yellow flower
J] *Codiaeum variegata*   x]   Orange red

116. Match the following :-
A] Laksha Ganga   i]   Grape
B] Arka Ajit   ii]   Tomato
C] Ruby   iii] Coconut
D] Angoorlata   iv]   Guava
E] Arka Vati   v]   Pomegranate
F] Arka Mridula   vi]   Pea
G] Avon   vii] Rose
H] Y2K   viii] Gladiolus
I]   Oscar   ix]   Chrysanthemum
J] Chitra   x]   Bougainvillea

True / False

117. Apricot is rich source of vitamin-A.   T/F

118. Black raspberry is considered as 'Bramble berry'.   T/F

Cucurbits
A large and diverse group of vegetable crops under Cucurbitaceae, used as vegetables (pumpkin, gourds, etc.), pickles (cucumber) and as desert fruits (muskmelon, watermelon, etc.).

Groundkeeper
An unwanted plant growing form seed or vegetative material that remains on a field form a preceding crop.

Pomology
The science of growing fruit crops is called pomology.

Drawn
A plant which is growing in a group that is too closely packed or in a poorly lighted position is inclined to become long and thin and is often pallid in colour.

119. Chirounji is commonly grown in Bundelkhand. T/F
120. CO-I is a hybrid variety of banana developed by TNAU. T/F
121. Datepalm is a monocotyledonous fruit crop. T/F
122. Full slip stage denotes the maturity of muskmelon. T/F
123. Indian Horticulture is published by IIHR. T/F
124. Low winter temp. in winter is limiting factor for banana in N. India. T/F
125. Pusa Drum Head is a variety of cauliflower. T/F
126. Ber flowers in the month of September. T/F
127. Epicotyle grafting is also known as 'Embryo grafting'. T/F
128. Fruits of *Fragaria vesca* are larger than *F. chiloensis*. T/F
129. In India, date is harvested at *Doka* stage. T/F
130. Olour is a polyembryonic variety of citrus. T/F
131. *Prunus dasycarpa* is hybrid of two species of apricot. T/F
132. Pusa Sawani is a popular variety of okra. T/F
133. Vegetative growth of determinate tomato is unlimited. T/F
134. Vitamin-A is precursor of carotene. T/F
135. A common product of date is raisin. T/F
136. Alphonso is an export quality apple. T/F
137. By 'Top working' inferior variety can be changed to superior one. T/F
138. Charmagz is self-incompatible variety of apricot. T/F
139. Florida Market brinjal is resistant to Phomopsis blight. T/F
140. *Hibiscus spectabilis* is an ornamental climber. T/F
141. Little leaf of brinjal is a viral disease. T/F
142. Stems of gooseberry are spiny/prickly. T/F
143. Asparagus & Pointed gourd both are perennial vegetables. T/F
144. Blanching is practised in banana. T/F
145. Botanically aonla is a berry. T/F
146. Kinnow is an inter generic hybrid. T/F

147. Pomegranate variety 'Jyoti' is selection from Bassein Seedless.　　T/F

148. Roksana (Rexona) is late ripening var. of apricot.　　T/F

149. Selection-I is a variety of doob grass.　　T/F

150. Tea is prepared from pods.　　T/F

151. 'Asphyxia' in apricot is caused due to waterlogging.　　T/F

152. Cabbage is a long day vegetable.　　T/F

153. Fruit growing in India is written by W.B. Haye.　　T/F

154. Fruit juice can be temporarily preserved by pasteurization.　T/F

155. In S. India, aonla flowers twice a year.　　T/F

156. Pusa Meghdoot is a variety of cauliflower.　　T/F

157. Tea grows well in acidic soils.　　T/F

158. Top working is successful in mango & custard apple.　　T/F

159. 'Turkey' apricot is suitable pollinizer for Charmagz.　　T/F

160. Acid content in under ripe fruit is very high.　　T/F

161. Amrit and Kanchan are varieties of aonla.　　T/F

162. Appropriate time for phalsa pruning in N. India is July.　T/F

163. Central Leader system is a method of 'pruning for form'.　T/F

164. Commercially tea is propagated by cuttings.　　T/F

165. I.I.V.R. is located at Bareilly.　　T/F

166. Micro irrigation technique is useful for banana.　　T/F

Fill in the Blanks

167. Grape is pruned during _____ in N.India .

168. Grape is pruned _____ a year in S. India.

169. _____ is called as 'queen of beverage crops'.

170. Kurrukan mango is _____ root stock.

171. Datepalm prefers 'head in fire and foot in _____ .

172. _____ is limiting factor for papaya cultivation.

173. The highest oil producing tree is _____.

174. India occupies _____ position in banana production.

175. _____ is salt resistant root stock of citrus.

Hydrophyte
Aquatic plant that normally grows in water e.g., Nelembium, Victoria amazonica, etc.

Understory trees
Small trees that grow well under larger ones. Common understory trees in a birch-poplar-spruce forest include ironwood, mountain maple, and serviceberry. Other common understory trees in a mixed maple-hardwood forest are witch hazel and black cherry.

Mesophyte
A plant that grows normally under conditions of moderate humidity, that neither requires nor survives culture in water or extreme drought.

Mature
A later phase of growth characterized by flowering, fruiting, and a reduced rate of size increase.

Gemma
A bud or out - growth of a plant which develops into a new individual; a leaf bud, a small cellular body in mosses, etc. that separates from mother plant and starts a new life.

Burl
A knot or woody growth of very irregular grain.

ABA
Acronym for the plant growth regulator Abscisic Acid, which inhibits growth and promotes abscission and dormancy. It is anti-gibberellic in mode of action.

176. _____ is common pollinator of mango.

177. _____ is vigourous root stock of pear.

178. Pear decline is caused by _____.

179. _____ variety of apple is indigenous to India.

180. Bael has _____ leaves.

181. Bael is propagated during _____.

182. Pineapple exhibits _____ parthenocarpy.

183. _____ is major breeding hindrance in pineapple.

184. Kandhari & Alandi are _____ seeded varieties of pomegranate.

185. Phalsa is commonly propagated by _____ .

186. Phalsa can tolerate temperature upto _____.

187. Seedlessness in litchi is due to _____.

188. Red Rot is disorder of _____.

189. Swarn Roopa is a variety of _____.

190. Footrot is a serious disease of _____.

191. Improper finger filling in banana is caused to _____ deficiency .

192. 'Shobha' is a mutant of _____ gladiolus.

193. Calcium Carbide releases _____ gas.

194. India ranks _____ in turmeric production.

195. The idea of wild garden was expounded by _____.

196. White colour is considered as _____ colour.

197. Red colour is _____ colour.

198. Taj Garden was built by _____.

199. Baradari is a typical feature of _____ garden.

200. _____ is situated in Lucknow.

As a young cartoonist, Walt Disney faced many rejections from newspaper editors, who said he had no talent. One day a minister at a church hired him to draw some cartoons. Disney was working out of a small mouse-infested shed near the church. After seeing a small mouse, he was inspired. That was the start of Mickey Mouse.

Script your success story

Successful people don't do Great things, they only do Small things in a great way.

Step-Nine *Losers say, "Something must be done"; Winners say, "I must do something".*

Multiple Choice

1.   Origin of Guava is
      A] America   B] China
      C] Australia   D] India

2.   For production of ' gynodioecious' lines in papaya, the most economic cross combination is-
      A] Male x Female   B] Male x Hermaphrodite
      C] Hermaphrodite x   D] Hermaphrodite x
         Female   Hermaphrodite

3.   Which is an anti -haemorrhagic vitamin-
      A] Vit-A   B] Vit-E
      C] Vit-K   D] Vit-C

4.   J.H. Hale is self-unfruitful variety of-
      A] Peach   B] Apple
      C] Pear   D] Strawberry

5.   Chromosome number reduced to half in-
      A] Meiosis-I   B] Meiosis-II
      C] Mitosis   D] Cytokinesis

6.   A serious handicap in citrus breeding programme is-
      A] Self-incompatibility   B] Seedlessness
      C] Viral diseases   D] Nucellar embryony

7.   Vanilla is harvested at which stage
      A] Fully ripe   B] Immature
      C] Tender   D] Ripe

8.   'Auto polyploidy' has been developed in-
      A] Bottle gourd   B] Watermelon
      C] Litchi   D] Banana

9.   Which vegetable has high calorie value
      A] Fenugreek   B] Amaranthus
      C] Potato   D] Tomato

Rejuvenlization
Stimulation of new growth of old plants usually accomplished with prunning.

Proliferation
Abnormal bush-like growths. A rapid and repeated production of new parts as the formation of leafy parts from flora parts.

Pollard
A tree whose crown has been cut back to induce production of shoots from the top.

Indexed plants
Plants that have been tested by pathological methods and found to be free of kown pathogens. Plants may be indexed for a single.

Variety

Subdivision of a species having a distinct though often inconspicuous difference, and breeding true to that difference. More general, also refers to clones.

Pungent

With a sharp, hard point; sharp and acid to taste or smell.

Line

A group of individuals having common parents or ancestors.

Haploid

An organism or cell having only one complete set of chromosomes.

Climber

Group of plant having weak items and ability to climb up the support with the help of modified organs.

10. State which has largest geographical area
A] Rajasthan   B] MP
C] UP   D] Bihar

11. Fruit which has maximum area under cultivation in India is
A] Mango   B] Banana
C] Citrus   D] Guava

12. The rubber has primary centre of origin
A] Malaya   B] Western ghats
C] Asia   D] China

13. Which nutrient has role in chlorophyll synthesis
A] N   B] p
C] Ca   D] Fe

14. Which state has the largest area under litchi-
A] U.P.   B] Bihar
C] West Bengal   D] Punjab

15. Which plant have aphrodisiac property
A] Rauwolfia serpentina B] Curcuma amada
C] Datura   D] All

16. Which cucurbit have highest vitamin A
A] Bitter gourd   B] Cucumber
C] Pumpkin   D] All

17. A disease having regular occurrence at a particular area is called as
A] Epidemic   B] Endemic
C] Sporadic   D] None

18. The tree having specific shape is
A] *Mellingtonia hortinsis* B] *Sida cordifolia*
C] *Polyalthia longifolia*   D] None

19. 'Hawk Moth' is a serious pest of -
A] Fig   B] Papaya
C] Jack fruit   D] Durian

20. Seed requirement of one hectare radish is-
A] 5-9 kg   B] 10-12 kg
C] 15-20 kg   D] 20-25 kg

21. Complete genetic heterozygosity & heterogenity is found in
    A] F_1 hybrids   B] Pureline
    C] Multilines   D] Synthetics

22. When two heterozygous plants are crossed, segregation is earliest seen in
    A] F_1 generation   B] F_2 generation
    C] Test cross   D] None

23. 'High Gate' banana is a dwarf bud sport of-
    A] Dwarf Cavendish   B] Monthan
    C] Poovan   D] Gross Michel

24. 'High Gate' is a mutant variety of-
    A] Chrysanthemum   B] Banana
    C] Rose   D] Pineapple

25. 'High Gate' variety of banana is a mutant of-
    A] Poovan   B] Cavendish
    C] Gross Michel   D] Harichhal

26. Jelly with high sugar & low acid is called as
    A] Weeping Jelly   B] Crystallized Jelly
    C] Soft Jelly   D] Cloudy Jelly

27. State having highest geographical area
    A] Madhya Pradesh   B] Uttar Pradesh
    C] Maharashtra   D] Tamil Nadu

28. Vegetable containing highest sodium content is-
    A] Spinach   B] Lettuce
    C] Cabbage   D] Brinjal

29. 'Hillock Head' orchard situated in Mashobra (H.P.) accommodates the plants of-
    A] Kinnow mandarin   B] Walnut
    C] Apple   D] Grape

30. In Nilgiri hills mandarin crop is harvested during -
    A] Dec-Sept   B] Aug- Oct
    C] March   D] July

31. Homozygocity and homogeneity is maximum in
    A] Pure line   B] Inbred line
    C] Land races   D] Synthetic varieties

Solidity
An approximate measure of head firmness of cabbage, chinese cabbage or head lettuce which is the head weight (g) divided by volume of head (cc).

Cone
A coniferous fruit, having a number of woody, leathery, or fleshy scales, each bearing one or more seeds, and attached to a central axis.

Silencer
A DNA sequence that helps to reduce or shut off the expression of a nearby gene.

Superdominance
A phenomenon which relates heterosis accruing from heterozygotes which carry divergent alleles $(A_1 A_4 > A_1 A_3 > A_1 A_2)$.

Caruncle
An outgrowth near the hilum of the seeds as seen in dolichos bean.

Resistant
Possessing qualities that hinder the development of a given pathogen or damaging factor.

Hoary
With a close white or whitish pubescence.

Homogeneon
A genetically and morphologically homogeneous species in which apomixis is absent and all members are interfertile.

Inbred line
A nearly homozygous line developed through continued inbreeding usually selfing, accompanied by selection.

32. Cell elongation hormone is
    A] Auxin   B] Ethylene
    C] Gibberelin   D] MH

33. Fruit ripening hormone is
    A] Auxin   B] GA
    C] Cytokinin   D] Ethylene

34. John Innes Horticultural Institute (JIHI) is located-
    A] Kent (England)   B] Rome (Italy)
    C] Shanghai (China)   D] New Delhi (India)

35. University of Horticulture & Forestry is located at
    A] Lucknow   B] Solan
    C] Pune   D] Bangalore

36. The book HORTICULTURE SCIENCE is written by
    A] Willium gadd   B] Swaminathan
    C] Jennick   D] Randhawa

37. Potato contain how much % of protein
    A] 1.8   B] 2.2
    C] 1.4   D] 3.2

38. The mango hybrid developed by IIHR
    A] Arka hans   B] Arka jyoti
    C] Arka aruna   D] Arka vati

39. A multiple hybrid variety 'Ruby' is a variety of-
    A] Bael   B] Brinjal
    C] Pomegranate   D] Grape

40. CH- I var of chilli is developed at
    A] TNAU, Coimbatore   B] PAU, Ludhiana
    C] IIHR, Bangalore   D] IARI, New Delhi

41. 'Panniyur- I' is a high yielding variety of-
    A] Bell pepper   B] Black pepper
    C] Cococa   D] Arecanut

42. In banana if score is above 16 then which will be its genome
    A] AA   B] AB
    C] A AB   D] ABB

43. For population improvement in any crop, best method is
    A] Pureline selection   B] Pedigree selection
    C] Mass selection   D] Recurrent selection

44. Which is useful for improving 'berry size' in grape-
    A] NAA   B] 2,4-D
    C] GA$_3$ D] IAA

45. Chromosome number in a diploid cowpea is-
    A] 2n=22   B] 2n= 14
    C] 2n=18   D] 2n=24

46. June drop in apple is due to -
    A] Lack of pollination   B] Moisture stress
    C] Ca deficiency   D] Alternate bearing

47. The browning in apple is due to
    A] Poly phenol oxydase   B] Tyrosene
    C] Hydrogenase   D] None

48. Bitter principle in bitter gourd is-
    A] Cucurbitacine   B] Momordicin
    C] Solanin   D] None

49. Isolation distance in brinjal for seed production is
    A] 50-100 m   B] 10-20 m
    C] 200-300 m   D] 500 m

50. Which type of flowers in brinjal produces much fruits-
    A] Yellow styled   B] Medium styled
    C] Pseudo short styled   D] True short styled

51. CO$_2$ accepter in C$_3$ plants is
    A] PEP   B] RUBP
    C] Both   D] None

52. 'Crown bud' in chrysanthemum appears in the month of-
    A] January   B] March
    C] May   D] September

53. Self incompatability in cole crops is predominantly is of-
    A] Sporophytic   B] Gametophytic
    C] Pin & thrum type   D] None

54. Main constraint in commercial cultivation of medicinal plants
    is
    A] Lack of planting   B] Lack of market
       material
    C] Lack of knowledge   D] All

Bulb forcer
A green house operator who produces pot plants and cut flowers form bulbous species using artificial growing conditions.

Putative
A taxon presumed of hybrid origin.

Rikka
A style of ' Ikebana' referring to larger, ornate and upright reproductions of the landscape by means of flowers and plants.

Torula
A group of yeasts important for the fermentation of fruit juice, fermenting yeasts, which produce undesirable changes in fruit products.

Uncongeniality
A phenomenon often encountered in a graft with plants which are not compatible; symptoms exhibited by uncongenial unions include swellings, overgrowths, severe dwarfing and ill-health.

Vein
a vascular tube in a leaf, typically referring to any within the blade except for the midrib

Decurrent
Continued down the stem in a ridge or wing, as applied to leaf-bases.

Agrochemicals
Biologically active chemicals used in agriculture, which include insecticide, fungicide, herbicide, growth substances, etc.

55. Advance form of sex in cucurbits is-
A] Monoecious   B] Dioecious
C] Hermaphrodite   D] Andromonoecious

56. For increase in flower size which practice is followed
A] Pinching   B] Deblossoming
C] Staking   D] All

57. Pod colour in Forestro variety of cocao is
A] Purple   B] Green
C] Blue   D] White

58. The spur in grape is
A] New growth of branch   B] Old shoot twig
C] Current season growth   D] None

59. Protein content in groundnut is
A] 22%   B] 25%
C] 40%   D] 43%

60. 'Mrig Bahar' in guava gives fruits in the month of-
A] June-July   B] August-September
C] May-June   D] October-November

61. A refrigerant commonly used in Indian Cold Storage is-
A] Ammonia   B] Carbide
C] Ethylene   D] Sodium benzoate

62. Which occupies maximum acreage in India-
A] Tomato   B] Onion
C] Cole crops   D] Potato

63. Crystal formation in jelly is due to
A] Excess of sugar   B] Less cooking
C] Both   D] None

64. Edible part in Jerusalem artichoke is-
A] Seed   B] Cutting
C] Layers   D] Tubers

65. NPK requirement in legumes is
A] 3:1:0   B] 4:2:1
C] 3:1:1   D] 1:2:2

66. Little leaf in Mango is due to
A] Zn deficiency   B] Mo deficiency
C] Zn toxicity   D] Mo toxicity

67. Spongy tissue in mango is-
    A] Bacterial disease   B] Viral disease
    C] Physiological disorder D] Fungal disease

68. For hybrid seed production in muskmelon, emasculation is necessary due to-
    A] Monoecious sex form B] Dioecious sex form
    C] Andromonoecious sex D] Gynoecious sex form
       form

69. Which cucurbit is dioecious in nature-
    A] Cucumber   B] Muskmelon
    C] Pointed gourd   D] Watermelon

70. Bolting problem in onion occur if temperature goes below
    A] 25°C   B] 20°C
    C] 15°C   D] 10°C

71. Deficiency symptoms of magnesium in plant usually appear first in-
    A] Tip of leaf   B] New leaves
    C] Older leaves   D] Whole plant

72. Nitrogen deficiency in plants leads to
    A] Excessive growth   B] Profuse flowering
    C] Lush green colour   D] Chlorosis

73. Completely tuberization in potato is stopped at temperature of
    A] 20 °C   B] 25 °C
    C] 30 °C   D] 23 °C

74. For foundation seed production in radish, isolation distance should be-
    A] 600 m   B] 900 m
    C] 1200 m   D] 1600 m

75. PGR used in root formation
    A] IBA   B] NAA
    C] ABA   D] MH

76. Blue colour in rose is incorporated by
    A] Gene transfer   B] Hybridization
    C] Selection   D] None

Thinning
1.Felling made in an immature tree stand for the purpose of improving the growth and form of trees that remain, without permanently breaking the canopy. 2. The removal of crop plants in a raw to reduce crowding and provide enough space between plants.

Frutescent
Becoming shrub-like in appearance or habit.

Aggregate flower
A single flower heaped or crowded into a dense flower cluster.

Agrestal
Applied to plants growing in arable land, uncultivated, growing wild.

Filler Trees
Trees placed in between permanent trees at the time of planting for getting some returns from them up to the stage, when permanent trees start production.

Cleistogamy
A built- in breeding mechanism where flowers remain closed at the time of pollination which favour self-pollination as seen in lettuce.

Theobromine
An important alkaloid present in cocoa (*Theobroma cacao*) and tea having medicinal properties.

Water-logging
A condition in consequence of inadequate drainage where the soil pores get filled with water by the exclusion of air.

77. Highest production in terms of million tonnes per year in India in that of following
A] Sugarcane   B] Potato
C] Pulses   D] Groundnut

78. Self compatibility is helpful in the development of-
A] Synthetic variety   B] Hybrid variety
C] Composite variety   D] Pure line variety

79. The major factors considered in the time of planting of trees on road side are-
A] Shade & Timber   B] Flowering only
C] Shape & Height   D] Oil content in leaves

80. Excessive chloride in tobacco lead to
A] Bushing inhibition   B] Difaced leaves
C] Reduce the storage life   D] All of above

81. Yellow colour in tomato is due to-
A] Lycopene   B] Carotene
C] Tomatin   D] Alline

82. Preservative used in tomato ketchup
A] Sodium benzoate   B] SO_2
C] Both   D] None

83. Which is extensively grown in tropical regions-
A] Date palm   B] Coconut
C] Walnut   D] Pomegranate

84. 'Yellow colour' in turmeric is due to-
A] Crucumin   B] Carotene
C] Turmarin   D] Lycopene

85. When petals are arranged in two or more than two whorls, flower is known as-
A] Apetalous   B] Single flower
C] Double flower   D] Bisexual flower

86. First hybrid in vegetables was developed in-
A] Tomato   B] Brinjal
C] Bottle gourd   D] Watermelon

87. First hybrid in vegetables was released in-
A] Bottle gourd (1971)   B] Brinjal (1976)
C] Tomato (1965)   D] Chilli (1970)

88. Major constraint in vegetative propagation of cardamom is
    A] Katte disease   B] Thirps
    C] Capsule Borer   D] None

89. The growing of tree in very dwarf form (5-20cm) is known as-
    A] Cascade Bonsai   B] Hanken Bonsai
    C] Mame Bonsai   D] Topiary

90. The Jelly in which fruit pieces are suspended is called as
    A] Jam   B] Jelly
    C] Marmalade   D] Candy

91. A cross in which order of male & female parent in reversed is called as
    A] Reciprocal cross   B] Back cross
    C] Test cross   D] Top cross

92. Horticultural work in which plants are given different shape is called as
    A] Training   B] Pruning
    C] Topiary   D] None

93. The plants in which staminate, pistillate and perfect flowers are borne on the same plant are known as-
    A] Andromonoecious   B] Androecious
    C] Trimonoecious   D] Dioecious

94. ' Inbreeding depression' is found in-
    A] Onion, Cabbage   B] Bottle gourd, Brinjal
    C] Tomato, Chilli   D] Muskmelon, Cucumber

Matching

95. Match the following :-
    A] Langra   i]   69.8% perfect flowers
    B] Chausa   ii] 42.9% perfect flowers
    C] Dashehari   iii] 30.6% perfect flowers
    D] Fazli   iv] 14.9 % perfect flowers
    E] Bombay Green   v]   9.2% perfect flowers

96. Match the following :-
    A] Lime   i]   *Citrus limon*
    B] Sour orange   ii] *Citrus aurantifolia*

Isogenic lines
Lines identical in genotype except for one gene.

Mosaic
A typical phenotypic variation in green and yellow tissue like a mosaic, caused either by "pattern" genes, which continue to reproduce that pattern or by certain viruses rather than by mutation.

Mesotherm
A plant that grows in intermediate condition of alternating high and low temperatures.

Row
A line of plants in line sowing system, sown with a desired spacing in field. Also a single file of containers lengthwise in the transport vehicle or in the store.

Tomatine
A steroidal glycoalkaloid present in *Lycopersicon* and *Solonum* species, which exhibits an antibiotic activity against a number of organisms.

Clustering/ Jhoomka
A disorder in mango characterized by a cluster of fruitlets at the tip of the panicles due to poor pollination and fertilization.

Diosgenin
A steroid derivative present in many species of yam (*Dioscroea* spp.) from which a number of steroid hormones including active birth control "pills" are manufactured.

C] Lemon   iii] *Juglans regia*
D] Pomegranate   iv] *Punica granatum*
E] Walnut   v] *Citrus aurantium*

97. Match the following :-
A] Limiting factor for pear cultivation   i] Salt tolerant rootstock of peach
B] Limiting factor for peach   ii] Vigourous rootstock of peach
C] Quince   iii] High humidity, Spring frost
D] De Bale   iv] Dwarf rootstock of pear
E] St. Julien-1   v] Fire blight, Quick decline

98. Match the following :-
A] Little leaf /Yellow mottling/Frenching   i] 'B' deficiency in Citrus
B] Green Mottling on leaves, Pale rind   ii] 'Fe' deficiency in Citrus
C] Exanthema, Bowing up; S-shaped shoots   iii] 'Zn' deficiency in Citrus
D] Water soaked flecks gum pockets in core   iv] 'Cu' deficiency in Citrus
E] Interveinal chlorosis, inability to mature   v] 'Mn' deficiency in Citrus

99. Match the following :-
A] Loose skinned orange   i] *C. grandis*
B] Tight skinned orange   ii] *C. sinensis*
C] Sour orange   iii] *C. medica*
D] Citron   iv] *C. reticulata*
E] Pummelo   v] *C. aurantium*

100. Match the following :-
A] Loquat   i] Native of South Asia
B] Mango   ii] Native of Malaya
C] Mangosteen   iii] China Origin
D] Passion fruit   iv] Japan-China Region
E] Peach   v] Native of Brazil

101. Match the following :-
A] Lucknow   i] Potato

B] Saharanpur   ii] Banana
C] Allahabad   iii] Mango
D] Nasik   v] Chirounji
E] Nagpur   v] Litchi
I] Kufri   v] Onion
G] Bhusawal   vii] Pomegranate
H] Bundelkhand   viii] Guava
I] Jalgaon   ix] Santra
J] Calicut   x] Black-pepper

102. Match the following :-
   A] Lycopene synthesis   i] Polygenic inheritance
   B] Yellow colour in   ii] 10-25°C
      tomato
   C] Unripened (green)   iii] K_2O deficiency
      tomato
   D] 'Grey Wall' in tomato   iv] 30-40°C
   E] Fruit Cracking in   v] 40-45°C
      tomato

103. Match the following :-
   A] *Lycopersicon*   i] Convolvulaceae
      esculentum
   B] *Lagenaria siceraria*   ii] Cabbage
   C] *Ipomoea batata*   iii] Bottle gourd
   D] Phylloxera   iv] Grape
   E] Head   v] Tomato

104. Match the following :-
   A] Lye solution   i] 20-55% sugar
   B] Brine solution   ii] Golden coloured enamel
   C] Syrup   iii] 1-3 %commonsalt
   D] Lacquering   iv] 15-25% salt
   E] Pickling   v] 1-2% caustic soda

105. Match the following :-
   A] $M_1m \times mm$ i] Hermaphrodite : Female
                                   (1:1)

   B] $M_2m \times mm$ ii] Hermaphrodite : Female
                                   (2:1)

Reproductive
Ability to increase in numbers. Development of organs capable of giving rise to new plants.

Gall
A pronounced localised swelling in plant parts with greatly modified tissue structure as a response to irritation by a foreign organism such as rootgall formation by nematode attack.

Floral biology
Study of flower in relation to its opening, dehiscence of anther, pollen viability, stigma receptivity, etc.

Chemotropism
A bending or turning in response to chemical stimulus.

C] $M_2m \times M_1m$ iii] Male : Female (1:1)
D] $M_2m \times M_2m$ iv] Male : Female (2:1)
E] $M_1m \times M_1m$ v] Male : Hermaphrodite : Female (1:1:1)

106. Match the following :-
A] M-9   i] Unsuitable for H.D.O.
B] M-27   ii] Widely used dwarf root stock
C] M-20   iii] Vigorous, Zigzag stem
D] Northern Spy   iv] Ultra dwarf root stock
E] M-25   v] Resistant to wooly aphid

107. Match the following :-
A] Magness (pear)   i] Self-unfruitful, Good for processing
B] Keiffer (pear)   ii] Susceptible to Fireblight & Stony Pit Virus
C] Flemish Beauty (pear)   iii] Mid season; High chilling requirement
D] Bartlett (pear)   iv] Male-sterile, Resistant to fireblight
E] Bosc (pear)   v] Self-fruitful, Good pollinizer

108. Match the following :-
A] Male & hermaphrodite flowers   i] Mango
B] Male, perfect & neutral flowers   ii] Cashew nut
C] Female-hermaphrodite-male   iii] Banana
D] Pistillate & perfect flowers   iv] Strawberry
E] Perfect & staminate flowers   v] Rambutan

109. Match the following :-
A] Maleic hydrazide   i] Seedlessness
B] Gibberellic acid   ii] Vase life
C] Indole butyric acid   iii] Male sterility
D] Ethylene   iv] Ripening
E] 8-H. Q. C.   v] Rooting

110. Match the following :-
     A] Malling -27   i]   Susceptible to tristeza
     B] Flying dragon   ii]   Most dwarfing apple
                                   rootstock
     C] Sour orange   ii]   Most dwarfting mango
                                   rootstock
     D] Vellaikolamban   iv]   Most dwarfing fortunella
     E] St. George, Gloria   v]   Phylloxera resistant grape
111. Match the following :-
     A] *Malus sylvestris*   i]   Early bearing rootstock
     B] *Malus baccata*   ii]   Ancestor of cultivated apple
     C] M-27   iii] Most vigorous rootstock
     D] M-IX   iv]   Crab apple
     E] M-XII   v]   Ultra dwarf rootstock
112. Match the following :-
     A] Mango   i]   Allo-octaploid
     B] Banana   ii]   Auto-tetraploid
     C] Cultivated Strawberry iii] Monosomic
     D] European Plum   iv]   Amphidiploid
     E] Aonla   v]   Aneuploid
     F] Persimmon   vi]   Auto-triploid
     G] Seedless Guava   vii] Allo-hexaploid
     H] 2n + 1   viii] Auto-hexaploid
     I] 2n-1   ix]   Trisomic
     J] 2n- 2   x]   Tetrasomic
113. Match the following :-
     A] Mango   i]   Auto-triploid
     B] European Plum   ii]   Amphidiploid
     C] Cultivated Strawberry iii] Allo-octaploid
     D] Persimmon   iv]   Allo-hexaploid
     E] Cultivated banana   v]   Auto-hexaploid
114. Match the following :-
     A] Mangosteen   i]   *Garcinia mangostana*
     B] Rambutan   ii]   *Nephelium lappaceum*
     C] Kiwi fruit   iii] *Eriobotrya japonica*
     D] Macadamia nut   iv]   *Actinidia deliciosa*
     E] Loquat   v]   *Macadimia ternifolia*

Heat delay
A delay in the initiation of a flower dub due to an abnormally high temperature.

Siliqua
A modified bicarpillary capsule in which a replum separates the locules into two locules which are opened by marginal sutures, pod of plants of mustard family.

Assartage
A type of agriculture in which crops are raised on the same land until it becomes uneconomical due to erosion or disease incidence; the land is then abandoned to produce forest or grassland and a new land is cleared and cultivated for crops.

Glossary

Freeze
1. Weather condition in which the air temperature falls to 32°F or 0°C or lower. 2. To preserve food products by rapidly reducing the temperature to 0°F and maintaining the temperature well below 32°F.

Microbudding
A method of budding in which the bud piece is reduced to very small size but it is not done in aseptic conditions.

Apospory
A form of apomixis in which the embryo sac develops from a vegetative cell of the ovule.

Haptogamy
A form of geitonogamy in which the flower touch one another.

---

115. Match the following :-

A]	Marcottage	i]	Stone-grafting
B]	Shield-budding	ii]	Chrysanthemum
C]	Epicotyle grafting	iii]	Air-layering
D]	Appertizing	iv]	Tea
E]	Dis-shooting	v]	T -budding
F]	Slipping stage	vi]	Canning
G]	Bower system	vii]	Mound layering
H]	Skiffing	viii]	Gladiolus
I]	Stooling	ix]	Blanching
J]	Scalding	x]	Pergola

116. Match the following :-

A]	Mattocking	i]	Guava
B]	Wintering	ii]	Chrysanthemum
C]	Bending	iii]	Citrus
D]	Flame peeling	iv]	Canna
E]	Disbudding	v]	Grape
F]	Indexing	vi]	Banana
G]	Dwarfing	vii]	Rose
H]	Cane pruning	viii]	Cauliflower
I]	Propping	ix]	Onion
J]	Blanching	x]	Bonsai

True / False

117. Riverbed cultivation is suitable for cucurbits.   T/P

118. Tomato is a day neutral plant.   T/F

119. Ashgourd (petha) shows very high rate of respiration.   T/F

120. Banana is seedless due to embryo abortion.   T/F

121. Bunchy top in banana is transmitted through honey bee.   T/F

122. Dashehari mango is originated from Malihabad.   T/F ·

123. In South India, phalsa is pruned twice a year.   T/F

124. Pea is rich source of pectin.   T/F

125. Premature die-back syndrome 'Apoplexy' occurs in apricot.   T/F

126. Zinc deficiency is a limiting factor for tea plantation.   T/F

127. *Cocos nucifera* is an important plantation crop. T/F
128. Edible part of cauliflower is flower. T/F
129. Sigatoka is a serious disease of citrus. T/F
130. Snowball is a variety of dahlia. T/F
131. Tart mangoes are not suitable for squash making. T/F
132. Vegetables showing low respiration rate have long storage life. T/F
133. World Coconut Germplasm Collection Centre (WCGCC) is located at Seepighat (Andaman) T/F
134. 'Early Rivers' cherry is resistant to cracking. T/F
135. Alphonso variety of mango is widely grown in A.P. T/F
136. High density planting of banana causes 'finger tip' disease. T/F
137. Pomegranate fruit is a Balausta. T/F
138. Pusa Delicious is a gynodioecious variety of apple. T/F
139. Rains in Feb.-March are ideal for blossoming in coffee. T/F
140. Vector of Yellow Vein Mosaic in okra is virus. T/F
141. Weeping jelly is formed due to excess of acid. T/F
142. 'Mridula' is a hybrid variety of pomegranate. T/F
143. Amrapalli is a vigorous variety of mango. T/F
144. Banana & date are rich source of energy. T/F
145. Datepalm fruit is one seeded berry. T/F
146. *Gardenia lucida* is a deciduous shrub. T/F
147. *Gomphrena* is a rainy season flowering annual. T/F
148. Sugarbaby is a variety of watermelon. T/F
149. Vrindavan Garden is located in Mysore. T/F
150. Yellow disease of pineapple occurs due to excess of Mn. T/F
151. Arka Jyoti variety of watermelon is released from IIVR. T/F
152. Edible part of pomegranate and litchi is aril. T/F
153. Florida Sun is a subtropical variety of pear. T/F
154. Heleid midge is main pollinator of cocoa. T/F

Dormancy
The period of inactivity in buds, bulbs, and seed when growth stops. Some change in environment usually required before growth will be resumes.

Refrigerator
An insulated cool chamber in which edibles, medicines, etc. are kept to preserve their freshness. It has major parts like, receiver, compressor, evaporator and condensor. The motor compressor compresses a suitable gas (refrigerant) which liquifies and expands in a closed system and in this process it absorbs the latent heat from the surroundings resulting in cooling of the chamber.

155. Pusa Bedana is a variety of grape.   T/F
156. Shoot and fruit borer is a serious insect-pest of mango.   T/F
157. Special taste (Tarrapan) in arecanut chewing is due to polyphenol (Catechine).   T/F
158. Spongy tissue is common disorder of Amrapalli.   T/F
159. Sweet Stella cherry is self-fruitful.   T/F
160. Walnut is the richest source of fat.   T/F
161. 'Compact Stella' is universal pollinizer of cherry.   T/F
162. Carnation is a winter season foliage plant.   T/F
163. *Citrus reticulata* is the botanical name of mandarin.   T/F
164. India is the largest producer of banana & lime.   T/F
165. Kathi & Sloh almonds are resistant to gummosis.   T/F
166. Low temperature is limiting factor for cocoa plantation.   T/F

Fill in the Blanks

167. Kochia is a _____ season foliage annual.
168. _____ is an example of English style.
169. *Wisteria chinensis* is the best climber for _____ place.
170. _____ is scented flowered climber.
171. Rose hips contain _____ Vit-C per 100 g part.
172. _____ is foliage climber.
173. _____ was first rose variety developed by Dr.B.P.Pal.
174. Pigment responsible for blue colour in roses is _____.
175. _____ is family of Bougainvillia.
176. _____ is a variety of Bougainvillia.
177. *Hibiscus* requires _____ situation.
178. _____ is a variety of chrysanthemum.
179. _____ is a single flowered dahlia.
180. *Canna speciosa* is native of _____.
181. Gladiolus belongs to _____ family.
182. Rajat Rekha is an improved variety of _____.

183. Swarna Rekha is a _____ flowered variety of tuberose.

184. *Nymphaea* is a _____ plant.

185. _____ has the largest size leaves.

186. _____ produces the smallest seeds.

187. _____ is an ornamental foliage bulbous plant.

188. *Jasminum humile* is known as _____ .

189. Arka Surabhi is an improved variety of _____.

190. *Poinciana regia* is the new botanical name of _____.

191. *Erythrina indica* (Rakta Madar) flowers during _____.

192. _____ flowers are ideal for Gajra & Veni making.

193. _____ variety of radish is tolerant to high temperature.

194. _____ is effective gametocide to induce male sterility in tomato.

195. Mughal gardens are replica of _____.

196. Papaya exhibits _____ sex form.

197. Datepalm has _____ seeds.

198. Little leaf is a serious disease of _____.

199. Amrapalli is a dwarf variety of _____.

200. Ikebana is a Japanese art of _____.

Systemic
Usually used to indicate that a pesticide could be translocated throughout the plant. Generally longer lasting than a spray application.

Immune
Can not be infected by a given pathogen.

Linkage
Association of characters in inheritance due to the fact that the genes determining them are physically located closely to each other in the same chromosomes.

Antispetic
A chemical substance that inhibits or prevents the growth of bacteria without killing them.

Monkey-hunters use a box with an opening at the top, big enough for the monkey to slide its hand in. Inside the box are nuts. The monkey grabs the nuts and now its hand becomes a fist. The monkey tries to get its hand out but the opening is big enough for the hand to slide in, but too small for the fist to come out. Now the monkey has a choice, either to let go off the nuts and be free forever or hang on to the nuts and get caught. Guess what it picks every time? You guessed it. He hands on to the nuts and gets caught.

We are no different from monkeys. We all hang on to some nuts that keep us from going forward in life. We keep rationalizing by saying, "I cannot do this because.." and whatever comes after "because" are the nuts that we are hanging on to which are holding us back.

[
Successful
people may
Analyze but
never
Rationalize.
]

Step-Ten

Losers are apart from the team;
Winners are a part of the team.

||||||||||| **Multiple Choice** |||||||||||

1. Heteromorphic self -incompatibility is found in-
    A] Carambola   B] Sweet Potato
    C] Brinjal   D] All of them

2. Sporophytic self -incompatibility was first reported by-
    A] Hughes & Babcock   B] East & Mangelsdorf
       (1950)   (1925)
    C] Freidlander (1977)   D] None of these

3. Choose the incorrect one in the group
    A] *Heuia brassilensis*   B] *Ficus elastica*
    C] *Ficus glomerata*   D] *Acecia mangium*

4. Pride of India is variety of -
    A] Cabbage   B] Cauliflower
    C] Radish   D] Turnip

5. The National fruit of India is-
    A] Aonla   B] Jackfruit
    C] Mango   D] Apple

6. The National flower of India is-
    A] Rose   B] Lotus
    C] Marigold   D] Lily

7. In North India, flowering in aonla occurs during-
    A] June-July   B] Sept.-Oct.
    C] Dec.-Jan.   D] March-April

8. In North India, generally pomegranate flowers during-
    A] Sept-Oct.   B] March-April
    C] Jan.-Feb.   D] May-June

9. ' Indian Horticulture' is published by-
    A] CSIR   B] IIHR
    C] ICAR   D] IARI

10. Which are indigenous to India-
    A] Brinjal & Apple   B] Mango & Aonla
    C] Chili & Brinjal   D] All of them

Necrosis
Symptom of pant injury, caused by spray damage, insect or disease injury or other causes, which is characterized by dead, discolored cells and tissues.

Infection
The attack of insect-pest on the host plant.

Interstock
A piece of stem inserted between stock and scion with a view to combining more than two kinds of plants together in a vertical arrangement by grafting.

Di-
A prefix indicating two.

Light flux
The light intensity times, the duration of light.

Foot
The lowering most internode of the floral scape of the tulip.

Exon
Any segment of an interrupted gene represented in the mature RNA product.

Connate
Like parts fused together into one, fused in to a tube.

Amorphophyte
Any plant which bears irregular flowers.

Soil ph
Basically, pH is a measure of the amount of lime (calcium) contained in your soil. A soil with a pH lower than 7.0 is an acid soil, a soil pH higher than 7.0 is alkaline soil. Soil pH can be tested with an inexpensive test kit.

11. Which is indigenous to South Africa-
A] Rose
B] Carnation
C] Gadiolus
D] None

12. *Bemesia tabaci* is an insect vector of-
A] Little leaf of brinjal
B] Leaf curl of tomato
C] Y.V.M of okra
D] All the above

13. The Indian Institute of Vegetable Research is located at
A] Ranchi
B] Delhi
C] Varanasi
D] Bangalore

14. The Central Institute which do research on oil palm is
A] NRCOP
B] NRCP
C] CPCRI
D] CTCRI

15. ' Internal Fruit Necrosis' is a serious disorder of-
A] Guava
B] Potato
C] Pineapple
D] Aonla

16. Which has not come into enforcement till now-
A] Tea Board
B] Cashew Board
C] Coconut Board
D] Spices Board

17. Dahlia flowers are classified into-
A] 5 groups
B] 8 groups
C] 11 groups
D] 15 groups

18. Bacteria are involved in production of
A] Nectar
B] Vinegar
C] Jam
D] Squash

19. Garlic crop is harvested in-
A] 80-100 days
B] 100-115 days
C] 120-160 days
D] 170-180 days

20. Photosynthesis occur in
A] Chloroplast
B] Mitochondria
C] Golgi bodies
D] None

21. How much irrigated area was under wheat cultivation in India during 1998-99
A] 35 mh
B] 32 mh
C] 22 mh
D] 18 mh

22. *Citrus myrtifolia* is a bud mutant of-
    A] *Citrus aurantium*   B] *Citrus sinensis*
    C] *Citrus maxima*   D] *Citrus nobilis*

23. 'Leaf cutting' is a commercial method of propagation in-
    A] Croton   B] Bryophyllum
    C] Dahlia   D] Pothos

24. 'Rome Beauty' is a commercial variety of-
    A] Pear   B] Apple
    C] Grape   D] Strawberry

25. 'Powdery mildew' is a common disease of-
    A] Loquat   B] Pomegranate
    C] Ber   D] Peach

26. 'Spongy tissue' is a common malady of-
    A] Pomegranate   B] Banana
    C] Mango   D] Cherry

27. 'Clonal selection' is a common method applied in-
    A] Guava   B] Cashewnut
    C] Banana   D] Peach

28. 'Fruit drop' is a common problem of-
    A] Mango   B] Litchi
    C] Banana   D] All of these

29. 'Root bulging' is a common process in-
    A] Carrot   B] Onion
    C] Sugar beet   D] Radish

30. 'Sweet Potato' is a crop of-
    A] Kharif season   B] Rabi season
    C] Zaid season   D] Spring season

31. Cultivated strawberry is a cross of-
    A] *Fragaria chiloensis* x   B] *Fragaria virginiana* x *F*
       F. vesca   *chiloellsis*
    C] *Fragaria virginiana* x   D] None of these
       F. vesca

32. 'Honey Dew' is a cultivar of-
    A] Coconut   B] Papaya
    C] Pomegranate   D] Pineapple

Vector
Any organism that is able to transmit a pathogen.

Fruit Juice Concentrate
This is a fruit which has been concentrated by removal of water either by heat or by freezing.

Syconus
A fruit type which is developed from a hollow, pear-shaped, fleshy receptacle enclosing a number of minute, male and female flowers and the fleshy receptacle grows to form the fruit.

Cover crop
A crop which is planted in the absence of the normal crop to control weeds and add humus to the soil when it is plowed in prior to regular planting.

Diaphragmed
Said of pith which is solid with transverse bars of denser tissue at intervals between the nodes (tulip tree) or at the nodes (grape).

Continuous
Said of pith which is solid; not spongy, chambered or interrupted by cavities.

Pyridine
A nitrogenous base which is the nucleus of many organic compounds, for example, nicotine.

Sport
A mutant that is inherited and transmitted to progeny.

Ephemer
An introduced plant which is unable to survive and soon disappears.

33. Red rust is a disease of-
 A] Apple　　　　　 B] Guava
 C] Litchi　　　　　 D] Mango

34. Water core is a disorder of
 A] Pear　　　　　　 B] Peach
 C] Quince　　　　　 D] Apple

35. *Celosia cristata* is a flowering-
 A] Shrub　　　　　 B] Annual
 C] Tree　　　　　　 D] Climber

36. *Cassia alata* is a flowerning-
 A] Tree　　　　　　 B] Bush
 C] Fera　　　　　　 D] Annual

37. 'Arka Vati' is a hybrid variety of-
 A] Bottle gourd　　 B] Chilli
 C] Banana　　　　　 D] Grape

38. 'Arka Shyam' is a hybrid variety of-
 A] Jamun　　　　　 B] Brinjal
 C] Grape　　　　　 D] Mulberry

39. Alternate bearing is a major problem in
 A] Banana　　　　　 B] Mango
 C] Citrus　　　　　 D] Grape

40. Spongy tissue is a major problem in
 A] Banana　　　　　 B] Apple
 C] Mango　　　　　 D] Orange

41. 'Cabbage Head' is a modification of-
 A] Leaves　　　　　 B] Stem
 C] Root　　　　　　 D] Flower

42. Sugarcane photoperiodically is a plant of
 A] Short day　　　　 B] Long day
 C] Day neutral　　　 D] None

43. *Ficus elastica* is a popular
 A] Flowering tree　　 B] Indoor plant
 C] Cacm　　　　　　 D] Climber

44. Pedigree selection is a principle of-
 A] Plant breeding　　 B] Method of breeding
 C] Rootstock　　　　 D] Scion

45. 'Panama wilt' is a serious disease of-
    A] Papaya   B] Guava
    C] Banana   D] Cucumber

46. 'Wooly aphis' is a serious insect pest of-
    A] Litchi   B] Mango
    C] Apple   D] Jack fruit

47. 'Berry borer' is a serious pest of-
    A] Grape   B] Tomato
    C] Strawberry   D] Coffee

48. 'Fruit cracking' is a serious problem in-
    A] Litchi   B] Banana
    C] Walnut   D] Date palm

49. 'Little leaf' is a serious problem in-
    A] Okra   B] Egg plant
    C] Tomato   D] Potato

50. 'Tree culture' is a synonym of-
    A] Arboriculture   B] Silviculture
    C] Horticulture   D] Floriculture

51. "Marcottage" is a synonym of-
    A] Ground layering   B] Air layering
    C] Patch budding   D] Bridge grafting

52. 'Green China' is a trade name of-
    A] Spice   B] Hukka Tobacco
    C] Insecticide   D] Tea

53. Pusa chikni is a var of -
    A] Bottle gourd   B] Ridge gourd
    C] Ash gourd   D] Sponge gourd

54. 'Pant Rituraj' is a variety of-
    A] Ber   B] Egg plant
    C] Tomato   D] Cauliflower

55. 'Pusa Naveen' is a variety of-
    A] Cauliflower   B] Bottle gourd
    C] Tomato   D] Bitter gourd

56. 'Pusa Ratnar' is a variety of-
    A] Chilli   B] Onion
    C] Garlic   D] Broccoli

Autophyte
A plant which manufactures its own food by photosynthesis and is not parasitic or saprophytic.

Carbonation
A method of fruit juice preservation with CO_2. It helps to make mist atmosphere inside the bottled juice by displacing oxygen which ultimately prevents microbial spoilage and fermentation of the juice.

Pectin
Any of the fruit juice substances which form a colloidal substance with water and are derived from pectose in the process of ripening. Derived from citrus fruits and apple wastes, used in jelly-making.

Cambium
The thin membrane located just beneath the bark of a plant.

Borecole
An alternative name for Kale, *Brassica olearacea* L. var. *acephala* DC.

Union
1. The proper joining of a stock with a scion. 2. The place on the stem where such a union has taken place.

Ketchup
Also called catchup, katsup, catsup. A concentrated juice or pulp without seed and skin where spices, salt, sugar, vinegar, etc. are added to the extent that it contains not less than 12 per cent fruit or vegetable solids and 28 per cent total solids.

57. 'Narendra Gobhi' is a variety of-
A] Early Cauliflower   B] Late Cauliflower
C] Early Cabbage   D] Late Cabbage

58. 'Pusa Seedless' is a variety of-
A] Grape   B] Papaya
C] Banana   D] Apple

59. 'Bombay Green' is a variety of-
A] Mandarin   B] Pomegranate
C] Banana   D] Sapota

60. 'Arka Amulya' is a variety of-
A] Mango   B] Chilli
C] Guava   D] Tomato

61. Dwarf Cavendish is a variety of-
A] Mango   B] Grape
C] Tomato   D] Banana

62. 'Pusa Narangi' is a variety of-
A] Marigold   B] Gladiolus
C] Sweet Orange   D] Rose

63. Arka Pragati is a variety of
A] Okra   B] Onion
C] Palak   D] Lettuce

64. Pusa ruby is a variety of
A] Potato   B] Brinjal
C] Tomato   D] Chilli

65. 'White Star' is a variety of-
A] Rose   B] Chrysanthemum
C] Canna   D] Dahlia

66. *Gomphrena globosa* is a _____ season annual.
A] Rainy   B] Winter
C] Spring   D] Summer

67. Which part of citrus is affected by 'Exanthema'-
A] Root   B] Stem
C] Leaf   D] Fruit

68. T budding is also referred as
A] Patch budding   B] Ring
C] Shield   D] Annular

69. 'Japanese garden' is an example of-
    A] Formal style
    B] Informal style
    C] Persian style
    D] Picturesque garden

70. Element which is an important constituent of organic compound is
    A] Carbon
    B] Phosphorous
    C] Potassium
    D] Nitrogen

71. 'Pusa Mukta' is an improved variety of-
    A] Cauliflower
    B] Cabbage
    C] Carrot
    D] Cucumber

72. 'Pusa Komal' is an improved variety of-
    A] Pea
    B] Cow pea
    C] Kidney bean
    D] Cucumber

73. Cryo preservation is associated with
    A] Liquid nitrogen
    B] Liquid oxygen
    C] Liquid potassium
    D] Liquid carbon di-oxide

74. Edible portion of cauliflower is a-
    A] Modified stem
    B] Modified Inflorescence
    C] Modified leaves
    D] Modified flower buds

75. Which crop is best suited for Green manuring
    A] Moong
    B] Dahincha
    C] Cowpea
    D] All

76. Citrus canker is caused by
    A] Fungus
    B] Bacteria
    C] Virus
    D] Mycoplasma

77. Damping off is caused by
    A] Pythium spp
    B] Fusarium spp
    C] Alternaria spp
    D] All

78. Powdery mildew of cucurbits is caused by-
    A] *Sphaerotheca fuiginea*
    B] *Colletotrichum* sp.
    C] *Pythum* sp.
    D] *Pseudomonos cubensis*

79. 'Root Knot' is caused by-
    A] Virus
    B] Nematode
    C] Mycoplasma
    D] Algae

Zen-ei- ka
A Japanese flower arrangement in which straight plant materials of uneven height are used in combination with any other inert material that a person can conceive of but the arrangement is pleasing and in agreement with the surrounding of the room.

Fungicide
A chemical that kills fungi or prevents them from infecting healthy plant tissue.

Herbicide
A chemical that kills or retards plant growth. Herbicides may kill the entire plant; or they may kill only the aboveground plant parts, leaving the roots alive.

80. Citrus dieback is caused by-
    A] Water stress   B] Virus
    C] Fungi & other   D] None

81. 'Rickets' disease is caused due to deficiency of-
    A] Vit-A   B] Vit-B
    C] Vit-C   D] Vit-D

82. 'Stone grafting' is commercially practiced in-
    A] Mango   B] Litchi
    C] Aonla   D] All of above

83. Passion fruit is commercially propagated by
    A] Cutting   B] Seed
    C] Grafting   D] Budding

84. Triploid variety is commercially used in
    A] Muskmelon   B] Cucumber
    C] Watermelon   D] All

85. 'Epicotyl grafting' is commercially useful in-
    A] Mango   B] Ber
    C] Grapes   D] Guava

86. Male sterility is common in
    A] carrot   B] Bean
    C] Okra   D] None

87. Tissue culture is common in
    A] Rose   B] Orchid
    C] Gladiolus   D] Palms

88. Indian Gooseberry is common name of
    A] Ber   B] Cherry
    C] Aonla   D] Banana

89. 'Double working' is commonly followed on-
    A] Peach   B] Apple
    C] Mango   D] Ber

90. Mango malformation is controlled by
    A] IAA   B] NAA
    C] 2, 4-D   D] ethophon

91. Back cross is crossing with
    A] Homoczygous recessive female parents
    B] Either of parents
    C] Recurrent parents
    D] None

92. Which tree is deciduous in nature
    A] Litchi
    B] Avacado
    C] Loquat
    D] Pear

93. Which vegetable is dioecious in nature and propagated by stem cutting
    A] Spinach
    B] Sweet potato
    C] Tobacco
    D] Pointed gourd

94. Pectin test is done by
    A] Jelmeter
    B] Alcohol
    C] Acid
    D] None

Matching

95. Match the following :-
    A] McIntosh, Crimson Gold   i] Triploid var.of apple
    B] Crispin, Baldwin   ii] Pollinizer of apple
    C] Discovery, Cox's Orange   iii] Dessert apple
    E] Golden Noble, Newton Wonder   iv] Diploid var. of apple
    E] Golden Delicious   v] Culinary apple

96. Match the following :-
    A] Metaxenia   i] Cho-Cho, Pointed gourd
    B] Gynoecium in cucumber   ii] Muskmelon
    C] Vegetative propagation   iii] Pointed gourd. Snake gourd
    D] Triploidy   iv] Cucumber, Bottle gourd
    E] Vivipary   v] Cho-Cho
    F] Monoecious lines (M-I, M-2)   vi] Muskmelon, Cucumber, Pumpkin
    G] Maleness in cucurbits   vii] Single dominant gene

Plot

A small piece of land, usually rectangular in shape and of definite size, used in comparing yields of crop varieties, testing different application of fertilizers, comparing different methods of tillage, etc. in agriculture or horticulture research.

Garden

A plot or land devoted to the growing of flower, shrubs, flowering and shade trees, creepers, herbs, other ornamental plants, fruit trees and vegetables in certain manners.

Arboretum

A garden with a large collection of trees and shrubs cultivated for scientific or educational purposes.

Annuals
Plants whose life cycle lasts only one year, from seed to blooms to seed.

Dsibudding
The removal of lateral flower buds on stems of plant such as carnation and chrysanthemum.

Stolons
Surface roots that travel laterally and root at the joints, growing new plants. An example is ground ivy (*Glechoma hederacea*, also called creeping Charlie).

Filiform
An extreme condition of leaf in certain plants which develops into a sinuous ribbon or thread shape, due to disease.

H] Pollination & fruit set in morning (6-8 AM)   viii] Ridge gourd, Bottle gourd

i] Pollination & fruit set in mid-day   ix] Water melon

J] Pollination & fruit set in early   x] High temp, Long day, Higher N_2 morning

97. Match the following :-
    A] *Mirabilis jalapa*   i] Self-sterility
    B] *Impatiens balsamina*   ii] Homogamy
    C] Orchids   iii] Dioecious
    D] Mulberry   iv] Pruning in October
    E] Potato   v] Self-poisoning

98. Match the following :-
    A] Monochromatic   i] White, Grey, Black
    B] Analogous scheme   ii] Opposite colours
    C] Contrast colour scheme   iii] Centre of lawn
    D] Primary colours   iv] Blue, Green, Violet
    E] Secondary colours   v] Red, Orange, Yellow
    F] Neutral colours   vi] Red, Yellow, Blue
    G] Cool colours   vii] Related colours
    H] Hot colours   viii] One side of lawn
    I] One way herbaceous border   ix] Violet, Green, Orange
    J] Double fronted herbaceous border   x] Single colour

99. Match the following :-
    A] Monoecious tree   i] Pistachionut; Datepalm
    B] Dioecious plant   ii] Pistachionut, Kiwi fruit
    C] Polymorphic species   iii] Strawberry
    D] Polyploid series of spp.   iv] Pecan, Chestnut
    E] 'Xenia' & 'Metaxenia'   v] Olive

100. Match the following :-
    A] Monoecious   i] Cucumber
    B] Andromonoecious   ii] Rambutan
    C] Gynomonoecious   iii] Pumpkin

D] Androgynous iv] Cashew nut

E] Dioecious v] Muskmelon

F] Polygamomonoecious vi] Papaya

G] Androdioecious vii] Coconut

H] Gynodioecious viii] Mango

I] Polygamous ix] Pointed gourd

J] Inbreeding depression x] Onion

101. Match the following :-

A] Monoecious i] Pomegranate

B] Andromonoecious ii] Pointed gourd

C] Dioecious fruit iii] Citrus

D] Dioecious vegetable iv] Banana

E] Heterostyly v] Cucurbits

F] Polyembryony vi] Date palm

G] Mono-carpism v] Loquat

H] Self-incompatibility vi] Muskmelon

I] Stenospermocarpy ix] Onion

J] Male-sterility x] Grape

102. Match the following :-

A] Monogenic fruits i] Bael, Aonla

B] Polyembryonic fruits ii] Guava, Litchi

C] Indigenous fruits iii] Trifoliate orange, Bael

D] Exotic fruits iv] Date palm, Papaya

E] Dioecious fruits v] Jamun, Citrus, Wood apple

103. Match the following :-

A] Monty & Tamuri i] Low Chilling var. of Peach

B] High Gate & Low ii] Male cultivars of Kiwi fruit
    Gate

C] Tribute & Tristar iii] Ornamental var. of Mango

D] Croton & Chitla Afag iv] Day neutral var. of
    Strawberry

E] Florda Sun & Sharbati v] Mutants of Banana

104. Match the following :-

A] Mosaic Mottling i] *Macrosteles divisus*
    Virus in Suran

B] Y.V.M.V in Okra ii] 'N$_2$' deficiency

Opposite
Said of leaves or leaf scars which are paired on opposite sides at each node.

Cadjan
A plaited coconut- leaf split from the midrib.

Interveinal chlorosis
Yellowing of leaf tissues between the veins.

Odd pinnate
A compound pinnate leaf with a terminal or odd numbered leaflet at the tip.

Digitate
Applied to a compound leaf, of which all the leaflets are borne on the apex of the petiole *e.g.,* Salmalia malabrica etc.

Cladophyll
A stem-like leaf.

Stipel

A stipule of a leaflet.

Bipinnate

Twice compound with the leaflets along each side of a common axis. The leaflets are further divided into pinnules.

Palmate

three or more leaflets radiating fan-like from a common base (for example, Horsechestnut); may also refer to lobes of some leaves (for example, Sugar Maple)

Needle

a slender, leaf usually in the shape of a needle or awl commonly found in some conifers, such as Junipers; contrast with scale-like

C] Hypocotyle Necrosis   iii] *Myzus persicae*

D] Yellow Mosaic Virus   iv] 'Mo' deficiency
in Turnip

E] Thin Leaf Virus in   v] 'B' deficiency
Carrot

F] Bigvein Virus in   vi] *Bemisia tabacii*
Lettuce

G] Buttoning in   vii] Leaf hopper
Cauliflower

H] Browning in   viii] French bean
Cauliflower

I] 'Whiptail' in   ix] Flea beetle
Cauliflower

J] Vector of Little Leaf   x] Seed borne

105. Match the following :-
A] Mottled leaf   i] Spike length
B] Gibberellic acid   ii] King of fruits
C] Indoor plant   iii] Sweet Orange
D] Stone weevil   iv] 'Zn' deficiency
E] Washington Naval   v] Caladium

106. Match the following :-
A] Mudigere, Allepey   i] Fenugreek
B] Sadhana, Swati   ii] Cinnamon
C] Prabha, Rajendra Kanti   iii] Cumin
D] Arka Mohini, Arka   iv] Black pepper
Gaurav
E] Nityashree, Navashree   v] Cardamom
F] Subhakara, Subhahara   vi] Turmeric
G] Suprabha, Suruchi   vii] Capsicum
H] Pratibha, Suroma   viii] Coriander
I] S- 79, PF-35   ix] Fennel
J] MC-43, S-404   x] Ginger

107. Match the following :-
A] Mughal Gardens   i] New Delhi
B] Japanese Gardens   ii] Jodhpur
C] Vrindavan Garden   iii] Informal style
D] Buddha Jayanti Park   iv] Bangalore

E] Rose Garden
F} Maitri Bagh
G] Lal Bagh
{h) Taj Garden
I] Mandor Garden
J] Sikander Bagh

v]   Mysore
vi]   Formal style
vii] Bhopal
viii] Chandigarh
ix]   Lucknow
x]   Agra

108. Match the following :-
A] Multibract Bougainvillea
B] Triploid Bougainvillea
C] Double flowered Hibiscus
D] Large flowered Chrysanthemum
E] Floribunda Rose

i]   Perfection
ii] Suryodaya
iii] Mahara
iv] Chandrama
v]   Mahatma

109. Match the following :-
A] Multiple Crown/ Fasciation
B] Fruit cracking
C] Dry neck
D] Bitter pit
E] Core breakdown

i]   Pomegranate
ii] Pear
iii] Apple
iv] Avocado
v]   Sapota

110. Match the following :-
A] Mutation bred rose
B] Mutation bred bougainvillea
C] Variegated bougainvillea
D] Double flowered bougainvillea
E] Hibiscus hybrids
F] Large flowered chrysanthemum
G] Fragrant gladiolus
H] Inter-specific hybrid of okra

i]   Pusa Bedana
ii]   Abhisarika, Madahosh
iii] Punjab Padmini
iv]   Thimma, Archana
v]   Lucky Star, Spice
vi]   Ashirwad, Geetanjali
vii] Varsha Uphar
viii] Partha, Jay Laxmi

Curling
Abnormal bending of the leaf blade downwards along the main vein.

Leaker
A can with visible leakage of the contents through the seams, perforations or nail holes. The leak may appear due to faulty seam, faulty lock seam or pinholes as a result of corrosion from the inside of the can or rusting from the outside.

Aromatic
Fragrantly scented, at least when broken or crushed.

Decussate
situation where opposite leaves at any two consecutive nodes on a stem are oriented at 90 degree angles to one another.

Leaf scar

Scars from which leaves have fallen. They usually occur characteristically either singly (alternate) or paired (opposite) or in groups of more than 2 (whorled) at each node. Leaf scars differ greatly in size and shape, and offer some of the best winter characteristics. The points where woody strands of vascular tissue passed up into the leaf are usually evident, and are called bundle scars or bundle traces. Typical leaf scars are essentially at the level of the stem; but they are raised o a pronounced base or leaf cushion in some cases, or the buds are covered by an articular membrane in other.

I] Inter-varietal hybrid of okra

J] Triploid Watermelon

ix] Beauty, Chandrama

x] Mahara, Cherry Blossom

111. Match the following :-

A] Mutation bred tomato   i] Arka Vardan, Punjab Chhuhara

B] Determinate tomato   ii] Pusa Sheetal

C] Indeterminate tomato   iii] H.S.-102

D] Tomato 'tolerant to high temp   iv] Sel-12,CO-I,CO-3

E] 'Cold tolerant' tomato   v] Arka Saurabh, Best of All

112. Match the following :-

A] N.H.B.   i] Abohar

B] CIPHET   ii] Bangalore

C] CTCRI   iii] Calicut

D] IIVR   iv] Kasaragod

E] IIHR   v] Thiruvananthpuram

F] IISR   vi] Bikaner

G] CISH   vii] Srinagar

H] CPCRI   viii] Varanasi

I] CIAH   ix] Gurgaon

J] CITH   x] Lucknow

113. Match the following :-

A] National Tree   i] Mango

B] National Flower   ii] Cardamom

C] King of Fruits   iii] Ber

D] Queen of Flowers   iv] Coconut

E] King of Arid fruits   v] Lotus

F] King of Spices   vi] Dahlia

G] Queen of Spices   vii] Pansy

H] King of Species   viii] Ashok

!] King of Flowers   ix] Black pepper

J] King of Annuals   x] Rose

114. Match the following :-

A] Nematode resistant grape R.S.   i] *V. arizonica, V. monticola*

B] Phylloxera resistant   ii] *V. doaniana* (Salt Creek), *V.*
    grape R.S.   *champini* Dogridge)

C] Pierce's disease   iii] *V. simpsoni* (Blue Lake), *V.*
    resistant R.S.   *smalliana*

D] Drought resistant   iv] Muscat Humburg; Bangalore
    grape R.S.   Blue

E] Anthracnose resistant v] *V. rupestris* (St.George),
    grape R.S.   *V. riparia* (Riparia Glorie)

115. Match the following :-

A] New Parlette,   i] Seedless
    Niagara, Robin-Cardinal   (Stenospermocarpic)
                               grapes

B] Perlette, Delight,   ii] Seedless (Parthenocarpic)
    Himrod   grape

C] Marvel Seedless,   iii] Seeded grapes
    Perle, Walli's Giant

D] Cordinal, Emperor.   iv] Tetraploid grapes
    Bhokri

E] Black Cornith   v] Mutation bred grapes

116. Match the following :-

A] Non-acidic   i] Mango, Tomato

B] Medium-acidic   ii] Wood apple, Lemon

C] Acidic   iii] Pea, Potato

D] Highly acidic   iv] Non-acidic products

E] Sulphur resistant cans v] Pumpkin, Turnip

True / False

117. Pusa Alankar & Pusa Sanyog are varieties of pumpkin.   T/F

118. Pusa Kesar is a variety of carrot.   T/F

119. Stone weevil is an insect-pest of mango   T/F

120. *Hevea brasilliensis* is botanical name of cocoa.   T/F

121. Kinnow is a cross of King orange x Willow leaf.   T/F

122. Kristin variety of cherry is self-infertile.   T/F

123. *Phoenix dactylifera* has dioecious sex form.   T/F

124. Pusa Ratnar and Punjab Padmini are varieties of onion.   T/F

Daughter bulb
Scales and leaves initiated by, and developing below and around the new daughter apex. This apex arises form a bud in the axil of a scale subtending the old or mother axis.

Fascicled
Clustered, like the leaves on a spur of barberry.

Involucre
Ring of small leaves or bracts just beneath the flower.

Alternate
the arrangement of leaves or buds one at a time along a stem at the nodes

Chlorophyll
The green pigment in leaves. When present and healthy usually dominates all other pigments.

Freckle
A dark-coloured lesion of a plant.

Hypobaric storage
Storage at less than atmospheric pressure for long term holding of flowers and planting material.

Light compensation point
That light intensity at which respiration and photosynthesis of the entire plant are balanced.

Lux
A measure of light spectrum.

Glossy
Shining, reflecting more light than if lustrous.

Lignified
Woody, hardened.

Ellipsoid
Elliptical in section, like a football.

125. Rataul mango is extensively grown in western U.P.   T/F
126. Spinach is the rich source of Vit-A.   T/F
127. Vaishali, Rupali & Naveen are hybrids of tomato.   T/F
128. Younger mango plants are more prone to malformation than older ones.   T/F
129. 'Bharat' is a hybrid of capsicum.   T/F
130. 'Longan' prefers cool tropical climate.   T/F
131. *Amaltas* flowers in rainy season.   T/F
132. Arka Rajhans is a hybrid variety of brinjal.   T/F
133. CCC is a plant growth inhibitor.   T/F
134. Cocoa bears hermaphrodite flowers.   T/F
135. Collar. rot in papaya is caused by *Fusarium* sp.   T/F
136. Kinnow is a commercial variety of pummelo.   T/F
137. Kochia is a rainy season annual.   T/F
138. Prunus mahaleb is also known as perfumed cherry.   T/F
139. Aggregate fruit develops from single ovary.   T/F
140. Botanically mango fruit is drupe.   T/F
141. *Delonix regia* flowers in April-May.   T/F
142. Ethylene is a gaseous hormone.   T/F
143. *Ficus elastica* belongs to family Euphorbiaceae.   T/F
144. Ganesh Gole is a hybrid of cabbage.   T/F
145. Jackfruit belongs to family Juglandacea.   T/F
146. Kagzi lime is highly tolerant to tristeza disease.   T/F
147. Sugar acts as a preservative.   T/F
148. 'Tapping' is practised in rubber plant.   T/F
149. C.I.S.H. is located at Lucknow   T/F
150. Guava wilt is more common is alkaline soils.   T/F
151. Jelmeter is used for estimation of sugar in squash.   T/F
152. Madhu & Milan are F_1 hybrids of cucumber.   T/F
153. Marmalade is prepared from citrus fruits.   T/F
154. Stella alld Vista cherries are universal donor.   T/F

155. Tahiti lime is a triploid and produces seedless fruits.   T/F
156. Valencia is a variety of grapefruit.   T/F
157. 'Brown blast' in rubber plants is caused due to more tapping.   T/F
158. 'Medjool' is an early variety of datepalm.   T/F
159. 'Petha' is prepared from *cucumis melo*.   T/F
160. Almond kernel is rich source of high energy.   T/F
161. Jack fruit is a typical monoecious.   T/F
162. Pruning is essential in banana.   T/F
163. Refrigeration slows down respiration of vegetables.   T/F
164. Sodium benzoate is used for preserving phalsa squash.   T/F
165. Tea is classified under spices crops.   T/F
166. Vapour Heat Treatment (VHT) in mango is mandatory for export.   T/F

Fill in the Blanks

167. Sanjose scale is serious insect-pest of _____.
168. _____ is known as 'King of fruits'.
169. _____ is 'National Flower' of Japan.
170. Chrysanthemum is propagated by _____ stem cuttings.
171. Stooling is commercial method of _____ propagation.
172. _____ is known as 'King of Arid Zone fruits'.
173. Ber is propagated by _____ budding.
174. _____ is summer season foliage annual.
175. Fruit splitting is common problem in _____.
176. 'Fruit Growing in India' was written by _____.
177. Floating vegetable gardens are found in _____ state.
178. Art of growing trees on a small & shallow pot is known as _____.
179. 'Nature in miniature' is a theme of _____ garden.
180. Leaf curl is a serious disease of _____.
181. _____ is chief pollinator of mango.

GLOSSARY

Claw
The constricted petiole- like base of petals and sepals of some flowers.

Hesperidium
A fleshy berry-like fruit with a hard rind and definite longitudinal partitions. Ex. Orange

Gelmeter
A long pipette-like glass apparatus used for determining the concentration of pectin present in fruit extract. It gives direct reading or indication of the amount of sugar needed for one litre of extract.

Knees
Pointed or dome-like outgrowths from cypress roots rising above the water.

Protocorm
A round or tuber - like structure that is the first stage of growth from the embryo to the adult plant.

Lenticels
Very small wart - like structures, breaking through the bark of most young twigs. Corky in texture and made of loosely packed cells, providing gaseous exchange between the inner tissues of the stem and the atmosphere.

Verrucose
Having a wart- like surface.

Palmate
Radiating, fan- like from a common point; as of leaflets of a palmately compound leaf or veins of a palmately-veined leaf. Digitate.

182. Post-harvest sprouting is a problem of _____.
183. 'Root Forking' is common problem of _____.
184. Biennial bearing is common in _____.
185. _____ is the richest source of Vit-C among arid fruits.
186. Cashewnut belongs to family _____.
187. Loquat belongs to family _____.
188. In South India, grape is pruned _____ a year.
189. _____ is suitable month for rose pruning.
190. Kiwi fruit has _____ sex form.
191. 'Two leaf and one bud' term is related to _____.
192. Amaltas flowers during _____.
193. *Poinciana regia* is the botanical name of _____.
194. *Pothos* is shade loving _____ climber.
195. *Allamanda cathartica* flowers are _____ in colour.
196. *Jacaranda mimosaefolia* flowers are _____ in colour.
197. Bending is practiced in _____.
198. Muskmelon exhibits _____ sex form.
199. _____ crops are most suitable for riverbed cultivation.
200. Brinjal is native of _____.

In 1914, Thomas Edison, at age 67, lost his factory, which was worth a few million dollars, to fire. It had very little insurance. No longer a young man, Edison watched his lifetime effort go up in smoke and said, "There is great value in disaster. All our mistakes are burnt up. Thank god we can start anew." In spite of disaster, three weeks later, he invented the phonograph.

Ask yourself after every setback: What did I learn from this experience? Only then will you be able to turn a stumbling block into a stepping stone.

Script your success story

Every success story is also a story of **G**reat failure.

Step-Eleven

Losers see the pain;
Winners see the gain.

Multiple Choice

1. Pusa komal is the var of
    A] French bean   B] Cluster bean
    C] Cow pea   D] Goa bean

2. Winter banana is the var of
    A] Mango   B] Banana
    C] Apple   D] Pear

3. Swarna rekha is the var of
    A] Pointed gourd   B] Bitter gourd
    C] Bottle gourd   D] None

4. First red is the var of-
    A] Rose   B] Aster
    C] Dahalia   D] Carnation

5. Cricket ball is the variety of
    A] Banana   B] Sapota
    C] Mango   D] Amla

6. Pusa yamdagni is the variety of-
    A] Cabbage   B] Carrot
    C] Spinach   D] Radish

7. Washing Navel is the variety of
    A] *Citrus reticulate*   B] *Citrus sinensis*
    C] *Musa paradisica*   D] *Mangifera indica*

8. Arka Krishna is the variety of
    A] Grape   B] Pomegranate
    C] Guava   D] Mango

9. Red gold is the variety of
    A] Marigold   B] Rose
    C] Daisy   D] Petunia

10. Winter banana is the variety of
    A] Pear   B] Peach
    C] Apple   D] Pineapple

Tuft
A small cluster, like that of grass, with the roots intertwined.

Bundle scar
Small dots or lines on the surface of the leaf scar marking the point of original departure of the vascular conducting strands into the leaf. Also called bundle trace.

Sex-linkage
Association or linkage of a hereditary character with the sex because gene controlling the character is on the sex chromosome.

Foliar feeding
Fertilizer applied in liquid form to the plants foliage in a fine spray.

Rostellum
A gland, literally a small beak, on the orchid stigma.

Prune

1. To remove live or dead branches or multiple leaders from standing trees for the improvement of the tree or its timber.
2. Any of the horticultural varieties of the plum (*Prunus* sp.) whose fruit can be preserved by drying and without fermenting.
3. Dried plum, colour of its juice, dark reddish purple.

Parasitic plant

A plant which lives on, and acquires it's nutrients from another plant. This often results in declined vigor or death of the host plant.

Biennial

A plant that lives through two growing seasons. It bears fruit and dies the second year.

11. Pusa narangi is the variety of-
    A] Rose   B] Carnation
    C] Jasmine   D] Marigold

12. 'Test weight' is the weight of-
    A] 100 seeds   B] 1000 seeds
    C] 10000 seeds   D] 10 seeds

13. Which crop is unable to produce seeds in plains-
    A] Radish   B] Cabbage
    C] Broccoli   D] All the above

14. Which vegetable is underground stem
    A] Potato   B] Raddish
    C] Carrot   D] Spinach

15. Which chemical is used as an aid to mechanical harvesting-
    A] Ethylene   B] GA_3
    C] Cycocel   D] ABA

16. 'Tetrazolium test' is used for the determination of-
    A] Seed germination   B] Dead seedlings
    C] Abnormal seedlings D] Seed viability

17. Which part of chrysanthemum is used for the extraction of 'pyrethrum'-
    A] Stem   B] Leaves
    C] Flowers   D] Root

18. Micro propagation is used for
    A] Orchid   B] Gladiolus
    C] Rose   D] Gerbera

19. The term 'Appertizing' is used for-
    A] Syruping   B] Sterilization
    C] Canning   D] Dehydration

20. Which plant is used in treatment of asthma & whooping cough
    A] Dill   B] Davana
    C] Ambrette seed   D] Adlusa

21. Lactic bacteria is used in
    A] Pickle   B] Chatny
    C] Murraba   D] Jelly

22. Which CHOs is used to enhance vase life of flowers
    A] Glucose   B] Maltose
    C] Sucrose   D] None

23. Which hormone is useful for berry elongation and thining in grape-
    A] GA_3 B] Ethephon
    C] CCC   D] NAA

24. Whlch fruit is useful for curing 'diabetes'-
    A] Aonla   B] Jamun
    C] Bael   D] Guava

25. Which annual is useful for shady situation
    A] Balsam   B] Clianthus
    C] Salvia   D] Aster

26. 'Flame Peeling' is useful for-
    A] Onion & Garlic   B] Mango & Guava
    C] Pea & Tomato   D] Apple & Carrot

27. Pusa ruby is var of-
    A] Tomato   B] Potato
    C] Brinjal   D] Pea

28. A phenomenon is which a gene has more than one phenotypic effect
    A] Polymorphism   B] Penetrance
    C] Expressivety   D] Pleiotropism

29. Edible portion of apple is -
    A] Embryo   B] Endosperm
    C] Ovary   D] Thalamus

30. Extension method is
    A] Learning process   B] Teaching process
    C] Learning & teaching   D] None
       process

31. According to ISI tomato has been specialized into
    A] 4 grades   B] 3 grades
    C] 2 grades   D] Only one grade

32. Carbohydrate content in potato is-
    A] 10%   B] 15%
    C] 22%   D] 25%

Entire
Neither toothed nor lobed, as applied to leaves.

Sinus
The space between two lobes, segments, or divisions; as of leaves or flower parts.

Canker
A definite, relatively localized, necrotic lesion primarily on the bark and cambium.

Mesophyll
Large parenchyma cells located within the epidermis layers of a leaf.

Filiform
Thread-like, long and very slender.

Ranked
Foliage is arranged in longitudinal planes around the stem.

33. Dehydration temperature for grape is-
    A] 45-55 °C   B] 67-71 °C
    C] 88-90 °C   D] 100-116 °C

34. The nature of auxins is-
    A] Acidic   B] Alkaline
    C] Neutral   D] Volatile

35. Seedless variety of Mango is-
    A] Amrapali   B] Sindhu
    C] Ratna   D] Neelam

36. Pulsing rate is
    A] Chemical insertion   B] Withholding the pot
       in flower
    C] Both   D] All

37. Thinning out is
    A] Complete removal of   B] Tip removal of shoots
       few branches
    C] Apical bud removal   D] None of these

38. Yellow flowering avenue tree is-
    A] Gulmohar   B] Amaltas
    C] Pride of India D] Fountain Tree

39. The origin of peach is-
    A] India   B] China
    C] Europe   D] South America

40. The origin of cauliflower is-
    A] Italy   B] India
    C] U.S.A.   D] China

41. Indigenous variety of apple is-
    A] Jonathan   B] Ambri
    C] Red Delicious   D] Baldwin

42. Man made fruit crop is-
    A] Kinnow   B] Annona
    C] Banana   D] Phalsa

43. *Citrus latifolia* is
    A] Triploid   B] Diploid
    C] Hexaploid   D] Octaploid

44. Anti-sterility vitamin is-
   A] Vit-K   B] Vit-E
   C] Vit-P   D] Vit-B$_2$

45. TSS of Jam should be minimum at time of end point
   A] 68.5%   B] 70%
   C] 50%   D] 80%

46. ' Jam' can be prepared from-
   A] Single kind of fruit   B] Two kinds of fruits
   C] Three kinds of fruits   D] All of these

47. Pigment in jamun is due to
   A] Anthocynin   B] Carotene
   C] Xanthophyll   D] None

48. 'Syneresis' of jelly occurs due to-
   A] Excess of acid   B] Over cooking
   C] Over cooling   D] Too much sugar

49. In fruit Jelly pectin and sugar is present in ratio of
   A] 1 : 65   B] 65 : 1
   C] 65 : 33   D] 33 : 65

50. How much juice is required in making of fruit jelly
   A] 10%   B] 20%
   C] 25%   D] 30%

51. Poornima, Kamini variety of Anjir is released by
   A] IIHR-Bangalore   B] Lucknow
   C] New Delhi   D] Chennai

52. ' Kera Ganga', 'Kera Shree' and 'Laksh Ganga' are vanetles'of-
   A] Muskmelon   B] Guava
   C] Coconut   D] Cashewnut

53. ' Kinnow' mandarin was bred by-
   A] H.B. Frost (1935)   B] Swingle (1932)
   C] Osbeck (1926)   D] J.H. Hale (1933)

54. Which is known as 'Heaven for orchids'
   A] Cherapunji   B] Kalimpong
   C] Shillong   D] Toklai

Symodial
Plants with a main stem or axis that cease growth each year and new growth arises form the base i.e. Cattleya orchid.

Herbaceous
Plants that are mainly soft and succulent, forming little or no woody tissue.

Experiment
A systematic procedure of making observation under controlled conditions in such a way that they can be used for drawing general conclusions regarding the population under study.

Tapka
A maturity index for mango harvesting indicated by fall of few ripe fruits naturally from that tree.

Double flower
A flower with many overlapping petals which gives it a very full appearance.

Undulate
wavy along a margin

Sinuate
With a strongly wavy margin.

Rhombic
a four- margined leaf that is diamond-shaped, having three prominent tips, two on the side and one at the top

Stipule scars
A pair of marks left after the stipules fall off, to either side of the leaf scar.

Suberin
A water-proof material secreted by cork cells.

55. Which is known as 'queen of night'-
A] *Cestrum nocturnum*   B] *Cestrum diurnum*
C] *Cestrum aurantiacum*   D] *Ixora coccinea*

56. Which is known as "9 O'clock" plant-
A] Balsam   B] Portulaca
C] Nastursium   D] Rose

57. Seeds without endosperm are known as-
A] Monoembryonic seeds B] Polyembryonic seeds
C] Exalbuminous seeds   D] Apomictic seeds

58. A well known cardiotonic is-
A] Cinchona   B] Digitalis
C] Periwinkle   D] Eucalyptus

59. ' L-49' variety of guava was bred by-
A] Cheema & Dhani   B] G.S.Cheema & Deshmukh
   (1920)   (1927)
C] G.S. Randhawa   D] None of these
   (1929)

60. ' La France' was the first variety of-
A] Floribunda rose   B] H.T. Rose
C] Polyantha rose   D] Rambler rose

61. Which is lare seed spices-
A] Fennel   B] Ferugreek
C] Cumin   D] All of them

62. Spice crop largely exported from India
A] Black pepper   B] Cardamom
C] Canine   D] Turmeric

63. Country having largest per capita availability of land is
A] Canada   B] Australia
C] India   D] Japan

64. The state leads in fruit production is
A] Maharashtra   B] Tamil Nadu
C] Andhra Pradesh   D] Orissa

65. The state leads in vegetable production is
A] Bihar   B] Punjab
C] U.P.   D] W.B.

66. For measuring leaf area following instrument is used
    A] Alanometer   B] Anemometer
    C] Porometer   D] Barometer

67. Insect vector of ' leaf curl' in chillies is-
    A] Honey bee   B] Thrips
    C] Khapra beetle   D] Aphid

68. Type of leaf of Bauhinia purpuria
    A] Cordate   B] Reniform
    C] Oblong   D] Optocordate

69. One bud and two leaf picking method is applied in-
    A] Coffee   B] Rubber
    C] Tea   D] Cocoa

70. ' Leaf Scorch' in mango occurs due to-
    A] Deficiency of Calcium B] Excess of Chloride
    C] Deficiency of Sodium   D] Excess of Magensium

71. A beverage containing at least 20% fruit juice and 15% T.S.S with 0.3% acid, is known as-
    A] Squash   B] Nectar
    C] R.T.S   D] Blended juice

72. Fruit beverage containing at least 25% juice, 30% T.S.S and 0.25% barley water is known as-
    A] Fruit juice powder   B] Carbonated beverage
    C] Fruit juice concentrate D] Barley water

73. Type of leaves of Gulmohar
    A] Monopinnate   B] Bipinnate
    C] Oblong   D] None of these

74. Interveinal chlorosis in younger leaves of plant occurs due to deficiency of-
    A] Boron   B] Zinc
    C] Iron   D] Sulphur

75. Stomata in leaves open when guard cells are
    A] Fully turgid   B] Reflexed
    C] Both   D] None

76. Amongst vegetable leaving potato, maximum production is of-
    A] Onion   B] Brinjal
    C] Tomato   D] Cucumber

Peat
Partially degraded vegetable matter found in marshy areas. Peat is commonly used as asoil amendment.

Annatto
A highly soluble colouring matter of vegetable origin derived from the pericarps of the seed of *Bixa orellana* (Annatoo plant) used as a colouring matter of cheese and table butter.

Leaf mold
Partially decomposed leaf matter, used as a soil amendment.

Homogamy
1. The simultaneous maturation of male and female elements in the same flower. 2. Bearing one kind of flowers throughout.

77. The ploidy level of jack fruit is—
    A] Diploid
    B] Triploid
    C] Tetraploid
    D] Hexaploid

78. The FPO licence is given by
    A] Govt of India
    B] State Govt.
    C] Agri. university
    D] None

79. Correlation coefficient lies between
    A] 0 to 1
    B] -1 to +1
    C] 1
    D] 0 to 1

80. Bending of plants towards light is called-
    A] Hydrotropism
    B] Geotropism
    C] Chemotropism
    D] Phototropism

81. Dry fruits like cashew & almond are richest source of
    A] vitamin B
    B] fat
    C] vitamin K
    D] carbohydrates

82. The strap like leaves with undeveloped leaf blade, resulting undeveloped curds in cauliflowers show the symptoms of-
    A] Buttoning
    B] Whiptail
    C] Blindness
    D] Damping off

83. ' Liliput' is a variety of-
    A] Rose
    B] Dahlia
    C] Chrysanthemum
    D] Jasmine

84. Male sterile line is maintained by a
    A] Crossing with heterozygous
    B] Nale fertile line
    C] Both
    D] None

85. The edible part of litchi is-
    A] April
    B] Endosperm
    C] Placenta
    D] Mesocarp

86. Mango seeds loose their viability within-
    A] 4-5 weeks
    B] 6-8 weeks
    C] 1-2 years
    D] 1-2 weeks

87. Seeds of which crop lost their viability if exposed to hot sun-
    A] Guava
    B] Citrus
    C] Muskmelon
    D] Ridge gourd

88. ' Lotus' is propagated by-
    A] Seeds   B] Layering
    C] Runners   D] Budding

89. In grafting, the lower part of stem containing roots is called as-
    A] Stion   B] Scion
    C] Stock   D] Graftage

90. Plantation and spices are main crops of-
    A] Eastern plateau &   B] Western Plateau & Hill zone
        Hill zone
    C] Southern Plateau &   D] Central Plateau & Hill zone
        Hill zone

91. Which is/are main symptoms produced by virus-
    A] Stunted growth   B] Curling of leaf
    C] Yellowing   D] All of these

92. Pure line selection is mainly applicable in-
    A] Self pollinated crops   B] Cross pollinated crops
    C] Often cross   D] All of them
        pollinated crops

93. 'Coleus' is mainly grown for its-
    A] Colourful flower   B] Colourful leaves
    C] Colourful fruits   D] All of them

94. Tikka is major disease of
    A] Tur   B] Sugarcane
    C] Cotton   D] Groundnut

Matching

95. Match the following :-
    A] Non-climacteric fruits   i] Papaya, Banana
    B] Climacteric fruits   ii] Kiwi fruit
    C] Single sigmoid curve   iii] Litchi, Strawberry
    D] Double sigmoid curve   iv] Bael, Cranberry
    E] Triple sigmoid curve   v] Walnut; Fig

96. Match the following :-
    A] Northern Spy   i]   Resistant to Fire blight
    B] Robusta No-5   ii]   Sensitive to Water logging

Tensiometer
A device for measuring the negative pressure (tension of water) in soil *in situ*. It is a porous, permeable ceramic cup connected through a tube to a manometer or vacuum gauge.

Refractometer
An instrument used for measuring the refractive index of liquids. Total soluble solids are estimated by this instrument.

Salometer
An instrument used for measuring the salt concentration of any substance (usually liquid) in terms of degree salometer.

Sac
A bag-like membrane-enclosed cavity in fruit or vegetable, etc.

C] Novale   iii] Resistant to Crown rot
D] M-7   iv] Tolerant to Drought
E] MM-106   v] Resistant to Wooly aphid

97. Match the following :-
A] Nutmeg and Mace   i] *Elettaria cardamomum*
B] Clove   ii] *Myristica fragrans*
C] Cardamom   iii] *Piper nigrum*
D] Cinammon   iv] *Syzygium aromaticum*
E] Black pepper   v] *Cinnamomum verum*

98. Match the following :-
A] Onion   i] Jamnagar
B] Garlic   ii] Sree Pallavi
C] Okra   iii] Sree Nandini
D] Sweet Potato   iv] Arka Kirtiman
E] Colocasia   v] Arka Abhay

99. Match the following :-
A] Open central core   i] Sweet orange
B] Solid central core   ii] Citron
C] Monogenic citrus   iii] Grape fruit
D] Indigenous citrus   iv] Sweet lime
E] Semi-hollow core   v] Poncirus sp.

100. Match the following :-
A] Pairy   i] Gynomonoecious
B] Cucumber   ii] Apple
C] Heterosis   iii] Mango
D] Malling   iv] Flower arrangement
E] Ikebana   v] Bottle gourd

101. Match the following :-
A] Papain   i] GA_3
B] Growth stimulant   ii] Anticaking agent
C] Liquid nitrogen   iii] Freezing agent
D] Bleaching agent   iv] Enzyme
E] Calcium silicate   v] Benzyl peroxide

102. Match the following :-
A] Papaya & Rubber   i] Native of Central Asia
B] Pistachionut & Apricot   ii] Central America origin

C] Bael & Jamun   iii] Native of Asia Minor
D] Guava & Pineapple   iv] Native of India
E] Apple & Chestnut   v] South America origin

103. Match the following :-
A] Parthenocarpic grapes i] Multiple genes interaction
B] Stenospermocarpy   ii] Seedlessness due to pollen stimulus
C] Seedlessness in grape iii] Seedlessness due to embryo abortion
D] Resistance to frost   iv] Multiple allele at a single locus
E] Sexuality in grape   v] Governed by recessive gene

104. Match the following :-
A] Partially Self-fruitful apple i] Crispin, Jona Gold
B] Cross-incompatible apple   ii] Jonathan, Golden Delicious
C] Partially Self-compatible   iii] Self-fruitful
D] Diploid cvs. of apple   iv] Green Sleeves, Kent
E] Triploid cvs. of apple   v] Cox, Suntan

105. Match the following :-
A] Pea   i] Juice & Squash
B] Tomato   ii] Dehydration & Bottling
C] Ginger   iii] Preserve & Candy
D] Radish   iv] Souce & Soup
E] Muskmelon   v] Pickle & Canning

106. Match the following :-
A] Pear Decline   i] *Erwinia* sp.
B] Fire blight   ii] MLOs
C] Pear Psylla   iii] 'B'deficiency
D] Green Stain & Cork Spot   iv] Vector of Quick decline MLO's
E] Blossom Blast & Pitting   v] 'Ca'deficiency

107. Match the following :-
A] Pecan   i] Chilling 500-600 hrs at 7.2°C

Floater-sinker
A method used for separating healthy bulbs for those with basal rot.

Hectare
Area measure in the metric system, equal to 10,000 square metres or 2.471 acres.

Micro nutrients
Mineral elements which are needed by some plants in very small quantities. If the plants you are growing require specific 'trace elements' and they are not available in the soil, they must be added.

Virus
A sub-microscopic obligate parasite consisting of nucleic acid and proteins.

B] Pistachionut   ii] Chilling 700-1000 hrs at 7.2°C

C] Blueberry   iii] Chilling 1000-1600 hrs at 7.2°C

D] Almond   iv] Chilling 250-600 hrs at 7.2°C

E] Apple   v] Chilling 100-700 hrs at 7.2°C

108. Match the following :-
A] Pecan   i] Hybrid of two species
B] Pistachionut   ii] Heterodichogamy
C] Chestnut   iii] Monoecious
D] Olive   iv] Dioecious
E] Cultivated strawberry   v] Evergreen tree

109. Match the following :-
A] Pectin content in jelly   i] 65%
B] Acid content in jelly   ii] 0.5-1%
C] Total soluble solids in jelly   iii] 0.5 -0.6%
D] Acidity in jam   iv] 0.5-0.75% (pH 3.2)
E] Total soluble solids in jam   v] 68%

110. Match the following :-
A] Peterss Chico   i] Pollinizers of Raspberry
B] Daviana, Woodford   ii] Pollinizers of Almond
C] Malling Jewel, Glen Clova   iii] Pollinizers of Pistachionut
D] Matua, Tomuri   iv] Pollinizers of Kiwi fruit
E] Non-Pareil, Marcona   v] Pollinizers of Hazelnut

111. Match the following :-
A] *Petrea volubilis*   i] Winter season annual
B] *Passiflora edulis*   ii] Light climber
C] *Antigonon leptopus*   iii] Annual climber
D] *Lathyrus odoratus*   iv] Deciduous climber
E] *Althea rosea*   v] Heavy climber

112. Match the following :-
A] pH 3.0
B] pH 3.2
C] pH 3.4
D] Immature fruits
E] Over ripe fruits

i] Jelly of 65° brix
ii] Protopectin
iii] Pectic acid
iv] Jelly of 60° brix
v] Jelly of 70° brix

113. Match the following :-
A] Phomopsis blight
B] Frog eye leaf spot
C] Cat faces
D] Leaf curl
E] Little leaf.

i] Virus
ii] Chilli
iii] M.L.O's
iv] Tomato
v] Brinjal

114. Match the following :-
A] Pineapple

B] Ginger
C] Rubber

D] Clove
E] Coconut
F] Litchi
G] Mango
H] Apple

I] Onion
J] Sapota

i] Thailand > Indonesia > Malaysia
ii] Karnataka > A.P. > T.N.
ii] Indonesia > Zangibar > Madagascar
v] India > Thailand > Japan
v] U.P. > Bihar > A.P.
vi] J&K > H.P. > Uttaranchal
vii] M.S. > Orissa > Gujarat
viii] Thailand > Sri Lanka > Hawaii
ix] Bihar > W.B. > U.P.
x] Indonesia > India > Philippines

115. Match the following :-
A] *Poinciana regia*
B] Male sterility
C] Cauliflower
D] Dioecious
E] Salad

i] Onion
ii] Date
iii] Avenue tree
iv] Browning
v] Lettuce

Pollen
The male cells or microspores produced by the stamens.

Obovate
widest above the middle of the leaf or leaflet

Homeoclimactic
Regions that may be miles or continents apart but share similar climates. Lauren Springer writes, for example, "My northern Colorado garden mirrors the extreme climate of the steppes and dry mountains of Central Asia and the eastern Mediterranean."

Tapestry hedge
A row of mixed species of shrub.

Growth habit
General appearance or mode of growth.

116. Match the following :-
    A] *Porana paniculata*   i] Scarlet flower
    B] *Jasminum humile*   ii] White flower
    C] *Pyrostergia venusta*   iii] Orange flower
    D] *Thunbergla grandiflora* iv] Yellow flower
    E] *Clerodendron splendens* v] Mauve flower

True / False

117. 'Apple colour' guava is tolerant to anthracnose. T/F
118. 'Brown blast' in rubber is a bacterial disease. T/F
119. 'Two leaf and one bud' is picking method of coriander. T/F
120. Anthocyanin is responsible for red colour in tomato T/F
121. Gynodioecious papaya produces inferior quality fruits than dioecious. T/F
122. In N. India, aonla flowers twice a year. T/F
123. Jaldhup pineapple is the most suitable for canning. T/F
124. Potassium metabisulphite is unsuitable for coloured juices. T/F
125. Southern slopes are best for almond cultivation. T/F
126. 'Internal Breakdown' is a physiological disorder of mango. T/F
127. Banarasi Karaka is a variety of aonla. T/F
128. Beheading in cashewnut is done in June-July. T/F
129. Bitterness in chilli is found in seeds. T/F
130. Coconut prefers hot & dry climate. T/F
131. First crop (*breba*) of San Pedro fig produces parthenocarpic fruits. T/F
132. *Lagerstroemia spaciosa* is the botanical name of Sawani. T/F
133. Lye peeling is suitable for oranges. T/F
134. Spring frost is limiting factor for almond cultivation. T/F
135. Whiptail develops due to 'B' deficiency. T/F
136. Cabbage seeds can be produced in plains. T/F
137. Fig is second richest source of sugar next to date. T/F
138. *Jacaranda mimosaefolia* is blue flowered Gold mohar. T/F

139. July to Sept. period is best for budding in almond.   T/F
140. Kerala is the highest producer of raw cashew.   T/F
141. Loquat can be grown successfully under tropical regions. T/F
142. Mango anthracnose is more prevalent in humid areas.   T/F
143. Rangpur lime is a dwarfing root stock of citrus.   T/F
144. Sugarbaby is an improved variety of sugarbeet.   T/F
145. 'Browning' in cauliflower occurs due to N_2 deficiency.   T/F
146. Apple seeds require 'after ripening' for germination.   T/F
147. Banana is an annual fruit tree.   T/F
148. Datepalm can be grown successfully at low temperature. T/F
149. Fountains & bridges are features of Japanese garden.   T/F
150. Red colour of jelly appears due to pectin.   T/F
151. Rose root stock is commonly propagated by cutting.   T/F
152. Soft wood grafting in mango was standardized by Dr. R.S. Amin.   T/F
153. *Ziziphus nummularia* is a dwarf root stock of ber.   T/F
154. 'Bunchy top' in banana is a viral disease.   T/F
155. Abscission layer is formed due to ethylene.   T/F
156. Amaranthus is a summer season vegetable.   T/F
157. Coconut is a typical tropical plant.   T/F
158. Flowering in almond occurs during May-June.   T/F
159. Karonda is also known as 'Christ's thorn'.   T/F
160. *Inga dulces* is the best for ornamental hedge.   T/F
161. Parbhani Kranti is a variety of okra.   T/F
162. Water suckers are better than sword suckers for banana planting.   T/F
163. Arecanut is highly susceptible to sun-scald.   T/F
164. Banana is propagated by roots.   T/F
165. Ber and Aonla prefers humid climate.   T/F
166. Exanthema is a viral disease of citrus   T/F

Androecious
A sex form in monoecious species where only staminate (male) flowers are produced, giving rise to supermale plant.

Autopolyploid
A polyploid containing more than two copies of the same genome of a single species *i.e.,* autotriploid (3x), autotetraploid (4x), autohexaploid (6x), *e.g.,* potato, sweet potato, etc.

Quassia
A tropical tree, mostly found in north-east Brazil and West Africa, of the Simarubaceae family, whose wood when boiled in water gives an extremely bitter extract; this extract is sometimes used as insect-repellant.

Gamete
A unisexual cell which must fuse with another gamete to produce a new individual.

Bud-mutation
A somatic mutation or variation, arising in bud, producing an abnormal branch.

IBA
Indole-3 -n- butyric acid, an auxin used to promote rooting of cuttings.

Coherent
Two or more similar parts or organs touching one another in very close proximity by the tissues not fused.

Composite
Compound; the common name of a member of the Compositae.

Fill in the Blanks

167. _____ variety of mango is susceptible to spongy tissue.
168. _____ is ultra dwarf root stock of Apple.
169. Datepalm has _____ sex form.
170. Umran is improved variety of _____.
171. _____ is essential Ingradlent for Jelly making.
172. 'Rose Scented' is a variety of _____.
173. There are _____ flushes of loquat in N. India.
174. Ikebana is _____ art of floral arrangement.
175. Flower adornment for condolence is known as _____.
176. Vrindavan Garden is situated at _____.
177. Mandor Garden is located at _____.
178. Sikander Bagh is situated at _____.
179. Red Fort Garden (N. Delhi) was built by _____.
180. Kachnar trees in Mughal Gardens symbolize _____.
181. Baradari is a typical feature of _____.
182. Shobha (gladiolus) is mutant of _____.
183. Geometrical style is also known as _____.
184. _____ are used for herbaceous borders.
185. Croton is an ornamental _____ plant.
186. Buddha Jayanti Park (N. Delhi) is an example of _____ style.
187. Mohini is a variety of _____ rose.
188. *Monslera deliciosa* is a _____ climber.
189. Arka Jay and Arka Vijay are varieties of _____.
190. Arka Ajit is a variety of _____.
191. Cluster bean (Guar) is generally grown in _____ climate.
192. Bhasinda is underground stem of _____.
193. Bolting and Buttoning are disorders of _____.
194. _____ root stock of apple is resistant to wooly aphid.

195. _____ is tetraploid root stock of mango.
196. Dominant beak of _____ mango is identified as genetic marker.
197. _____ is monotype genus of citrus.
198. Vesicles in citrus fruits are arranged in _____ fashion.
199. *Brassica caulorapa* is the botanical name of _____.
200. Club rot of cauliflower is serious in _____ soils.

A flood was threatening a small town and everyone was leaving for safety except one man who said, "God will save me. I have faith." As the water level rose a jeep came to rescue him, the man refused, saying "God will save me. I have faith." As the water level rose further, he went up to the second storey, and a boat came to help him. Again he refused to go, saying, "God will save me. I have faith." The water kept rising and the man climbed on to the roof. A helicopter came to rescue him, but he said, "God will save me. I have faith." Well, finally he drowned. When he reached his maker he angrily questioned, "I had complete faith in you. Why did you ignore my prayers and let me drown?" The Lord replied, "Who do you think sent you the jeep, the boat, and the helicopter?"

The only way to overcome the fatalistic attitude is to accept responsibility and believe in the law of cause and effect rather than luck. It takes action, preparation and planning rather than waiting, wondering or wishing, to accomplish anything in life.

Script your success story

[Luck favours those who Help Themselves.]

Step-Twelve

Losers see problems;
Winners see possibilities.

Multiple Choice

1. ' Makhana' is produced from-
   A] *Nymphaea* sp.　　　　B] *Vectoria regia*
   C] *Euryale ferox*　　　　D] All the above

2. In pointed gourd field male plants should be-
   A] 2-5%　　　　B] 10-15%
   C] 20-25%　　　　D] 25-50%

3. A ' male sterile' variety of peach is-
   A] Shan-e-Punjab　　　　B] J.H. Hale
   C] Elberta　　　　D] All of these

4. By crossing male x female papaya, progeny will be dioecious in ratio of-
   A] 25:75　　　　B] 75:25
   C] 50:50　　　　D] 60:40

5. ' Malling' rootstocks of apple are generally propagated by-
   A] Seed　　　　B] Air-layering
   C] Stooling　　　　D] Cutting

6. Seeds of mandarin orange is sown-
   A] Immediately　　　　B] After extraction
   C] After dormancy　　　　D] After stratification

7. Identify the mango cultivar suitable for canning
   A] Alphonso　　　　B] Amrapalli
   C] Neelum　　　　D] Bombay Green

8. The major constraint in mango cultivation is-
   A] Stone borer　　　　B] Spongy tissue
   C] Biennial bearing　　　　D] All of them

9. Family of mango is
   A] Rosaceae　　　　B] Musaceae
   C] Vitaceae　　　　D] Anacardiaceae

Ecosystem
Stable, though not necessarily permanent, community of plants that have developed interrelationships with each other and with native wildlife to form a distinct, self-sustaining system. A few examples of ecosystems are tallgrass prairie, boreal forest, estuary, and oak savannah. Though ecosystems are a useful concept, in real life a "pure" ecosystem is unusual; more common are areas in which several ecosystems overlap to various degrees.

Leaf scorch
Crescent shaped necrotic areas that develop along the margin and tips of leaves as a result of physiological imbalances.

Fumigun
A large 'hypodemic needle' device for hand-injection of fumigant chemicals in to the soil.

Propagation
Means production of new individuals.

Suckers
On woody plants, new stems that emerge from the roots. These can occur next to existing stems or many feet distant, depending on the species and how far the roots spread.

Purines
A group of nitrogenous organic compounds such as uric acid, xanthine and caffein. The compounds have hydrogen and nitrogen.

10. ' Mango malformation' in mango was first observed in-
   A] Uttar Pradesh (1900)
   B] Bihar (1891)
   C] Orissa (1905)
   D] Maharashtra (1875)

11. Longevity of mango seed is
   A] 1 month
   B] 2 months
   C] 3 months
   D] 4 months

12. Flower bud differentiation in mango takes place in-
   A] July-August
   B] October-December
   C] February-March
   D] May-June

13. The principal pollinator of mango tree is-
   A] Honey bee
   B] House sparrow
   C] Mealy bug
   D] House fly

14. Compared to mango, banana & guava, apple is most rich in-
   A] Vitamin A
   B] Calcium
   C] Phosphorous
   D] Potassium

15. Which is not correctly matched-
   A] Aonla-T-Budding
   B] Grape-Cutting
   C] Mango-Inarching
   D] Guava-Wedge grafting

16. Which is not correctly matched-
   A] Colocasia blight-. *Pythium* sp
   B] Kinnow-Mandarin
   C] Wilt-Guava
   D] Dieback-Cucurbits

17. Which is not correctly matched-
   A] Sweet sop- *A.squamosa*
   B] Sour sop-*A. muricata*
   C] Bullock's Heart- *A. reticulata*
   D] Custard apple-*A. atimoya*

18. The most suitable packaging material for cut flower is-
   A] Card board boxes
   B] Wooden boxes
   C] Plastic boxes
   D] Iron boxes

19. ' Mauritius' is a variety of-
   A] Grape
   B] Oil palm
   C] Pineapple
   D] Wood apple

20. State accounting maximum area & production of cardamom is-
    A] Tamil Nadu   B] Kerala
    C] Karnataka   D] Meghyala

21. Crop requiring maximum irrigation
    A] Maize   B] Gram
    C] Barley   D] Sugarcane

22. The simplest measure of variability in a data set is -
    A] Mean   B] Mode
    C] Range   D] Median

23. TSS is measured by
    A] Hand refractometer   B] Hydrometer
    C] Peperinometer   D] None

24. ' Meera', 'Nazrana' and 'Apsara' are varieties of-
    A] Rose   B] Gladiolus
    C] Chrysanthemum   D] Dahlia

25. ' Mercury' is found in-
    A] Beet root   B] Onion
    C] Garlic   D] Radish

26. Back cross method of breeding means-
    A] Mutation breeding   B] Polyploidy breeding
    C] Tissue culture   D] Croosing f_1 to desirable
                               parents

27. The propagation method of cashew nut is
    A] Patch budding   B] Softwood grafting
    C] Cutting   D] None

28. The quickest method of developing pureline variety of a heterozygous population is through
    A] Selfing   B] Submatirig
    C] Doubling of   D] Doubling chromosome of
       chromosome of haploids   diploids

29. Most suitable method of extraction of floral perfumes is-
    A] Enflurage   B] Distillation
    C] Maceration   D] Solvent extraction

Superficial
On the surface, not connected to inner tissues.

Tepal
A segment of perianth not differentiated into calyx or corolla. Ex.- tulip, magnolia.

Fleshy
For a stem, not hard and woody; for a fruit or bud scales, not dry. Succulent.

Twiner
A herb that does not possess special structures but climb over support by twining themselves spirally e.g., Honeysuckle, Madhavi, etc.

Included
Not protruding as stamens not projecting beyond a corolla; as opposed to exserted.

Subcontinuous pith
With occasional but not regular gaps in the pith.

Nitrogen fixer
Some plants (notably many legumes such as clover, peas, beans, and alfalfa) have the ability to transform nitrogen from the air into a form of nitrogen that plants can use. As nitrogen is a key nutrient that often limits growth, nitrogen fixers can influence the fertility of the surrounding soil and the growth of neighboring plants. The actual "fixing" or chemical transformation is performed by bacteria that live in nodules on the roots of the host plant.

30. Most economic method of irrigation of an orchard under water scarcity condition is
    A] Flood system   B] Ring system
    C] Sprinkler system   D] Drip system

31. In which method of planting, maximum number of plants per unit area can be accommodated
    A] Square   B] Hexagonal
    C] Contour   D] Diagonal

32. 'Tissue culture' is a method of................propagation.
    A] Sexual   B] Asexual
    C] Natural   D] Apomixis

33. Frontier is mid season variety of-
    A] Peach   B] Plum
    C] Apricot   D] Grape

34. Which non -mineral nutrient is essential for plant
    A] Carbon   B] Nitrogen
    C] Phosphorous   D] Potassium

35. Endosperm is missing in seed of
    A] Carnation   B] Rose
    C] Chrysnthemum   D] Orchid

36. Sweet potato is a modified form of-
    A] Potato   B] Root
    C] Stem   D] Rhizome

37. The residual moisture in dehydrated vegetable should be-
    A] More than 8-9%   B] Less than 6-8%
    C] More than 10-11%   D] Less than 9-10%

38. The ' monoecious form of sex' is found in-
    A] Date palm, & Papaya B] Pistachionut & Hazelnut
    C] Aonla & Coconut   D] Mango & Apple

39. Potato variety more suitable for chips making
    A] Kufri Jyoti   B] Kufri Naveen
    C] Kufri Badshah   D] Kufri Chipsona

40. Which is more useful in hulling of walnut
    A] GA_3 B] Kinetin
    C] IAA   D] CCC

41. 'Yellow Vein Mosaic' is a serious disease of-
    A] Okra   B] Lemon
    C] Bitter gourd   D] Chilli

42. 'Yellow Vein Mosaic' is a serious problem in-
    A] Okra   B] Brinjal
    C] Chilli   D] Tomato

43. Where the most intensive farming is prevalent
    A] China   B] India
    C] Japan   D] Indonesia

44. Which crop is the most suitable for 'river-bed cultivation'-
    A] Watermelon   B] Pumpkin
    C] Muskmelon   D] Bottle guurd

45. Which preservative is the most suitable for coloured juices-
    A] KMS   B] Sodium benzoate
    C] Hydrogen peroxide D] Sorbic acid

46. Vegetable crop mostly affected by thrips
    A] Tomato   B] Okra
    C] Chilli   D] Onion

47. In India mostly grown annual for loose flower
    A] Jasmine-   B] Marigold
    C] Chrysanthemum   D] None

48. 'Diamond Back Moth' is a serious pest of-
    A] Apple   B] Cauliflower
    C] Chilli   D] Cassava

49. ' Mughal gardens' are replica of-
    A] Italian style   B] English style
    C] Japanese style   D] Persian style

50. ' Muletail or Crazy Top' is a genetic disorder of-
    A] Peach   B] Pecan
    C] Walnut   D] Almond

51. A bud mutant cultivar of mango is
    A] Noorjahan   B] Rosica
    C] Creton   D] Ratna

52. B. N. of Rose geranium is
    A] *Pelargonium graneolens* B] *Rose damascena*
    C] *Rosa indica*   D] None

Grassy growth
Excessive and noticeable production of axillary branches such as on stem of snapdragon.

Heteroploid
An individual carrying chromosome numbers other than the diploid (2x) number.

Native plant
Any plant that occurs and grows naturally in a specific region or locality.

Alliin
A colourless, odourless, water soluble amino acid present in the uninjured bulb of garlic which, on crushing, breaks down in presence of the enzyme alliinase to allicin, the principal ingredient of the odoriferous diallyl disulfide.

53. National Horticulture Board (N.H.B.) was established in-
   A] 1975   B] 1984
   C] 1988   D] 1990

54. Movement of NAA and IAA in aerial parts of plants is-
   A] Non-directional   B] Acropetal
   C] Basipetal   D] Gaseous form

55. The botanical name of 'Brimha Kamal' (Lotus) is-
   A] *Nymphaea odorata*   B] *Nelumbo nucifera*
   C] *Euryale ferox*   D] *Vectoria regia*

56. The Botanical name of Indian Jujubee is
   A] *Zizyphus jotundifolia* B] *Z. mauritiana*
   C] *Z. mummularia*   D] None

57. Example of narcotic crop is-
   A] Poppy   B] Isabgol
   C] Vinca   D] Tea

58. Which crop needs maximum water
   A] Barley   B] Maize
   C] Sorghum   D] Sugarcane

59. ' Neelam' variety of mango is a_____ season variety.
   A] Early   B] Mid
   C] Late   D] Very late

60. The Headquarters of International Network for Improvement of Banana and Plantain (INIBAP) is situated in-
   A] Shillong (India) B] Italy (Rome)
   C] Nairobi (Kenya)   D] Montpelliere (France)

61. There is no commercial hybrid of which one-
   A] Tomato   B] Cabbage
   C] Cucumber   D] Peas

62. The 2n no. of cucumber is
   A] 14   B] 24
   C] 38   D] 18

63. Which is non bulb forming member of onion family
   A] Leek   B] Garlic
   C] Onion   D] None

64. Which is non climacteric in nature
    A] Pineapple   B] Mango
    C] Banana   D] Avacado

65. Which is/are non-climacteric fruit-
    A] Lime   B] Pineapple
    C] Litchi   D] All of these

66. Which is nonpungent vegetable when uncooked
    A] Lettuce   B] Raddish
    C] Onion   D] Chilli

67. Which is not a C_3 plant
    A] Maize   B] Barley
    C] Sugarbeet   D] All

68. Which is not a cytockinin
    A] BA   B] Zeatin
    C] Etheral   D] Kinatin

69. Which is not a green house gas
    A] CO_2 B] CH_4
    C] CO D] O_2

70. Which is not a legume crop
    A] Pea   B] Beans
    C] Wheat   D] Fenugreek

71. Which is not a pest of Apple
    A] Wolly aphids   B] San Josescale
    C] Tent caterpillar D] Gall shoot maker

72. Following is not a var of arecanut
    A] mangla   B] sumangla
    C] kaddyam   D] none

73. Who do not adorn the D.G. post of ICAR-
    A] M.S. Swaminathan   B] V.L. Chopra
    C] N .S. Randhawa   D] K.L. Chadha

74. Which crop not belongs to cole crop-
    A] Cabbage   B] Cauliflower
    C] Kale   D] Turnip

Calyx
1. The sepals of a flower as a group.
2. The outermost of the floral whorls, the external part of the flower.

Torus
1. The receptacle of a flower, part of the axis on which the flower parts are inserted. 2. The thickening in the centre of the membrane in bordered pits. 3. Modified end of stem.

Perianth
The two floral envelopes of a flower; a collective term embracing both corolla and calyx as a unit; often used when it is not possible to distinguish one series from the other (as in most monocots) and the parts then called tepals.

75. 'They are not hungry but are thirsty' explains to which type of soils-
    A] Temperate zone soils   B] Coastal soils
    C] Arid zone soils   D] Black cotton soils

76. Which variety of guava not requires 'bending'-
    A] Behut Coconut   B] L-49
    C] Allahabad Safeda   D] Apple colour

77. Which is not the variety of Brinjal
    A] Pant Samrat   B] Pant Bahar
    C] Pant Rituraj   D] Pusa ankur

78. Which is not used as preservative
    A] KMS   B] Sodium benzoate
    C] Acetic acid   D] Benlate

79. ' Nut' crops are generally produced in-
    A] U.P.   B] Kerala
    C] J & K   D] Karnataka

80. ' Nutmeg Act' was enforced during-
    A] 1960   B] 1967
    C] 1973   D] 1989

81. The metal nutrient present in the structure of chlorophyll is
    A] Fe   B] Zn
    C] Ca   D] Mg

82. Micro- nutrient required for photolysis of water in photosynthesis is
    A] Fe   B] Mn
    C] Bo   D] Zn

83. Seed variation is greatly observed in-
    A] Papaya   B] Jamun
    C] Mango   D] Brinjal

84. Metaxenia is observed in
    A] Root crops   B] Tuber crops
    C] Bulb crops   D] Cucurbits

85. Which country occupies first position in production of fruits
    A] Brazil   B] USA
    C] China   D] India

86. The browning occur in
    A] Cauliflower   B] Brinjal
    C] Okra   D] None

GLOSSARY

87. A naturally occurring growth inhibitor is-
    A] GA$_3$ B] Auxin
    C] CCC   D] ABA

88. Lack in reproduction capacity occurs due to deficiency of-
    A] Vitamin-A   B] Vitamin-C
    C] Vitamin-K   D] Vitamin-E

89. Failure in jelly setting occurs due to-
    A] Lack of acid   B] Lack of pectin
    C] Too much sugar   D] All of above

90. Flowering in *Cassia fistula* occurs during-
    A] February-March   B] April-May
    C] June-July   D] September-October

91. 'Browning' disorders occurs in-
    A] Cauliflowers   B] Apple
    C] Cabbage   D] Citrus

92. The concept of 'Bioaesthetic planning' was first propounded by-
    A] Persy Lancaster   B] Lancelot Hogben
    C] William Rabinson   D] G.S. Randhawa

93. Appropriate quantity of 'sodium benzoate' for squash preservation is-
    A] 250 ppm   B] 350 ppm
    C] 500 ppm   D] 600 ppm

94. The concept of 'totipotency' of plant cell was propounded by-
    A] Woodhouse (1950) B] Swingle (1914)
    C] Haberlandt (1902) D] Darwin (1820)

Matching

95. Match the following :-
    A] Pot marigold   i]   Helichrysum
    B] Foliage annual   ii] Nastursuim
    C] Dry flower   iii] Vernonia
    D] Hanging basket   iv] Calendula
    E] Foliage climber   v]   Iresine

Germplasm
A collection of genotypes of an organism.

Fruiting
1. The production of bearing of fruit by a plant. 2. Designating a branch, cane, etc. on which the fruit is borne.

Burning over
The practice of burning straw on fields not due to be lifted and this practice advance flowering.

Pinking
Abnormal pink coloration of caladium leaves due to low temperature.

Pathogen
An organism capable of causing a disease. (such as fungus, bacterium, or virus)

Chimeras

A plant consisting of cells of two or more genotype in the same part of a plant. It may result from mutation, irregular mitosis and somatic crossing over.

Squash

1. The fruit of certain tender, vine-like annual plants of the genus *Cucurbita*, which is used as vegetables.
2. Unfermented fruit juice beverage which consist essentially of strained juice containing moderate quantities of fruit pulp to which canesugar is added for sweetening. It should contain 25 pet cent juice and 45 per cent total soluble solids.

F]   Evergreen hedge
G]   Edging
H]   Bonsai
I]   Indoor plant
J]   Cut-flower

vi]   Kochia
vii]   Dieffenbachia
viii]   Clerodendron
ix]   Banyan
x]   Carnation

96.   Match the following :-
A]   Preserve
B]   Marmalade
C]   Squash
D]   Sauces
E]   Ketchup

i]   40-50% T.S.S.
ii]   30% T.S.S.
iii]   68% T.S.S.
iv]   28% T.S.S.
v]   65% T.S.S.

97.   Match the following :-
A]   Propyl gallate (PG)
B]   Indigo carmine
C]   Lycopene
D]   Pelargonidin
E]   Catechin

i]   Carotenoid
ii]   Tannin
iii]   Antioxidant
iv]   Anthocyanin
v]   Synthetic colour

98.   Match the following :-
A]   Protopectin
B]   Pectin
C]   Vit-A&D
D]   Dextrose
E]   Pectic acid

i]   Fat soluble
ii]   Water insoluble
iii]   Inverted sugar
iv]   Water soluble
v]   Inverted pectin

99.   Match the following :-
A]   *Prunus behini*
B]   *Prunus davidiana*
C]   Fuzziness
D]   Smooth Skin
E]   Veteran

i]   Hybrid of Almond x Peach
ii]   Peach
iii]   High chilling peach
iv]   Ornamental peach
v]   Nectarine

100.   Match the following :-
A]   *Prunus domestica*
B]   *Prunus salicina*
C]   *Prunus cerasifera*
D]   *Prunus insililia*
E]   *Prunus armeniaca*

i]   Myrobalan Plum
ii]   Apricot
iii]   Japanese Plum
iv]   European Plum
v]   Damsons

101. Match the following :
     A] Pungency in chilli   i] Clove
     B] Pungency in garlic   ii] Corn
     C] Pungency in onion   iii] Placenta
     D] Edible part of taro   iv] Pericarp
     E] Edible portion of   v] Scale leaves
        watermelon

102. Match the following :-
     A] Pungency in ginger   i] Methyl iso-thio cyanate
     B] Pungency in Chilli   ii] Quercetin
     C] Pungency in Radish   iii] Allyle propyle disulphide
     D] Bitterness in Potato   v] Calcium oxalate
     E] Pungency in Onion   v] Gingerol
     F] Pungency in Garlic   vi] Capsainthin
     G] Colour in Onion   vii] Capsaicin
     H] Red colour in Tomato   viii] Diallyle-disulphide
     I] Acridity in Colocasia   ix] Solanin
     J] Red colour in Chilli   x] Lycopene

103. Match the following :-
     A] Purple colour of new   i] 'B' deficiency
        leaves in mango
     B] Leaf scorch in mango   ii] Mango
     C] Internal necrosis in   iii] 'K' deficiency
        mango
     D] Tetraploid   iv] Genetic marker
     E] Amphidiploid   v] Bael

104. Match the following :-
     A] Pusa Komal   i] Mango
     B] Arka Neelmani   ii] Watermelon
     C] Arka Anmol   iii] Guava
     D] Arka Amulya   iv] Cowpea
     E] Pusa Jyoti   v] Grape

105. Match the following :-
     A] Pusa Purple Long   i] Winter Squash
     B] India   ii] Grape
     C] Cucurbitaceae   iii] Brinjal
     D] Bonville   iv] Garden Pea
     E] Thompson Seedless   v] Mango

Loam
A rich soil composed of clay, sand, and organic matter.

Hypocarp
A fleshy modified peduncle of certain fruits as cashew apple.

Formalin
A 40% solution of commercial formaldehyde used as preservative of plant tissues.

Strain
A group of plants of common lineage though not taxonomically distinct from others of the species or variety, are distinguished on the basis of some character or characters such as productiveness, vigour, resistance to cold, diseases, etc. or of other ecological or physiogical characteristic.

GLOSSARY

Ploidy
Refers to number of complete sets of chromosome.

Double cooling
A type of cooling in which precooled bulbs are planted an returned to cold storage.

Blackout system
A means of covering plants to shorten the photoperiod, to promote flowering in short day plant such as Chrysanthemum inorifolium. Black sateen clothe or black polyethylene film generally used.

Denuded
Naked through the loss of covering.

Artificial long days
Interruption of dark period or extension of natural day length to prevent flower

---

106. Match the following :-

A] Pusa Rasraj (Muskmelon)   i] Monoecious line (M-3) x Durgapur Madhu
B] Punjab Hybrid (Muskmelon)   ii] PSPR x Sel-11
C] Punjab Rasila (Muskmelon)   iii] Japanese Long Green < Green Long aples
D] Pusa Bedana (Watermelon)   iv] IHR-20 x Crimson Sweet
E] Pusa Sanyog (Cucumber)   v] WMR-29 x Hara Madhu
F] Pusa Alankar (Summer squash) vi] IIHR-54 x IIHR-24
G] Pusa Meghdoot (Bottle gourd) vii] EC-2070SO x 51-1-8
H] Arka Sumeet (Ridge gourd)   viii] Male sterile line(MS₁) x Hara Madhu
I] Arka Jyoti (Watermelon)   ix] P.S.P.L. x Sel-2
J] Pusa Manjari (Bottle gourd) x] Tetra-2 x Pusa Rassal

107. Match the following :-

A] Pusa Red Plum   i] Inter-varietal hybrid of tomato
B] Pusa Ruby   ii] Ancestor of cultivated tomato sp.
C] Rarkuna First   iii] Inter-specific hybrid of tomato
D] *L. peruvianum*   iv] Parthenocarpic tomato
E] *L. pimpinellifolium* v] Resistant to Root Knot nematode

108. Match the following :-

A] Pusa Snow Ball   i] Apple
B] Sharbati   ii] Peach
C] Pusa Chetki   iii] Papaya
D] Honey Dew   iv] Cauliflower
E] Red Delicious   v] Radish

109. Match the following :-
    A] Queen Alizabeth, Crrimson Glory   i] Hibiscus
    B] Mahara, Thimma   ii] Dahlia
    C] Aikata, Shanti   iii] 'Single' Tuberose
    D] Chandrama, Snowball   iv] Rose
    E] White Star, Tamtam   v] 'Double' Tuberose
    F] Radio, Ambassador   vi] Bougainvillea
    G] Eurovision, Friendship   vii] Gladiolus
    H] Jhumka, Vibhuti   viii] Chrysanthemum
    I] Rajat Rekha, Shrinagar   ix] Portulaca
    J] Suvasini, Swaran Rekha   x] Canna

110. Match the following :-
    A] Queen of nuts   i] *Juglans cardiformis*
    B] Butter fruit   ii] Pecan
    C] Heart nut   iii] *Malus sikkimensis*
    D] Butter nut   iv] Avocado
    E] Apomictic sp. of apple   v] *Juglans cinerea*

111. Match the following :-
    A] R T S & Nectar   i] Asepsis
    B] Cleanliness   ii] Refrigeration
    C] Heating above 100°C   iii] Pasteurization
    D] Temp. 0 to 5°C   iv] Sterilization
    E] Temp. 18 to 40°C   v] Freezing

112. Match the following :-
    A] R.S. resistant/tolerant to 'Gummosis'   i] Sour orange
    B] R.S. tolerant/resistant to 'Foot rot'   ii] Sour orange, Trifoliate orange, Troyer Citrange
    C] R.S. tolerant/resistant to 'tristeza'   iii] Rough lemon, Mandarin, Orange, Rangpur lime
    D] R.S. resistant to 'Canker'   iv] Rangpur lime, Kumquat, Tangerine
    E] R.S. tolerant/ resistant to 'Salt'   v] Cleopatra mandarin, Rough lemon, Rangpur lime

Proto-epiphyte
A group of epiphytes that are compelled to acquire nourishment from the surface of the supporting host and from atmosphere e.g., certain ferns, *Peperomia*, etc.

Vascular bundle
A discrete group of conducting vessels.

Vase-life
Longevity of cut flowers.

Pseudo-epiphytes
A group of epiphytes that germinate in the soil but their stems gradually die from below upwards so that at maturity they exist like hemiepiphytes e.g., some Lianes.

Leaching
The removal or loss of excess salts or nutrients from soil. The soil around over fertilized plants can be leached clean by large quantities of fresh water used to 'wash' the soil. Areas of extremely high rainfall sometimes lose the nutrients from the soil by natural leaching.

Slipping of bark
Period of extreme cambial activity in spring when the bark- phloem section of a stem separates readily form the xylem at the cambial layer.

Flower bud aboption
Cessation of floral bud development at any stages of development.

113. Match the following :-
A] R.T.S., Cordial   i]   Sweetened juice
B] Cider, Champagne   ii]   Unfermented beverages
C] 85 % juice & 10% T.S.S.   iii] Ready to serve (RTS)
D] 10% fruit juice & 10% T.S.S.   iv] Fermented beverages
E] 20% fruit juice & 15% T.S.S.   v] Nectar

114. Match the following :-
A] Radish var. tolerant to high temp.   i]   Arka Abhay
B] Temperate type Radish   ii]   Pusa Chetki
C] Hybrid Carrot   iii] Pusa Jyoti
D] Garden beet resistant to bolting   iv] Imperator
E] Salinity resistant Turnip   v] Pran
F] Mutation bred French bean   vi] Pusa Himani
G] Mutation bred Palak   vii] Arka Pragati
H] A natural hybrid Onion   viii] Avon Early
I] Kharif Onion   ix] Pusa Kanchan
J] YVM resistant Okra   x] Pusa Parvati

115. Match the following :-
A] Radish   i]   Detroit, Crimson Globe
B] Carrot   ii]   Pusa Chandrama, Pusa Swarnima
C] Sugar beet   iii] Pusa Navbahar, Sharad Bahar
D] Turnip (Asiatic)   iv] Pusa Reshmi, Arka Nishant
E] Temperate type Turnip   v] Pusa Komal, Arka Garima
F] French bean (Rajmah)   vi] Pusa Meghali, Pusa Kesar
G] Cowpea (Lobia)   vii] Arka Jay, Arka Vijay
H] Dolichos bean (Sem)   viii] Pusa Kanchan. Pusa Sweti

J] Cluster bean (Guar) ix] Pusa Ratnar, Pusa Red
J] Red skinned Onion x] Pusa Parvati, Arka Komal

116. Match the following :-

A] Recurrent apomixis i] Mangosteen
B] Zygotic seed ii] Citrus
C] Nucellar budding iii] Papaya
D] Parthenogenetic seed iv] Banana
E] Crown v] Blackberry
F] Shoot tip culture vi] Cashew nut
G] Leaf bud cutting vii] Mango
H] Soft wood grafting viii] Jamun
I] Micro grafting ix] Apple
J] Polyembryony x] Pineapple

True / False

117. House fly is chief pollinator of almond. T/F

118. $K_2 S_2 O_5$ is the formulae of ethrel. T/F

119. Litchi is commercially propagated by seed. T/F

120. Pusa Reshmi variety of radish is free from pithiness. T/F

121. *Quisqualis indica* flowers are red-white mix. T/F

122. 'Tristeza' is a viral disease of citrus. T/F

123. Chilling requirement of almond is lower than peach. T/F

124. Fruit fly is serious in guava. T/F

125. Grape is deciduous under tropical conditions. T/F

126. Hollyhock is a tall bush. T/F

127. Incompatibility in coffee is gametophytic type. T/F

128. Mallisona cultivar of almond is self-fruitful. T/F

129. Sioux is a variety of tomato. T/F

130. Arkel is the earliest variety of pea. T/F

131. Banana & pineapple are monocot fruits. T/F

132. Early rains is a limiting factor in commercial cultivation of grape & date in N. India. T/F

133. Greening is a disease of banana. T/F

134. India ranks third in cashewnut production. T/F

Flower initiation
Visible organization of flower primodia (buds) at the stem apex.

Drying
A method of preservation of foods by reducing the moisture content to the point at which microorganisms fail to survive on them and the action of enzymes is also checked as well and in this situation osmotic pressure due to increased dissolved soilds act as active agent for preservation.

Greenhouse phase
That portion of forcing which encompasses the time form placing the pants in the greenhouse until flowering.

Terrarium
A closed type of glass container used to provide usual environment for the growth of plants.

Over Walling
This mode of healing is system mostly on prunning wounds.

Indeterminate
Said of those kinds of inflorescence whose terminal flowers open last, hence the growth or elongation of the main axis is not arrested by the opening of the first flowers.

Graftage
It is process of joining a part of plant with another in such a way that both will unite to work as a unit and unit will continue to grow.

135.	Jack fruit belongs to family Sapindaceae.	T/F
136.	Kochia is a summer season flowering annual.	T/F
137.	Mule Tail/Crazy Top is genetical disorder of almond.	T/F
138.	Wood apple belong to family Rosaceae.	T/F
139.	'Compierganj' is a culinary variety of banana.	T/F
140.	Barbados cherry is the richest source of Vit-C	T/F
141.	Coffee rust is caused by Algae.	T/F
142.	First International Symposium on Plantation Crops was held in India.	T/F
143.	Giant Kew is a variety of papaya.	T/F
144.	Indian Journal of Horticulture is published by ICAR.	T/F
145.	*Inga dulces* is suitable for protective hedge.	T/F
146.	Murraya *paniculata* is suitable for topiary work.	T/F
147.	Self-sterility in almond is due to genetic incompatibility.	T/F
148.	*Amrapalli* mango is suitable for kitchen garden.	T/F
149.	Bitterness in almond kernel is governed by single recessive gene.	T/F
150.	Coffee is short day plant.	T/F
151.	I.B.A. is a natural plant hormone.	T/F
152.	In determinate type tomato apical growth is unlimited.	T/F
153.	In northern India, litchi ripens in June.	T/F
154.	Inarching is an attached method of grafting.	T/F
155.	Journal of Ornamental Horticulture is published by IARI.	T/F
156.	Mangala & Sumangala are varieties of coffee	T/F
157.	Thermometer is used to determine the end point of jelly.	T/F
158.	*Aegle marmelos* is a tristeza tolerant root stock of citrus.	T/F
159.	Cauveri variety of coffee is resistant to leaf rust.	T/F
160.	Commercial varieties of banana are mostly diploids.	T/F
161.	Cracking of litchi and grapes in North India is mainly due to early rains.	T/F
162.	First book on horticulture was related to litchi.	T/F

163. Frame working is common in ber.   T/F
164. Laksha Ganga is a variety of guava.   T/F
165. Pectin is essential for marmalade preparation.   T/F
166. Seedlessness in guava occurs due to tetraploidy.   T/F

Fill in the Blanks

167. The term 'Blown' is associated to _____.
168. A prefloral fleshy apical meristem in cauliflower is known as _____.
169. _____ is leading tuber crop of India.
170. First seedless variety of watermelon was developed by _____.
171. _____ is a dioecious cucurbit.
172. Little leaf of brinjal is transmitted through _____.
173. *Pusa Anmol* is a hybrid variety of _____.
174. _____ is responsible for red colour in chilli.
175. Green buds & fleshy flower stalk is edible portion of _____.
176. Mature and unripe guava are suitable for _____ preparation.
177. Momtaze is an Iranian variety of _____.
178. Taj Garden is an example of _____ design of gardening.
179. _____ is loose skinned orange species.
180. _____ is monoembryonic species of citrus.
181. _____ is vector of 'Bunchy top' in banana.
182. Moko disease is serious problem in _____.
183. Giant Kew is a commercial variety of _____.
184. _____ is tetraploid variety of ber _____.
185. From fruit set to maturity, sapota takes about _____.
186. _____ is octaploid variety of ber _____.
187. Ripe mango is rich source of _____.

Tender
In the horticultural sense of not enduring winter conditions.

Spear
A sprout or shoot of plant, as a blade of grass, a young stalk of asparagus, etc.

Chemotaxonomy
A method of classification of plants where the chemical constituents of plants are used in taxonomy.

Phloem
A complex vascular tissue of plants which may include sieve tubes, companion cells, fibres, parenchyma and secretory cells. Portion of the bark immediately adjacent to the cambium in woody plants. Responsible for translocation of food in solution.

188. Dry Karonda is the richest source of _____.

189. _____ is the edible part of litchi fruit.

190. Kokrol is botanically known as _____.

191. *Colocasia esculenta* belongs to _____ family.

192. Pusa Navrang is a variety of _____.

193. Mango necrosis is also known as _____.

194. Mango tiprot (Koeli) disorder is prevalent in _____ India.

195. _____ is yellow colour due to the lack of light.

196. Over growth of plant due to increase in number of cells is termed as _____.

197. Yellowness on leaf due to deficiency of certain element is known as _____.

198. Heart rot of sugar beet occurs due to deficiency of _____.

199. Yellow disease of pineapple occurs due to excess of _____.

200. Gilas is a variety of _____.

Once someone asked a farmer, if he had planted wheat for the season. The farmer replied, "No, I was afraid it wouldn't rain." The man asked, "Did you plant corn?" The farmer said, "No, I was afraid of insects eating the corn." Then the man asked, "What did you plant?" The farmer said, "Nothing, I played it safe."

Take risks but don't gamble. Risk-takers go with their eyes open. Gamblers shoot in the dark.

Script your success story

[Success

involves Taking

Calculated

Risks.]

Step-Thirteen

Losers believe for them to win someone has to lose;
Winners believe in win/win.

Multiple Choice

1. The concept of 'wild garden' was propounded by-
    A] M.S. Randhawa   B] Persy Lancaster
    C] William Edward   D] William Robinson

2. Aleurone layer of a monocot seed has chromosome number.
    A] n   B] 2n
    C] 3n   D] 4n

3. Each fruitlet of a multiple fruit (pineapple) is a-
    A] Drupe   B] Berry
    C] Achene   D] Sorosis

4. The ability of germination of a seed is known as-
    A] Purity   B] Initiation
    C] Immunity   D] Viability

5. Imperial Council of Agricultural Research (1929) was renamed as present ICAR during-
    A] 1929   B] 1935
    C] 1942   D] 1947

6. Imperial Council of Agricultural Research is presently known as-
    A] IARI   B] IIHR
    C] ICAR   D] UPCAR

7. Imperial Council of Agricultural Research was renamed as present ICAR in-
    A] 1929   B] 1936
    C] 1947   D] 1950

8. The process of removal of air from cans is known as-
    A] Appertizing   B] Exhausting
    C] Brining   D] Syruping

9. The process of removal of air from cans is known as-
    A] Processing   B] Ventilation
    C] Brining   D] Exhausting

Acre-inch
A measure of quantity of water flow covering an acre to a depth of one inch, assuming no seepage, evaporation and run-off losses.

Air layering
A specialized method of plant propagation accomplished by cutting into the bark of the plant to induce new roots to form.

Jointed
Having nodes or points of real or apparent articulation.

Shrinkage
1. The process of reducing in dimension, weight or volume. 2. Loss of weight in fruit computed from harvest time to some later date.

Rooting
1. The production of root by a plant.
2. A root used for propagation. 3. The production of roots by a cutting.

Freezer-burn
A defect of stored frozen food which appears usually as dry, grainy and brownish spots due to dehydration of frozen foods by sublimation during air freezing and in all conditions of frozen storage. It irreversibly alters the colour, texture, flavour and nutritive value of frozen foods.

Pappus
An appendage or tuft of such appendages forming the top of an achene, helps in the wind dispersal of seed.

10. Botanical name of Amaltas is
    A] *Delonix regia*   B] *Cassia fistula*
    C] *Jacoaranda spp*   D] *Butea monosperma*

11. The presence of anthocyanin stripes in apple skin is controlled by-
    A] Single recessive gene   B] Single dominant gene
    C] Multiple gene   D] Jumping gene

12. Diploid varieties of apple are generally-
    A] Self-sterile   B] Cross-compatible
    C] Self-infertile   D] Self-fruitful

13. 'Fire blight' of apple is caused by-
    A] Virus   B] Fungus
    C] Nematode   D] Bacteria

14. Which variety of apple is indegenous to India-
    A] McIntosh   B] Ambri
    C] Maharaja   D] King of Pippins

15. Which variety of apple is susceptible to 'bitter pit'-
    A] Jonathan   B] Rymer
    C] Nothern Spy   D] King of Pippins

16. Dwarfing rootstock of Apple is
    A] M-12   B] M-13
    C] M-9   D] M-4

17. National Board of Aromatic & Medicinal Plants is located at
    A] Bangalore   B] Mumbai
    C] Lucknow   D] New Delhi

18. Bunchy top of Banana is caused by
    A] Fungus   B] Bacteria
    C] Virus   D] Physiological factor

19. Panama wilt of banana is disease of which origin-
    A] Bacterial   B] Fungal
    C] Mycoplasma   D] None

20. Storage temperature of banana is
    A] 2-3°C   B] 10-13°C
    C] 4-5°C   D] 15-20°C

21. Common blight of bean is caused by
    A] Virus   B] Physiological disorder
    C] Bacteria   D] Mycoplasma

22. 'Red colour' of beet-root is due to-
    A] Carotene   B] Xanthophyll
    C] Chlorophyll-b   D] Anthocyanin

23. 'Umran' variety of ber is-
    A] Triploid   B] Diploid
    C] Tetraploid   D] Octaploid

24. Botanical name of bitter gourd is:
    A] *Cucumis melo*   B] *Momordica charantia*
    C] *Citrullus lanatus* D] None

25. First variety of Bougainvillea raised in India (1920) was-
    A] Trinidad   B] Mrs. H.C. Buck
    C] Cherry Blossom D] Scarlet Queen

26. Which is a variety of Bougainvillea-
    A] Chandrama   B] Golden Ball
    C] Best of all   D] Thimma

27. Which is a variety of Bougainvillea-
    A] Super Star   B] The President
    C] Crimson Glory   D] Mahara

28. Ornamental value of bouganvillia lies in-
    A] Colourful bract B] Fruits
    C] Foliage   D] Seed

29. Little leaf of brinjal is caused due to
    A] Virus   B] Mycoplasma
    C] Bacteria   D] Fungi

30. Good source of carbohydrates, vitamins, fats, minerals & protein is
    A] Vegetables   B] Egg
    C] Milk   D] None

31. Edible part of cashew apple is
    A] Kernel   B] Endosperm
    C] Pericarp   D] Peduncle

Shoot
1. Young growth of the current season in fruit plant. 2. Stem of a plant including the roots. 3. A new leaf growth from the bud. 4. To develop new stems.

Caryopsis
The fruit of members of the grass family; not basically distinct from an achene.

Articular membrane
A membrane consisting of the thin, enlarged base of the petiole, on which the leaf scar occurs.

Ageing
A process involving increment of time which is generally accompanied by physiological changes.

32. Which variety of cauliflower is suitable for south Indian condition
    A] Pusa Subhra   B] Pusa Early synthetic
    C] K-3   D] All

33. Seed rate of cauliflower is
    A] 100 g/hac   B] 200 g/hac
    C] 350-600 g/hac   D] 1000 g/hac

34. Largest producer of cauliflower is
    A] China   B] Kenya
    C] India   D] Australia

35. The ability of cell to generate into a whole plant is known as
    A] Regeneration capacity   B] Totipotancy
    C] Autolysis   D] None

36. Who was the architect of Chandigarh city-
    A] Dr. M. S Randhawa B] Le Carbusier
    C] Albert Meyer   D] Dr. B. P Pal

37. The cracking of cherry fruits is related to-
    A] Boron deficiency   B] Copper deficiency
    C] Excess of calcium D] Excess of water

38. The pungency of chilli is due to
    A] Capsanthin   B] Capsacin
    C] Isothiocynate   D] None

39. 'Spreading decline' of citrus is caused by-
    A] Citrus nematode   B] *Xanthomonos citri*
    C] Burrowing nematode D] Nutritional deficiency

40. Green mould of citrus is caused by
    A] *Penicillium digititum* B] *P. italicum*
    C] *Lemon butterfly* D] None

41. 'Mottled leaf' of citrus is due to deficiency of-
    A] Sodium   B] Lithium
    C] Molybdenum   D] Boron

42. Which species of citrus is the most susceptible to cold injury-
    A] Sweet orange   B] Lemon
    C] Lime   D] Mosambi

43. Banana prefers which type of climate-
    A] Hot   B] Dry
    C] Cold   D] Humid

44. 2n number of cotton is
    A] 40   B] 32
    C] 52   D] 60

45. Which part of country contributes maximum apple production
    A] J & K$_2$ B] HP
    C] MP   D] UP

46. The hybrids of crop, mostly preferred in India are
    A] Brinjal   B] Cucumber
    C] Onion   D] Tomato

47. White rust of crucifer is caused by
    A] *Crostopus candida*   B] *Albugo cadida*
    C] *Peranospora*   D] *Cleviceps*

48. Storage temperature of cucumber is
    A] 0°C   B] 4-5°C
    C] 6- 7°C   D] 10°C

49. Seed rate of cucumber per hectare is
    A] 2.5-3.5kg   B] 5-6kg
    C] 1 -2kg   D] 6- 8kg

50. The most of cultivars of *A. comosus* are-
    A] Self-fertile   B] Cross-incompatible
    C] Self-sterile   D] Self-incompatible

51. Daria method of cultivation is followed in which vegetable crop
    A] Cucurbits   B] Cole crops
    C] Tubers   D] Root crops

52. Correct order of different subphases of Interphase
    A] G_1-G_2-S   B] S-G_2-G_1
    C] G_1-S-G_2 D] G_2-S-G_1

53. The removal of dried flowered shoots in canna is known as-
    A] Training   B] Deshooting
    C] Mattocking   D] Dehorning

Geotropic bending
Upward curvature of tips of spike flowers such as gladiolus when held horizontally on the ground surface.

Chimera
Plan part consisting of tissue of diverse genetic constitution, often observed in flowers.

Anisogamete
A gamete of either of two kinds, sexually differentiated in size and structure.

Compound
a leaf composed of two or more distinct leaf blades (leaflets), each attached to a petiole-like structure (rachis) or directly to the top of the petiole; compare with simple

54. Seed production of early cauliflower can be done in-
A] Hills only   B] Plains
C] Europe   D] None

55. 'Little leaf' of egg plant is caused by-
A] Virus   B] Mycoplasma
C] Nematode   D] Bacteria

56. Rich source of fat is
A] Cashew   B] Walnut
C] Almond   D] Pecanut

57. The pollination of fig is affected by
A] House fly   B] Wind
C] Honey bee   D] Wasp

58. For regulation of flowering in guava, which chemical is used
A] IAA   B] NAA
C] Ethephon   D] GA_3

59. Which one of following contain blue flower
A] *Jacaranda acutifolia*   B] *Delonix regia*
C] *Cassia fistula*   D] *Lagerstomia speciosa*

60. In preservation of food how much quantity of colour is permitted
A] 250 ppm   B] 300 ppm
C] 200 ppm   D] 500 ppm

61. 'Terminal bearing' of fruits is found in-
A] Mango   B] Fig
C] Papaya   D] Citrus

62. Which principle of gardening is ignored when the large tree is planted near the small building
A] Proportion   B] Balance
C] Rhythum   D] All

63. Alternate form of gene at the same locus are refered as-
A] Chromosome   B] Plastid
C] Allele   D] Dominant

64. Two varieties of gladiolus are treated with three concentrations of GA_3 and replicated thrice, the total treatment combination will be-

A] (2 x 3)+3   B] (2 x 3 x 3) = 18
C] (3 x 3) + 11   D] (3 x 3)-2 = 7

65. The most common method of grape preservation is-
A] Jam making   B] Jelly making
C] Dehydration   D] Pickling

66. 'Little leaf' of grape vine is caused due to deficiency of-
A] Cu   B] Zn
C] Fe   D] Mn

67. Seedless variety of guava bear following type of fruits-
A] All seedless   B] All seeded
C] Both seeded & seedless   D] None of above

68. L-49 variety of guava is also known as-
A] Cheema Sahebi   B] Sadar Guava
C] Arka Amulya   D] Pear Shaped

69. Fruit type of guava is
A] Pome   B] Berry
C] Drupe   D] Nut

70. For every 900 feet of height, the atmospheric temperature falls by-
A] 1°C   B] 2°C
C] 3°C   D] 5°C

71. International Institute of Horticulture (IIH) is located in-
A] U.S.A.   B] China
C] Brazil   D] South Korea

72. XYZ system of hybrid production is followed in which crop
A] Wheat   B] Maize
C] Soyabean   D] Rye

73. The most suitable method of identification of lack of certain element in a plant is-
A] Sand culture   B] Solution culture
C] Rock culture   D] Tensiometer

74. Food Corporation of India (FCI) was established in year
A] 1950   B] 1955
C] 1960   D] 1965

Sub-
A prefix often used to mean nearly, as in subacute, subsessile, etc.

Dead heading
The removal of old blossoms to encourage continued blooming or to improve the appearance of the plant.

Wart
A protube-rance on a plant part caused by some external agents like fungus.

Marketing phase
The movement of the plants and/or flowers form the forcing facilities to the whole seller and/or retailer at the proper stage or development so that the consumer receives the maximum possible enjoyment.

Aril
An additional covering formed on certain seeds by expansion of the stalk of the ovule after fertilization *e.g.,* litchi.

Dormant oil
Oil sprayed on deciduous trees while they are dormant. Dormant oils are used to kill overwintering insects or insect eggs on plant bark.\

Hypothesis
A speculative statement based on observation suggesting a possible cause and effect relationship and forming the beginning point of an experimental study.

Parasite
An organism living on or in another living organism and obtaining its food form that plant.

75. A popular delicious fruit of India is-
    A] Banana   B] Litchi
    C] Apple   D] Pineapple

76. National flower of India is
    A] Rose   B] Tulip
    C] Lotus   D] Carnation

77. Annual flower of Indian origin is
    A] Antirrhinum   B] Gaillardia
    C] Gomphrena   D] Statice

78. *Poplular climber* of Indian origin is
    A] *Tecoma catensis*   B] *Bigonia venusta*
    C] *Gloriosa superba* D] None

79. Chromosomal theory of inheritance was given by
    A] C. Darwin   B] Mendal
    C] Wesimann   D] Sutton & Boveri

80. National flower of Japan is
    A] Rose   B] Chrysanthemum
    C] Tulip   D] Carnation

81. pH value of jelly should be-
    A] 2.0   B] 3.0
    C] 4.0   D] 1.0

82. The availability of litchi fruits in market occurs in-
    A] January-February   B] March-April
    C] May-June   D] August-September

83. Botanically the mature fruit of litchi is-
    A] Nut   B] Berry
    C] Aril   D] Drupe

84. Botanical Survey of India (B.S.I.) was established by-
    A] J.D. Hooker   B] Charles Gray
    C] Larson   D] A. Mukhopadhyay

85. Horticultural Society of India (H.S.I) was established during-
    A] 1935   B] 1942
    C] 1950   D] 1955

86. The most commercial varieties of loquat are-
    A] Cross-incompatible   B] Self-sterile
    C] Self-incompatible   D] Self-pollinated

87. Scorching on the margins of lower leaves manifests the deficiency of-
    A] Nitrogen   B] Zinc
    C] Potassium   D] Copper

88. 'Red Rust' of mango is caused by-
    A] Bacteria   B] Algae
    C] Fungus   D] Virus

89. Black tip of mango is caused by-
    A] Fungi   B] Bacteria
    C] Virus   D] None

90. 'Black tip' of mango is caused by-
    A] Virus   B] Fungus
    C] Bacterium   D] None of these

91. Kinnow variety of mardarin is obtained by crossing of
    A] Citrus sinensis x C. reticulate
    B] C. nobilis x C. deliuosa
    C] C. deliuosa x C. nobilis
    D] None of the above

92. For inactivation of microbial growth which method is followed
    A] Freezing   B] Sulphuring
    C] Blanching   D] None

93. Young leaves of new plant found often distorted, small & abnormally dark green, is the symptom of
    A] Excess of Nitrogen   B] Deficiency of Calcium
    C] Deficiency of   D] Excess of Potassium
       Phosphorous

94. Maximum percentage of Nitrogen present in fertilizer is
    A] Urea   B] Anhydrous Ammonia
    C] CAN   D] Ammonium Sulphate

Phyllotaxy
Arrangement of leaves on stem.

Prickles
Slender, sharp outgrowths of the stem tissues beneath the epidermis.

Root ball
The clump consisting of the main roots of a plant and the soil (or other growing medium) clinging to them. The root ball should be kept intact when transplanting.

Bloom
A waxy coating found on stems, leaves, flowers and fruits, usually of a grayish cast and easily removed.

Cauliflory
Bearing of fruit directly on the branches or main stem.

Matching

95. Match the following :-
   A] Red Gold, Indira   i] Hybrids of Bougainvillea
   B] Nazrana, Apsara   ii] Hybrids of Rose
   C] Chitra Vati,   iii] Hybrids of Chrysanthemum
       Dr. H.B. Singh
   D] Tribal Queen, Aikta   iv] Hybrids of Hibiscus
   E] Raj Kumari, Jawahar   v] Hybrids of Gladiolus

96. Match the following :-
   A] Regular bearing   i] J.H.Hale
   B] Self-sterile   ii] Turmeric
   C] Y.V.M.   iii] Pusa Seedless
   D] Damping off   iv] Amrapalli
   E] Oleoresin   v] Kurukkun
   F] Little leaf   vi] Pseudoananas
   G] Polyembryony   vii] Okra
   H] Hermaphrodite flowers   viii] Satputia
   I] Self-fertile   ix] Brinjal
   J] Spur pruning   x] Papaya

97. Match the following :-
   A] Residual moisture in   i] Not more than 6-8 %
       dehydrated fruits
   B] Residual moisture in   ii] 66-71°C
       dehydrated vegetables
   C] Dehydration of   iii] 115°C for 40-75 mmutzs
       vegetables
   D] Dehydration of fruits   iv] Not more than 10 -20 %
   E] Processing of   v] 60-66°C
       vegetables

98. Match the following :-
   A] Root Forking   i] Cauliflower
   B] Post-harvest sprouting   ii] Sugarbeet
   C] Ricyness   iii] Radish
   D] Fruit splitting   iv] Litchi
   E] Zoning   v] Onion

99. Match the following :-
    A] Rooting   I] Carotene
    B] Citrus   ii] Polyembryonic
    C] Carrot   iii] Pollinizer
    D] Mangosteen   iv] I.B.A.
    E] Apple   v] Parthenocarpy

100. Match the following :-
    A] Rose   i] Malvaceae
    B] Bougainvillea   ii] Apocynaceae
    C] China Rose   iii] Rosaceae
    D] Chrysanthemum   iv] Euphorbiaceae
    E] Canna   v] Nyctaginaceae
    F] Gladiolus   vi] Amaryllidaceae
    G] Tuberose   vii] Compositae
    H] Poinsettia   viii] Irridaceae
    I] Oleander   ix] Scitaminae
    J] Amaltas   x] Caesalpinoidae

101. Match the following :-
    A] Rosette in Pecan   i] 'Zn' deficiency
    B] Mouse ear in Pecan   ii] 'Mn' deficiency
    C] 'Blank' fruits   iii] Blue berry
    D] 'Cat-faced' fruit in strawberry iv] Pistachionut
    E] 'Mummy' berry   v] 'B' deficiency.

102. Match the following :-
    A] Rubber   i] Maharashtra > Kerala > T.N.
    B] Coriander   ii] Assam > Kerala > W.B.
    C] Sweet potato   iii] Kerala > T.N. > Kanataka
    D] Banana   iv] Karnataka > T.N. > A.P.
    E] Papaya   v] A.P. > Orissa > Maharashtra
    F] Cocoa   vi] Orissa > U.P. > Bihar
    G] Chilli   vii] Kerala > Meghalaya > Orissa
    H] Cassava   viii] A.P. > Rajasthan > T.N.
    I] Pineapple   ix] Kerala > T.N. > A.P.
    J] Ginger   x] Kerala > Karnataka > T.N.

103. Match the following :-
    A] Runner   i] Raspberry
    B] Sucker   ii] Lily

Comparium
A group composed of one or more cenospecies that are able to intercross.

Multi-branched plant
One plant with several shoots and flowers, which is achieved by pincing.

Hemizygous
A diploid organism having only one copy of a particualr gene present in the genome. e.g., males are hemizygous for genes on the X chromosome (Sex chromosome).

Gynoecious
A sex from where only pistillate flowers are produced giving rise to a super female plant.

Unilateral
Arranged on one side only, as unilateral leaves, where leaves lean to one side of the stem.

Lawn
Lawn is an open area with green grass of the garden e.g. *Cynodon dactylon*.

Follicle
A capsular fruit which opens on one side only *e.g., Rouwolfia, Calotropis,* etc.

Variety
A group of strains or a single strain of closely related plants of common origin which have similar characteristics and can be differentiated on the basis of structural and functional characters from another group of plants.

C] Stolon   iii] Onion
D] Offsets   iv] Strawberry
E] Bulb   v] Potato
F] Tuber   vi] Bermuda grass
G] Stem cutting   vii] Gladiolus
H] Corm   viii] Banana
I] Tunicated bulb   ix] Datepalm
J] Rhizome   x] Pointed gourd

104. Match the following :-
A] Rutaceae   i] Fenugreek
B] Orchidaceae   ii] Ocimum
C] Papilionaceae   iii] Curry leaves
D] Iridaceae   v] Clove
E] Lauraceae   v] Vanilla
F] Labiaceae   vi] Cardamom
G] Myrtaceae   vii] Saffron
H] Alliaceae   viii] Black pepper
I] Piperaceae   ix] Cinnamon
J] Zingiberaceae   x] Garlic

105. Match the following :-
A] Sauerkraut   i] Acetic acid
B] Cauliflower   ii] Pickle
C] Vinegar   iii] Apple
D] Cider   iv] Tomato
E] Ketchup   v] Cabbage

106. Match the following :-
A] Scurf disease   i] Black pepper
B] Spice   ii] Bitter principle
C] Solanin   iii] Potato
D] Capsaicin   iv] Acidic soils
E] Dahlia   v] Tuber

107. Match the following :-
A] Sealing of Cans   i] 100°C
B] Cooling of processed   ii] At 74°C
    Cans
C] Processing of Tomato   iii] 121°C

D] Processing of Carrot   iv] 79°C to 82°C

E] Filling of Urine solution v] To 39°C

108. Match the following :-

A] Seed rate 2.5-3.5 kg/ha   i] Watermelon, Tinda, Ridge gourd

B] Seed rate 3.5-5.0 kg/ha   ii] Tropical Asia origin

C] Seed rate 5-6 kg/ha   iii] Carrot

D] Seed rate 7-10 kg/ha   iv] Snake gourd, Bitter gourd, Ash gourd

E] Pointed gourd, Cucumber   v] Cucumber, Sponge gourd

F] Muskmelon, Watermelon, Bottlegourd   vi] Central America origin

G] Ashgourd, Bittergourd   vii] Pumpkin, Summer Squash

H] Cho-Cho   viii] Japanese Radish

I] Male sterility   ix] Indian origin

J] Protandry   x] Tropical Africa origin

109. Match the following :-

A] Self-fertile pear   i] Flemish Beauty, Beurre Hardy

B] Self-unfruitful pear   ii] Pineapple, Patharnakh

C] Self-sterile pear   iii] Magness, Winter Nelis, Keiffer

D] Low chilling pear   iv] Fertility, Conference

E] High chilling var. of pear   v] Bartlett, Anjou

110. Match the following :-

A] Self-incompatibility   i] Avocado, Walnut

B] Heterostyly   ii] Pomegranate, Sapota

C] Self-sterility   iii] Jackfruit

D] Dichogamy   iv] Banana

E] Pin type styles   v] Mango, Loquat

F] Thumb type styles   vi] Mangosteen

G] Parthenocarpy   vii] Grape

H] Stenospermocarpy   viii] Litchi, Pomegranate

Bivalent
A pair of sinapsed or associated homologous chromosomes during meiosis.

Glaucous
Covered with a white or bluish bloom which can usually be rubbed off easily.

Primocane
The first year's shoot or cane of a biennial woody stem, ex. Rubus.

Vesicle
A small bladdery sac or cavity filled with air or fluid.

Single flower
A flower having only a minimum number of petals for that variety of plant.

Agameon
A species that contains only apomictic individuals.

I] Parthenogenesis   ix] Almond, Carambola
J] Vivipary   x] Peach, Pear

111. Match the following :-
A] Self-pollination   i] Garden beet, Spinach
B] Often-cross pollination   ii] Lettuce Papaya
C] Wind pollination   iii] Asparagus, Cowpea
D] Insect-pollination   iv] Chestnut
E] Cleistogamy   v] Protogynous
F] Chasmogamy   vi] Okra, Brinjal
G] Duodichogamy   vii] Pistaschionut
H] Dioecious plants   viii] Cucurbits, Turnip
I] Hetero-dichogamy   ix] Banana, Pineapple
J] Bird-pollination   x] Tomato, Chilli

112. Match the following :-
A] Self-sterile peaches   i] Dixigem, Florida Sun, Babcock
B] Clingstone peaches   ii] Baifeng, May Gold, Florida Queen
C] Freestone peaches   iii] Sun Red, Sun Rise, Annqueen
D] Peach suitable for canning   iv] J.H. Hale, Halberta, June Elberta
E] Nectarines   v] Elberta, Red Globe, Golden Bush

113. Match the following :-
A] September, Pride of India   i] Cauliflower
B] Pusa Deepali, Pusa Subhra   ii] Badger Shipper
C] White Vienna, Purple Vienna   iii] High temperature
D] Danish Prize, Early Dwarf   iv] Knol-khol
E] Yellows resistant cabbage   v] Acidic Soil

F] Salinity tolerant cabbage   vi] Cabbage

G] 'Black-rot' tolerant cauliflower   vii] 'B'deficiency

H] 'Club-rot' in cauliflower   viii] Pusa Drum Head

I] 'Riceyness' in cauliflower   ix] Brussel's Sprout

J] 'Browning' in cauliflower   x] Pusa Snowball K-1

114. Match the following :-

A] Sharp freezing   i] Freezing at 18°C & drying at 20°C

B] Quick freezing   ii] -15°C to 29°C for 3-72 hours

C] Cryogenic freezing   iii] -18°C

D] Storage of frozen guava   iv] 0 to-4°C for 30 minutes

E] Freeze drying   v] Below 60°C

115. Match the following :-

A] Sigatoka   i] Radish

B] Spongy Tissue   ii] Banana

C] Exanthema   iii] Carrot

D] Cavity Spot   iv] Mango

E] Root Bulging   v] Citrus

116. Match the following :-

A] Single dahlia   i] Masterpiece, Triumph

B] Star dahlia   ii] Giraff, Disneyland

C] Anemone dahlia   iii] Kokette, Frances

D] Collerate dahlia   v] Lovely Looker, Dorris Duke

E] Peony flowered   v] White Star

F] Decorative dahlia   vi] Jean Lister, Tamtam

G] Fancy dahlia   vii] Comete

H] Pompon dahlia   viii] Bishop of Llandoff

I] Miniature dahlia   ix] Standard, Model

J] Miscellaneous/orchid dahlia   x] Areoline, Scarlet Queen

Bunch
A cluster of plants or fruits.

Suture
A line of dehiscence or groove marking a face of union.

Mulch
A layer of organic or inorganic material on the soil surface. Mulches help to moderate the temperature of the soil surface, reduce loss of moisture from the soil surface, suppress weed growth, and reduce run off.

Leaf
An expanded or more or less flattened outgrowth of a stem or directly from root, usually green, which is primarily concerned with the manufacture of carbohydrate by photosynthesis.

Glossary

Sheath
Any elongated, more or less tubular sructure enveloping an organ or plant part.

Stalk
1. The stem or main axis of a plant especially of a herbaceous plant.
2. Any stem or plant part which supports leaves, flowers or fruits, a petiole, pedicel or peduncle.

Hypotriploid
A triploid lacking one or more chromosomes.

Torulose
Twisted or knobby, irregularly swollen at close intervals.

Cauliflorous
Flowering on the trunk or on specialized spurs from it or from the larger branches (redbud).

True / False

117.	Walnut is the richest source of Vit-B_6 among all nuts.	T/F
118.	Anemone is a class of dahlia.	T/F
119.	Asian Vegetable Research & Development Centre is located in Taiwan.	T/F
120.	Curding is an intermediate stage of cauliflower.	T/F
121.	Illaichi variety of mango is free from malformation.	T/F
122.	*Juglans regia* (Walnut) is native of Iran.	T/F
123.	Leaf curl is a serious disease of tomato.	T/F
124.	Oilpalm is the latest addition among plantation crops.	T/F
125.	Pusa Majesty is a dioecious variety of papaya.	T/F
126.	Tomato & Squash are day neutral vegetables.	T/F
127.	CaC_2 releases acetylene gas.	T/F
128.	Chandler and Waterloo are varieties of walnut.	T/F
129.	Conference is a variety of pear.	T/F
130.	Generally orchids are epiphytes.	T/F
131.	Improved Golden Yellow is a variety of litchi.	T/F
132.	In North India, Kamrakh ripens during summer.	T/F
133.	India is 'Home of Spices'.	T/F
134.	Kochia is a summer season foliage annual.	T/F
135.	Radish & beet are short day vegetables.	T/F
136.	Washington Naval is a lemon variety.	T/F
137.	Arka Ajit is a variety of pea.	T/F
138.	Asparagus is a perennial vegetable.	T/F
139.	California Advance is a variety of loquat.	T/F
140.	Duncan and Foster are cultivars of grape.	T/F
141.	Guava is propagated by stooling.	T/F
142.	In N. India, carambola flowers in rainy season.	T/F
143.	Inarching is the latest method of mango propagation.	T/F
144.	Mould is spoiling agent of jelly.	T/F
145.	Yellow leaf in arecanut is caused by MLOs.	T/F

146. Careless Love is a variety of rose.   T/F
147. Damping off of tomato seedlings is caused by virus.   T/F
148. Mango is difficult-to-root by cuttings.   T/F
149. Phalsa and papaya are propagated by seeds.   T/F
150. Rose and Grape, both are pruned in October.   T/F
151. Shalimar Bagh in Kashmir was built by Babur.   T/F
152. Tea is a monocotyledonous plant.   T/F
153. 2, 4-D at lower concentration acts as a hormone.   T/F
154. Amchur is raw mango's dried powder.   T/F
155. Arka Anmol is a variety of grape.   T/F
156. Basrai Dwarf is a variety of banana.   T/F
157. Cool and short growing season is limiting factor in walnut production.   T/F
158. Karonda is a rich source of Iron & Vit-A.   T/F
159. *Katte disease* is serious problem in cardamom.   T/F
160. Mosambi is incompatible with *Trifoliate* orange.   T/F
161. Pineapple is originated from Brazil.   T/F
162. Pumpkin produces bisexual flowers.   T/F
163. Appertizing is synonym of preserving.   T/F
164. *Ixora parviflora* is a flowering climber.   T/F
165. *Lawsonia alba* is a beautiful/flowering hedge plant.   T/F
166. Loquat fruits are harvested during Sept.-Oct..   T/F

Fill in the Blanks

167. Thompson is a variety of _____.
168. Spinach belongs to _____ family.
169. L-49 is synonym of _____.
170. Careless Love is a variety of _____.
171. Sathgudi is a cultivar of _____.
172. _____ is important for jelly setting.
173. *Vitis vinifera* is the botanical name of _____.

Nullisomic
An otherwise diploid cell or organisms lacking both members of a chromosome pair (2n-2).

Arboreal
Tree-like or pertaining to trees.

Alloplasmic
A cell or line or plant with nucleus of one species and cytoplasm of another species.

Bleeding
Loss of water or sap from the plants as result of wound or pruning.

Pneumatophore
A knee-shaped or spike-like upward projection of the roots of swamp trees enabling the submerged roots to obtain oxygen.

Sporocarp
A body containing sporangia or spores.

Bud scale
A modified leaf or stipule (there may be one, a few, or many) protective of the embryonic tissue of the bud.

Preformed
Already with definite shape or structure, as with leaves within a bud.

Cespitis
A turf horizon, or surface soil layer held together by a tangle of roots, mostly grass.

Hardy
Able to withstand unfavourable or various environmental factors. Hardy vegetables like, broccoli, cabbage, asparagus, peas, turnip, garlic, knolkhol, radish, spinach, onion, etc. are tolerant to frost.

174. Grape fruit belongs to _____ family.
175. Sugar baby is a variety of _____.
176. Phalsa belongs to _____ family.
177. _____ is the best grass for lawn making.
178. Cultivation of citrus is called as _____.
179. _____ is serious viral disease of okra.
180. Epilachna beetle feeds on _____.
181. *Phoenix dactylifera* has _____ sex form.
182. _____ is the botanical name of Karonda.
183. The placing of horticultural commodity in consumer size package is known as _____.
184. _____ is Japanese art of growing of trees & shrubs in miniature form.
185. _____ is maximum point of respiration of mature fruits.
186. _____ is responsible for red colour in tomato.
187. Shoot and fruit borer is serious pest of _____.
188. Marry Palmer is famous variety of _____.
189. _____ is causal organism of little leaf of brinjal.
190. Cell elongation is stimulated by _____ hormone.
191. _____ is edible part of pomegranate fruit.
192. _____ is a stem vegetable of cruciferae family.
193. Citrus seeds are viable for _____ only.
194. _____ is used for preparation of lye solution.
195. Molybdenum deficiency causes _____ in cauliflower.
196. Sioux is hybrid variety of _____.
197. Washington Naval is a variety of _____.
198. _____ is necessary for better fruiting in grape.
199. _____ is applied for uniform ripening of pineapple.
200. Ethrel releases _____ gas.

A polio victim at the age of five started swimming to regain strength. It was because of her desire to suceed that she went onto become a world record holder at three events and won the gold at the 1956 Olympics at Melbourne. Her name is Shelley Mann.

When people lack purpose and direction, they see no opportunity. If a person has the desire to accomplish something, the direction to know his objective, the dedication to stay focused, and the discipline required to put in the hard work, then other things come easy. But if you don't have them, it doesn't matter what else you have.

Script your success story

The **B**urning desire to **S**ucceed is the driving **F**orce of success.

Step-Fourteen

Losers see the past;
Winners see the potential.

Multiple Choice

1. The tendency of one cross over to reduce the chance of another cross over in its adjacent region is called as
    A] Interference   B] Coincidence
    C] Penetrance   D] Expressivity

2. The pungency of onion is due to
    A] Allyl propyl   B] Diallyl disulphide
        disulphide
    C] Both   D] None

3. Pink colour of onion is due to
    A] Qucertin   B] Anthocynin
    C] Carotene   D] None

4. Yellow colour of onion is due to
    A] Question   B] Myrcetin
    C] Allicin   D] Malvelin

5. Which variety of Onion is suitable for kharif season
    A] Pusa Red   B] Pusa Madhuri
    C] N-53   D] Early Goano

6. Contour system of orchard planting is generally followed in-
    A] Hills   B] Saline soil
    C] Punjab   D] UP

7. Largest genera of orchid is
    A] Dendrabium   B] Bulbophyllum
    C] Catteleya   D] Cymbidium

8. The centre of origin of cocao is
    A] Brazil   B] South America
    C] Asia   D] Africa

9. Commercial propagation of papaya is done by-
    A] Seed   B] Cutting
    C] Budding   D] Layering

Pistil
The seed-bearing organ of a flower, consisting of the ovary, stigma, and style.

Vine
A slender stemmed climbing or trailing plant.

Humus
The brown or black organic part of the soil resulting from the partial decay of leaves and other matter.

Transgenic
A term applied to organisms that have been altered by introducing DNA molecules into them.

Stipules
Small leaf-like organs occurring in pairs on either side of the leaves; occasionally each one extends half way around the twig, respectively.

10. Gynodioecious variety of papaya is
    A] pusa giant   B] Pusa dwarf
    C] Pusa delicious   D] Co 2

11. Which type of Parthencarpy is found in bread fruit-
    A] Stenospermocarpy B] Stimulative
    C] Vegetative   D] None of these

12. Which is suitable variety of pea for green pods-
    A] Boneville   B] Rachana
    C] Plant-P   D] None of these

13. Powdery mildew of pea is caused by-
    A] Fungi   B] Bacteria
    C] Virus   D] None

14. Which species of pineapple does not produce slips-
    A] *Annanas comosus* B] *A. bracteatus*
    C] *A. erectifolius* D] *Pseudoananas sagenarius*

15. Fruit type of pineapple is
    A] Berry   B] Sorosis
    C] Balusta   D] Amphisarca

16. National Bureau of Plant Genetic Resources (NBPGR) was
    established in-
    A] 1966   B] 1968
    C] 1972   D] 1976

17. Best time of planting for temperate fruits is
    A] June-July   B] Sept-Oct
    C] Dec-Jan   D] March-April

18. Binomial nomenclature of plants is written in which language-
    A] English   B] Roman
    C] Latin   D] French

19. Which variety of plum is suitable for plains of U.P.-
    A] Santa Rosa   B] Victoria
    C] Titron   D] Kelsey

20. Fruit type of plum is
    A] Drupe   B] Berry
    C] Pome   D] None

21. The ratio of pollinizer and main crop in apple orchard should be-
    A] 1 : 6   B] 1 : 20
    C] 1: 1   D] 9: 1

22. 'Late blight' of potato is caused by-
    A] *Alternaria solani*   B] *Phytophthora solani*
    C] *Phytophthora infestans* D] *Xanthomonos solani*

23. The most dangerous disease of potato is-
    A] Mosaic   B] Black heart
    C] Early blight   D] Late blight

24. Stooling method of propagation is generally used in-
    A] Walnut   B] Apple
    C] Litchi   D] Guava

25. 'Stooling' method of propagation is practiced in-
    A] Mango   B] Litchi
    C] Doobgrass   D] Guava

26. Rich source of Protein (> 12%) is
    A] Walnut   B] Cashew nut
    C] Pecanut   D] Mango

27. Proper time of pruning in ber is-
    A] July-August   B] October-November
    C] February-March   D] May-June

28. Best time of pruning of peach is-
    A] Mid summer   B] Mid winter
    C] Autumn   D] Spring

29. The colour of aril of ripened litchi is generally-
    A] White   B] Yellow
    C] Red   D] Brown

30. Main function of RNA is
    A] DNA replication   B] protein synthesis
    C] DNA transfer   D] None

31. Which var of rose is exported
    A] Raktagandha   B] Rirst red
    C] La france   D] None

Imbricate
Said of scales which overlap like shingles; the opposite of valvate in which the scales meet along a line without overlapping.

Imbricate
Said of scales which overlap like shingles; the opposite of valvate in which the scales meet along a line without overlapping.

Nucleus
A specialised spherical or ovoid protoplasmic body embedded in the cytoplasm of eukaryotic cell which is predominately composed of nucleoprotein and responsible for growth and reproduction of the cell.

Tetrasomic
An individual having one pair of chromosome in addition to the normal somatic chromosome complement (2n + 2).

Tunic
The dry, papery scales that surround the fleshy organs of a bulb or corm.

Lineate
Lined; bearing thin parallel lines.

Ovary
The ovule-bearing part of a pistil.

Core
1. The innermost part of an apple, pear, etc. which contains the seeds.
2. To remove the central seed-containing portion of fruits for canning or preserving.

32. Direct sowing of seed is done in-
A] Tomato   B] Cabbage
C] Fenugreek   D] Brinjal

33. Dioecious form of sex is found in-
A] Asparagus   B] Papaya
C] Date palm   D] All of these

34. Permissible limit of SO_2 in Jam is
A] 300 ppm   B] 40 ppm
C] 70 ppm   D] 100 ppm

35. For packing of spinach, cauliflower and sweet potato which type of can is used
A] AR cans   B] SR cans
C] Plain   D] All

36. Central Institute of Sub-tropical Horticulture (CISH) is located at-
A] Abohar (Punjab)   B] Lucknow (U.P.)
C] Hissar (Haryana)   D] Godhra (Gujarat)

37. The inflorescence of sugarcane is known as
A] Spike   B] Arrow
C] Cymose   D] None

38. In presence of sunlight, CO_2 & water (with help of chlorophyll) are converted into carbohydrates is called as-
A] Respiration   B] Metabolism
C] Photosynthesis   D] Solar radiation

39. Maximum producer of sweet orange in the world is
A] USA   B] China
C] Brazil   D] India

40. Which part of tea plant is harvested
A] A bud &young leaves   B] Mature leaves
C] Bark   D] All

41. The most of the banana fruits show tropism as-
A] Negative geotropism   B] Positive chemotropism
C] Photo tropism   D] Positive geotropism

42. Which part of the country is the highest producer of apple
A] HP   B] J & K
C] UP Hills   D] AP

43. In which of the following fruit tissue culture is followed
    A] Banana   B] Mango
    C] Citrus   D] Litchi

44. Which one of the following is highly cross-pollinated
    A] Okra   B] Bitter gourd
    C] Chilli   D] Cowpea

45. Which one of the following is non climacteric fruit
    A] Apple   B] Mango
    C] Pineapple   D] Banana

46. Which one of the following is the best maturity indice of Grape
    A] Colour   B] Shape
    C] Size   D] TSS

47. Which one of the following is used as growth retardant
    A] GAB   B] NAA
    C] Ethephon   D] MH

48. Which one of the following is used for preparing best Jelly
    A] Jamun   B] Guava
    C] Mango   D] Annona

49. Which one of the following is used to enhance more fruit set
    A] NAA   B] 2, 4-D
    C] 2, 4, 5- T   D] Ethylene

50. Which one of the following medicinal plant is used against asthama
    A] *Rauwolfia serpentina*   B] *Pongamia glabra*
    C] *Tylophora*   D] *Mesua feria*
       antiacthencatica

51. Which one of the following state is IInd largest in area
    A] MP   B] AP
    C] UP   D] Rajasthan

52. In which of the following underground root is edible
    A] Potato   B] Ginger
    C] Onion   D] Raddish

53. Which one of the following variety of cashew has bold nuts
    A] Kanaka   B] Dhana
    C] HS-517   D] H-1591

Vegetable
Herbacious plant or part of plant which are use for culinary purpose.

Segment
Any one of the parts into which an object an be naturally separated or divided; part cut off or separable from the other parts; a section of an orange, a flower, or an insect.

Jelmeter
An apparatus used for pectin test during the preparation of jam and jelly. It directly indicates the amount of sugar to be added per litre of extract.

Balling
A measurement of sugar percentage in simple syrup at fixed temperature. Also called brix.

Lysimeter
A device for measuring percolation and leaching losses from a column of soil under controlled conditions.

Polygamous
1. With both perfect and imperfect flowers present on the same plant. 2. Bearing both unisexual and bisexual flowers on the same or different plants of the same species e.g., mango, cashewnut, kola, etc.

Falls
The outer whorl of petals of an iris flower, often broader than the inner petals and often drooping or flexuous.

54. Which one of the following was founded first
A] KAU B] TNAU
C] PAU D] GBPAU

55. Which variety of tomato can be sown throughout year
A] Pusa ruby B] Pusa sadabahar
C] Sioux D] None

56. Rupali variety of tomato is developed by
A] Indo-American Seed Company B] Ankur seeds
C] Mahyco seeds D] None

57. The fruit of tomato is
A] Pome B] Stone
C] Berry D] Drups

58. 'Bending' type of training is applied in-
A] Apple B] Pomegranate
C] Guava D] Peach

59. Kniffin system of training is used in -
A] Mango B] Grape
C] Litchi D] Papaya

60. Oldest system of training of grape practiced in India
A] Head B] Bower
C] Kniffin D] Telephone

61. The most common system of training of vine crops is-
A] Head system B] Hedge system
C] Bower system D] Cordon system

62. The best 'system of training' in apple orchard is-
A] Espalier B] Central leader
C] Open centre D] Modified leader

63. Ploidy level of *Triticum asitivum* is
A] Diploid B] Triploid
C] Hexaploid D] Tetraploid

64. For measurement of TSS which instrument is used
A] Dendrometer B] Thermometer
C] Refractrometer D] None

65. Underground stem of tuberose is known as
    A] Rhizone   B] Tuber
    C] Corn   D] Bulb

66. Maximum life of Vanilla is
    A] 10 years   B] 15 years
    C] 30 years   D] 50 years

67. Square root of variance is referred to as
    A] Coefficient of   B] Standard error
       variantion
    C] Standard deviation   D] Range

68. Root knot of vegetable is caused by-
    A] Fungi   B] Bacteria
    C] Nematode   D] Virus

69. Daily requirement of vegetable per capita/day is
    A] 150 gram   B] 200 gram
    C] 285 gram   D] 400 gram

70. Indian Institute of Vegetable Research is located at
    A] New Delhi   B] Allahabad
    C] Varanasi   D] Mumbai

71. Soft rot of vegetables is caused by
    A] *Verticillium spp*   B] *Erwinia carotovora*
    C] *Fusarium*   D] None

72. During dehydration of vegetables, the temperature of dehydrator should be-
    A] 40-50 °C   B] 60-66 °C
    C] 86-90 °C   D] 100-115 °C

73. Richest source of vitamin A is
    A] Papaya   B] Bael
    C] Banana (Ripe)   D] Mango (Ripe)

74. Daily requirement of vitamin-D for a healthy adult is
    A] 200 I.U.   B] 400 I.U.
    C] 600 I.U.   D] 800 I.U.

75. The movement of water from root to leaves is due to
    A] High potential of water in roots than leaves
    B] Low potential of water in roots than leaves
    C] Equal potential of water
    D] No relation of water potential

Emarginate
Applied to leaves or petals which are notched at the apex, or having a notched margin.

Petals
Modified leaves forming the inner floral envelope.

Rachis
the stem or petiole-like central axis of a pinnately compound leaf, connecting the petiolules to the petiole

Piperine
Alkaloid, extracted from pepper by suitable chemical processes, used medicinally as carminative, febrifuge and as antiperiodic in malaria piperine, piperic acid and piperidine synthesized from it.

76. Maximum density of water is at temperature
    A] 10°C   B] 4°C
    C] 20°C   D] 80°C

77. Dough stage of wheat is a stage when
    A] Maximum tiller appear
    B] Flowering stage
    C] When white fluid in grain just begin to harden
    D] Just before harvesting

78. The fruit of wheat is known as
    A] Pod   B] Capsule
    C] Siliqua   D] Caryopsis

79. Loose smut of wheat is
    A] Externally seed borne   B] Internally seed borne
    C] Both   D] Soil borne

80. Loose smut of wheat is
    A] Externally seed borne   B] Internally seed borne
    C] Both   D] None

81. The seeds of which fruit lost their variability-
    A] Lime   B] Mango
    C] Papaya   D] Guava

82. Apple is a type of-
    A] Achene   B] Samara
    C] Caryopsis   D] False Fruit

83. Citrus is the native of-
    A] Africa   B] Australia
    C] India   D] China

84. Turmeric is the native of-
    A] Europe   B] India
    C] Peru   D] Mexico

85. All cytokinins are derivatives of-
    A] Guanine   B] Cytosine
    C] Adenine   D] Uracil

86. *In vitro* culture consists of-
    A] Pollen culture   B] Tissue culture
    C] Embryo culture   D] All of these

87. Vegetables are rich source of-
A] Vitamins　　　　B] Proteins
C] Minerals　　　　D] Carbohydrates

88. % of oil in mint
A] 0.1- 0.5%　　　B] 0.5- 1.0%
C] 1 -1.5%　　　　D] 2 -3%

89. ' Old leaves' of a plant is a site of synthesis of-
A] Ethylene　　　　B] GA_3
C] IAA　　　　　　D] Cytokinin

90. The best quality ' Oleoresin' is extracted from-
A] Onion　　　　　B] Castor
C] Chilli　　　　　D] Coconut

91. First book on horticulture in India was written by-
A] B.P. Pal　　　　B] W.B. Hayes
C] Persy Lancaster　D] M.S.Randhawa

92. Carrot seed rate for one hectare is-
A] 5-7 kg　　　　　B] 8-10 kg
C] 12-15 kg　　　　D] 16-20 kg

93. Replacement of one pair of chromosome of cultivated spp with those of wild doner spp
A] Alien substitution　B] Alien addition
C] Introgression　　　D] Somatic hybridization

94. Substitution of one purine by another purine is called as
A] Addition　　　　B] Deletion
C] Transition　　　D] Transversion

Matching

95. Match the following :-
A] Soft wood cutting　i] Rubber
B] Seed　　　　　　ii] Tea
C] Epicotyl grafting　iii] Cocoa
D] Forkert budding　iv] Turmeric
E] Rhizome　　　　v] Cashewnut

Pinnule
The leaflet of a pinna; a secondary leaflet of a pinnately decompound leaf.

Gynomonoecious
A sex form where pistillate and hermaphrodite flowers are separately produced in the same plant.

Unisexual flowers
Flowers having either pistils or anthers, but not both.

Fistulous
Hollow, with excavated pith.

Adventitious
Not in the usual place (buds that have remained undeveloped so that they are no longer evidently axillary, or that really originate elsewhere, as on a root).

Staminode
A structure appearing in place of a stamen but bearing no pollen.

Weed
Any plant in a place where it is a nuisence.

Juvenile
An early phase of plant growth, usually characterized by non-flowering, vigorous increase in size, and often thorniness.

Quadrangular
Four angled, or pith or a twig.

Heeling in
Temporarily setting a plant into a shallow trench and covering the roots with soil to provide protection until it is ready to be permanently planted.

96. Match the following :-
A] *Solanum xanthocarpum*   i] Highly tolerant to 'Rootknot nematode'

B] *S. integrifolium*   ii] Resistant to 'Phomopsis blight'

C] *S. elaegnifolium*   iii] Susceptible to 'Little leaf'
D] *S. torvum*   iv] Resistant to 'Bacterial wilt'
E] *S. auriculatum*   v] Tolerant to 'Epilachna beetle'

97. Match the following :-
A] Solitary Inflorescence   i] Pear
B] Raceme Inflorescence   ii] Ber, Sweet orange
C] Catkin   iii] Fig
D] Corymb   iv] Blackberry, Raspberry
E] Solitary cymose   v] Banana, Coconut
F] Fasicle   vi] Mulberry, Pecan nut
G] Panicle   vii] Gladiolus
H] Spadix   viii] Grape, Mango
I] Hypanthodium   ix] Almond, Peach
J] Spike   x] Sapota, Phalsa

98. Match the following :-
A] Sorosis   i] Peach, Ber
B] Pome   ii] Banana, Guava
C] Drupe   iii] Pomegranate
D] Nut   iv] Pineapple, Jack fruit
E] Etaerio of berry   v] Aonla, Carambola
F] Berry   vi] Loquat, Quince
G] Hesperidium   vii] Raspberry, Custard apple
H] Capsule   viii] Fig
I] Cyconus   ix] Orange, Grape fruit
J] Balausta   x] Litchi, Rambutan

99. Match the following :-
A] Spike shedding   i] Sugarbeet
B] Zoning   ii] Chilli
C] Bulging   iii] Papaya
D] Anthracnose   iv] Grape
E] Stoney fruits   v] Black pepper
F] Damping off   vi] Mango

G] Hollow heart   vii] Radish

H] Black tip   viii] Potato

I] Multiple crown   ix] Custard apple

J] Phylloxera   x] Pineapple

100. Match the following :-

A] Spinach   i] Beet leaf

B] Amaranthus   ii] Dioecious

C] Pusa Jyoti, Pusa Harit   iii] Fenugreek

E] Arka Kiran, Arka Kanchan   iv] Monoecious

E] Prabha, Rajendra Kanti   v] Chaulai

101. Match the following :-

A] Spongy Tissue   i] Acid treatment

B] Panniyur-1   ii] Black pepper

C] Bunchy Top   iii] Leafy vegetable

D] Seed Scarification   iv] Banana

E] Spinach   v] Alphonso

102. Match the following :-

A] Sree Roopa, Sree Kirthi   i] Okra

B] Sree Prakash, Sree Vishakham   ii] Dioscorea

C] Sree Pallavi, Sree Rashmi   iii] Cococasia

E] Sree Vardhini, Pusa Sundari   iv] Cassava

E] Arka Gaurav, Arka Anamika   v] Sweet Potato

103. Match the following :-

A] Sterculiaceae   i] Coffee

B] Rubiaceae   ii] Date

C] Theaceae   iii] Rubber

D] Palmaceae   iv] Cocoa

E] Euphorbiaceae   v] Tea

Stock

1. The rooted plant onto which a graft is made. 2. A plant or plant part that furnishes cutting for propagation. 3. The main stem of a plant stumps, butt, main trunk. 4. A rhizome or rootstock.

Senescence

Aging of the plant parts, such as the flower, usually the stage form full maturity to death.

Freak

An unusual type of plant sometimes comes across in breeding.

Ephemeral

A short-lived plant species; applied to plant movements, as expanding of buds; completing life-cycle within a brief period.

Host
An organism or a plant that harbours a parasite.

Specimen or specimen plant
A plant that is grown in relatively open ground with little competition and therefore develops an unnaturally (in most cases) broad spread and dramatic form. Contrast with masses, drifts, thickets, or groves of plants, in which individual specimens intermingle with each other and may even be hard to distinguish from each other.

Quantitative long day plant
Plant that is not inhibited form flowering by short-day treatment, but hastened by long day treatment.

104. Match the following :-
A] Strawberry   i] *M. acuminata x M. balbisiana*
B] Banana   ii] *F. chiloensis x F. virginiana*
C] Pineapple   III] King x Willow Leaf
D] Kinnow   iv] Self-sterile cultivars
E] Tangor   v] *C. sinensis x C. reticulata*

105. Match the following :-
A] Strawberry, Grape   i] Pome fruits
B] Filbert, Chestnut   ii] Temperate-small fruits
C] Banana, Pineapple   iii] Tropical-berry fruits
D] Avocado, Datepalm   iv] Herbaceous-perennial
E] Quince, Loquat   v] Temperate-nuts

106. Match the following :-
A] Sub-tropical, Deciduous   i] Rambutan
B] Tropical, Evergreen   ii] Banana
C] Temperate, Evergreen   iii] Olive
D] Vine, Deciduous   iv] Bael
E] Herbaceous, Evergreen   v] Grape

107. Match the following :-
A] Sulphur dioxide   i] 700 ppm SO_2
B] Fruit juices   ii] 1-8 g/ltr. CO_2
C] Squash & Cordial   iii] 350 ppm SO_2
D] RTS & Nectar   iv] Chemical preservative
E] Fruit juice beverages   v] 100 ppm SO_2

108. Match the following :-
A] Sundaram, Singara   i] Arecanut
B] Cauvery, Taferikela   ii] Coffee
C] Forstero, Criollo   iii] Cashewnut
D] Vengurla- 1, Ullal-2   iv] Tea
E] Mohitnagar, Sreemangla   v] Cocoa

109. Match the following :-
    A] Sweet orange
    B] Pummelo
    C] Grape fruit

    D] Sour lime

    E] Santra

    i] Thompson, Marsh, Ruby
    ii] Pramalini, Vikram, Kadayam
    iii] Pineapple, Sathgudi, Valencia
    iv] Coorg Mandarin, Ponkan, Nagpur, Mandarin
    v] Walter, Chakaiya, Red Fleshed

110. Match the following :-
    A] Symptoms of Tristeza disease
    B] Symptoms of Xyloporosis
    C] Symptoms of Greening
    D] Symptoms of Citrus Canker
    E] Symptoms of Granulation

    i] Peg like out growth, Wood pitting, Tissue gum
    ii] Granular lesions on leaves, (Cochexia) twigs & fruits
    iii] Multiple buds, Narrow leaves, Chlorosis of leaf
    iv] Stem pitting, Seedling yellows, Root decay
    v] Hard, Grey & enlarged juice sacs

111. Match the following :-
    A] Tangelo, Sweet lime, Kumquat
    B] Sweet orange (Dancy, Hamlin)
    C] Mandarin, Sweet orange
    D] Citrange (Troyer)
    E] Poncirus, Carizzo citrange

    i] Tolerant to 'citrus nematode'
    ii] Sensitive to 'Salt'
    iii] Susceptible to 'Granulation'
    iv] Susceptible to 'Psorosis'
    v] Susceptible to 'Xyloporosis' (Cochexia)

112. Match the following :-
    A] Tangor
    B] Tangelo
    C] Kinnow
    D] Lemonnage
    E] Citrange

    i] *C. reticulata x C. paradisi*
    ii] *C. limon x C. sinensis*
    iii] King x Willow leaf
    iv] *P. trifoliata x C.sinensis*
    v] *C.sinensis x C. reticulata*

113. Match the following :-

A] Tea   i] Theaceae
B] Cocoa   ii] Arecaceae
C] Cashewnut   iii] Sterculiaceae
D] Arecanut   iv] Euphorbiaceae
E] Rubber   v] Anacardiaceae

114. Match the following :-

A] Temple; Clementine; Monreal   i] Tangelo cultivars
B] Sampson, Minneola, Seminole   ii] Lemon, Sweet lime
C] Shield/Patch budding   iii] Lime, Mandarin
D] Apomictic seedling   iv] Tangor cultivars
E] Cutting/ Air layering   v] Sweet orange, Grape fruit

115. Match the following :-

A] Thermodormancy   i] Apple
B] Zygodormancy   ii] Radish
C] Stratification   iii] Strawberry
D] Spongyness   iv] Lettuce
E] Blossom end rot   v] Sugar beet
F] Heart Rot   vi] Tomato
G] Speckled Yellow   vii] Aonla
H] Marsh spot   viii] Pea
I] Rosettes   ix] Cabbage
J] Catfaced fruits   x] Pecan

116. Match the following :-

A] Thompson Seedless Grape   i] Budsport of Thompson
B] Red blush (Grape fruit)   ii] Synthetic chimeras of Citrus
C] Foster (Grape fruit)   iii] Bud sport of Walter
D] Sarah (Orange)   iv] Bud sport of Marsh (Grape fruit)
E] Bizzarria, Kobayashi Mikan   v] Bud sport of Shamouti

True / False

117.	Mango is a non-climacteric fruit.	T/F
118.	Saffron belongs to family Iridaceae.	T/F
119.	Taj Garden in Agra is laidout in geometrical style.	T/F
120.	Tea mosquito bug is serious pest of coffee.	T/F
121.	Walnut is a monoecious tree.	T/F
122.	Waxy scale is an insect pest of guava & ber.	T/F
123.	'Sir Prize' is a typical triploid variety of apple.	T/F
124.	Angoorlata is a variety of tomato.	T/F
125.	Artificial pollination can be practiced in walnut.	T/F
126.	Buddha Jayanti Park is a formal style garden.	T/F
127.	Chrysanthemum is propagated by root suckers.	T/F
128.	*Euphorbia splendens* is a flowering shrub.	T/F
129.	Indigenous tea was first discovered in Meghalaya.	T/F
130.	Maleic hydrazide is a growth inhibitor.	T/F
131.	Sub-tropical fruits require chilling period.	T/F
132.	Tapioca belongs to family Araceae.	T/F
133.	'Fish leaf plucking' is a standard picking in tea.	T/F
134.	Alka Nishant is a variety of radish.	T/F
135.	Arka Manik is a variety of watermelon.	T/F
136.	Coorg Honey Dew is an improved variety of banana.	T/F
137.	Cytokinin activates cell enlargement.	T/F
138.	Diploid cultivars are usually self-fruitful.	T/F
139.	Okra & Aonla are botanically capsule.	T/F
140.	Red Chief is a spur type variety of apple.	T/F
141.	Skiffing is practised in Tea.	T/F
142.	'Kaithli' ber is much susceptible to powdery mildew.	T/F
143.	'Two leaf, a bud plucking' is also known as *Janam* plucking.	T/F
144.	C.M.R.S. is located in Bihar.	T/F

Ground cover
A group of plants usually used to cover bare earth and create a uniform appearance.

Hydroponics
The growth of plant in water without soil and supplied nutrient in water is known as hydroponics.

Spiratism
An abnormal condition of plants wherein normally straight stems produce an irregular scraggly growth with a flat or spiral form.

Heterogen
A variable group of plants which arise as hybrids, sports, mutation, etc., types of which may or may not breed true.

Glossary

Race
A group of individual plants which have certain common characteristics because of ancestry, a sub-division of a species, a permanent variety, a particular breed.

Rouging
Elimination of undesirable plants, which might be diseased, inferior, or non-typical.

Gutta-percha
A tough plastic substance obtained from the latex of some trees used especially for electrical insulation.

Plumose
Feather-like, plumy.

Apical meristem
The growing point of the shoot or root.

145. Cacti are xerophytic plants. T/F
146. Chicken & Hen is a disorder of tomato. T/F
147. Climacteric peak is maximum point of respiration in mature fruit. T/F
148. Generally, Valleys collect cool air. T/F
149. Kinetin is useful for artificial fruit ripening. T/F
150. Pecan is rich source of phosphoric acid. T/F
151. The stone of peach is known as kernel. T/F
152. Arkel & Bonville are popular varieties of grape. T/F
153. Drupe fruits are developed from inferior ovaries. T/F
154. Flavour in tea is due to linalool. T/F
155. Hoppers are serious pests of mango. T/F
156. Mauritius is a mid season variety of pineapple. T/F
157. Peters and Chico are universal pollinizers of pistachionut. T/F
158. Pistachionut flowers during May-June. T/F
159. Seed rate of chilli is higher than brinjal seed rate. T/F
160. Wilt is common disease of guava. T/F
161. Winter Banana is a variety of diploid banana. T/F
162. 'Ringing' is a method of asexual propagation. T/F
163. Asha and Jwala are clones of tea. T/F
164. Bartlett is a self-fruitful variety of pear. T/F
165. Coconut is botanically fibrous drupe. T/F
166. Fruit cracking in litchi is more prevelant in orchard situated at lower elevations. T/F

Fill in the Blanks

_____ bearing is problem of mango.
168. Indian Gooseberry is rich source of _____.
169. Hoppers are serious pest of _____.
170. Okra belongs to family _____.
171. ICAR headquarters is situated at _____.
172. 'Indian Horticulture' is published by _____.

173. Olour is polyembryonic variety of _____.
174. _____ wilt is serious disease of banana.
175. Mycorrhizal association is found in _____ tree.
176. Granulation is physiological disorder of _____ fruits.
177. Pointed gourd is propagated by _____.
178. Muscat flavour is found in _____.
179. Greening is a common viral disease of _____ fruits.
180. Pusa Sawani is a variety of _____.
181. Bonevilla is a variety of _____.
182. Loquat belongs to family _____.
183. _____ induces rooting in cuttings.
184. Slips are propagating material of _____.
185. Leaf curl is a serious disease of _____.
186. Citrus canker is a _____ disease.
187. Lalit is a nematode resistant variety of _____.
188. Fruits are rich source of _____.
189. Phalsa is commercially propagated by _____.
190. _____ variety of mango is suitable for intensity orcharding.
191. Banana is rich source of _____.
192. Decomposition of protein is known as _____.
193. ICMR headquarters is situated at _____.
194. Niranjan variety of mango gives fruits in _____ month.
195. Romani variety of mango flowers during _____.
196. _____ is the oldest fruit crop.
197. Mango is an _____ plant in growth habit.
198. Dashehari mango accommodates _____ plants/ha.
199. Planting distance of Amrapalli mango is _____ meter.
200. Grape is considerd as a fruit of _____ region.

Synthetic variety
In cross pollinated species a variety obtained by mating in all possible combinations a number of lines that combine well with each other.

Intraovarian
A type of artifical pollination in which pollen is inserted directly into the ovary rather than on the stigma.

Polyhaploid
A haploid from a polyploid.

Agar
A gelatinous and complex polysaccharides obtained from sea weeds (*Gracilaria* sp. *Gelidium* sp., etc.) widely used as a substrate in aseptic cultures and in making jelly, ice-cream, soap, pastries as emulsifier.

Several years ago Lockheed introduced the L-1011 Tristar plane. In order to ensure safety and test the strength of the jetliner, Lockheed exposed the plane to the roughest treatment for 18 months, costing $ 1.5 billion. Hydraulic jacks, electronic sensors and a computer put the airplane through its paces for more than 36,000 simulated flights, amounting to 100 years of airline service, without one single malfunction. Finally after hundreds of tests the aircraft was given the seal of approval.

There is every reason to believe that this plane would be safe to fly, because of all the effort put into preparation.

Script your success story

Having a positive **A**ttitude without making the **E**ffort is nothing more than having a **W**ishful dream.

Step-Fifteen

Losers are like thermometers;
Winners are like a thermostat.

Multiple Choice

1. Pungency in onion is due to
    A] Allyl propyl disulphide
    B] Di allyl disulphide
    C] Capsacin
    D] None

2. In North India, onion nursery is sown in-
    A] June-July
    B] August-September
    C] October-November
    D] January-February

3. Bolting of onion takes place if temperature goes below
    A] 10°C
    B] 15°C
    C] 20°C
    D] 25°C

4. A rose variety bears only one flower bud on a single stem, that variety belongs to-
    A] Floribunda group
    B] Ramblers
    C] Hybrid Tea group
    D] Polyantha group

5. In Brinjal only those flowers set fruits which have
    A] Short style
    B] Long style
    C] Medium style
    D] Both b & c

6. The commercial orchards of grapefruit are found in-
    A] Southern India
    B] Western India
    C] Eastern India
    D] Jammu & Kashmir

7. Repotting in orchid is done
    A] Every year
    B] Every 2-3 year
    C] 5 year
    D] 10 year

8. Food Product Order (FPO) was passed in-
    A] 1945
    B] 1950
    C] 1955
    D] 1975

9. Centre of origin of rice is
    A] Indo-Malaya region
    B] China
    C] USA
    D] Turkey

Annuals
Are the annual plants that live for one year or less, that is, the plant makes its vegetative growth flowers, and produces seed within one year from the sowing date then plants die.

Sample
A part of the population selected as a representative to the whole.

Heading back
Removal of a portion of a stem leaving another portion for promoting new growth.

Genus
A group of species possessing fundamental traits in common but differing in other lesser characteristics.

Mat watering
Irrigation of potted plants by capillary. Mats are composed of fabric, cellulose or other water-absorbing materials.

Parts per million (ppm)
It is equivalent to milligram per liter.

Spring
1. The season preceding summer.
2. Refers to the spring season in which a crop is harvested or a flower blooms, as a spring crop, a spring bloom, etc.

Caducous
Falling away early or prematurely.

Crush
A drink or pulp prepared by crushing fruit *e.g.*, orange crush.

10. Gladiolus is originated from-
    A] Europe　　　　　B] U.S.A.
    C] China　　　　　D] South Africa

11. Pineapple is originated in-
    A] India　　　　　B] Brazil
    C] New-zealand　　D] China

12. 'Journal of Ornamental Horticulture' is released from-
    A] IIHR, Bangalore　　B] IARI, New Delhi
    C] CIMAP, Lucknow　　D] ICAR, New Delhi

13. Cultivated rice *Oryza sativa* has following no. of chromosomes
    A] 32　　　　　B] 24
    C] 20　　　　　D] 18

14. A mature and ripened ovary containing seed is termed as-
    A] Embryo　　　　B] Stone
    C] Fruit　　　　　D] Grain

15. Modified atmospheric packaging (map) of fruit & vegetable prevent the building up of the following-
    A] Sugar　　　　　B] Protein
    C] CO_2 & acetylene　　D] None

16. ' Panama' is a serious disease of-
    A] Apple　　　　　B] Ber
    C] Banana　　　　D] Guava

17. Origin of Papaya is
    A] Central America　　B] South America
    C] China　　　　　D] India

18. The edible part of citrus is-
    A] Mesocarp　　　　B] Juicy placenta
    C] Endocarp　　　　D] Fleshy receptacle

19. The economic part of sandalwood tree is
    A] Leaves　　　　　B] Heartwood & roots
    C] Inflorescence　　D] Bark

20. ' Parthenogenesis' is found in-
    A] Mangosteen　　　B] Grape
    C] Banana　　　　　D] Jamun

GLOSSARY

21. Idea of particulate nature of inheritance was given by-
    A] Mendel   B] Jones
    C] Darwin   D] Bateson

22. Softness of pea is measured by
    A] Derometer   B] Refractometer
    C] Alanometer   D] Tendrometer

23. In jelly preparation, pectin acts as-
    A] Preservative   B] Stabilizer
    C] Buffers   D] Flavoring agent

24. The quickest method of peeling is-
    A] Steam peeling   B] Flame peeling
    C] Lye peeling   D] Mechanical peeling

25. The botanical name of perennial chrysanthemum is-
    A] *C.segetum*   B] *C.carinatum*
    C] *C. hybrida*   D] *C. morifolium*

26. Which crop perfume has high demand in market
    A] Jasmine   B] Chirata
    C] Kalmegh   D] All

27. ' Persian Style' of gardening was introduced into India by-
    A] Shahjehan   R] Babur
    C] Noorjehan   D] Akbar

28. Rate of photosynthesis is maximum in
    A] Red light   B] Black light
    C] Green light   D] Yellow light

29. ' Phylloxera' resistant rootstock of grape is-
    A] Harmony   B] Freedom
    C] St. George   D] Anab-e-Shahi

30. Whiptail is physiological disorder of-
    A] Cabbage   B] Cauliflower
    C] Tomato   D] Potato

31. In tea crop, picking of leaves is started after-
    A] 1-2 year   B] 2-3 year
    C] 4-5 year   D] Within 6 months

32. Vinegar concentration in finished pickle should not be-
    A] Less than 10%   B] Less than 5%
    C] Less than 2%   D] Less than 50%

Marmalade
A fruit jelly generally prepared from citrus fruits like oranges and lemons in which the slices of fruits or peels are suspended.

Saponine
A biologically active substance present in a number of vegetable crops like spinach, tomato, etc. that promotes intensive growth and development of the accompanying plants.

Kolanin
A heart stimulant glucoside present in the seeds or nuts of cola (*Cola nitida*).

Unarmed
Without either spines or prickles, though the leaves may have sharp teeth at the margins or tip like holly.

33. In fruit canning and pickling, salt acts as a/an-
    A] Oxidant   B] Antioxidant
    C] Reductant   D] Enzyme

34. Which is a famous place for mango orcharding-
    A] Nasik   B] Pune
    C] Allahabad   D] Lucknow

35. National Research Centre on Plant Bio-technology is situated at-
    A] Bangalore   B] New Delhi
    C] Mumbai   D] Nagpur

36. The concept of ' plant ideotype' was first propounded by-
    A] C.M. Donald   B] Meyer
    C] G.J. Mendal   D] J.D.Eastim

37. Which medicinal plant is effective for control of blood pressure
    A] Rauwolfia serpentina   B] Withania sommifera
    C] Catharanthus rosens   D] All

38. Which of the horticultural plant is grouped under 'C$_4$' plant-
    A] Mango   B] Banana
    C] Amaranthus   D] All of above

39. Which medicinal plant is used in controlling blood cancer
    A] *Rauwolfia serpentine*   B] *Catharanthus rosens*
    C] *Withania somnifera*   D] None

40. The foliage plant is
    A] *Cassia fistula* B] *Bauhinia spp*
    C] *Jacoranda spp*   D] *Aucaria cokii*

41. The origin of egg plant is-
    A] India   B] China
    C] Africa   D] Mexico

42. Shade loving plant is
    A] Pancy   B] Diffenbachia
    C] Anthurium   D] Both b & c

43. Which is referred as plantation crop-
    A] Grape   B] Potato
    C] Rubber   D] Pineapple

44. The beauty of trees planted along the water canals get enhanced due to-
    A] Refraction   B] Reflection
    C] Dispersion   D] Colourful stem

45. Appropriate time for grape planting is-
    A] September-October B] June-July
    C] March-April   D] December-January

46. Most popular planting method for mango is
    A] Square   B] Hexagonal
    C] Contour   D] Diagonal

47. Most popular planting method for mango is
    A] Square   B] Hexagonal
    C] Contour   D] Quincunx

48. The best planting time of Bougainvilia in Northern plains is-
    A] July-Sept.   B] Nov.-Dec.
    C] March-April   D] May-June

49. The best planting time of rose in north India is-
    A] Jan-Feb.   B] March-April
    C] Sept.-Oct.   D] July-Aug.

50. Leaves of which ornamental plants are much poisonous when eaten or chewed-
    A] Marigold   B] Chrysanthemum
    C] Dieffenbachia   D] Hollyhock

51. The maximum number of plants can be planted in-
    A] Square system   B] Contour system
    C] Hexagonal system D] Rectangular system

52. Respiration in plants is essentially a process related to the following-
    A] Oxidation   B] Evaporation
    C] Transpiration   D] None of these

53. 'Green house plants' are also known as-
    A] Sun loving plants   B] Shade loving plants
    C] Xerophytic plants   D] Climbers

54. The chromosome ploidy in the endosperm of coconut will be
    A] n   B] 2n
    C] 3n   D] 4n

Infusion
A method of seed priming where the seeds are immersed in acetone and dichloromethane solution for 1 to 4 hours for the chemicals to be infused into the seeds following removal of the solvent by evaporation. The incorporated chemical is then absorbed directly into the embryo with subsequent soaking in water.

Dehydro-brining
A storage procedure of perishables where the foods are at first dried partially and then salted which helps to tie up the free water present in food. These foods have good shelf life without any refrigeration.

55. " Plough Crop" is common name of the following fruit crop
   A] Walnut   B] Cashewnut
   C] Ber   D] Datepalm

56. ' Polygamomonoecious' form of sex is found in-
   A] Coconut   B] Date palm
   C] Papaya   D] Cashew nut

57. ' Polygamous' flowers are found in-
   A] Mango & Litchi B] Guava & Loquat
   C] Papaya & Coconut D] Cucumber & Chlii

58. Botanically, pomegranate fruit is a-
   A] Drupe   B] Berry
   C] Balausta   D] Pome

59. Which is popular as 'Pride of tropical S.E. Asia'-
   A] Litchi   B] Durian
   C] Carambola   D] Rambutan

60. Which is popular climber of Indian origin
   A] Indian ivy creeper B] Railway creeper
   C] Glory lily   D] All

61. The edible portion of Jerusalem artichoke is
   A] Leaf stalk   B] Tuber
   C] Fruits   D] Bulb

62. Fruit of portulaca (A summer leafy vegetable) is
   A] Capsule   B] Berry
   C] Siliqua   D] None

63. Chemical formulae of ' Potassium meta-bisulphite' is-
   A] KMS   B] K_2SO_4
   C] KSO_2 D] $K_2S_2O_5$

64. Greening in potato is due to
   A] High temperature B] High light intensity
   C] Low temperature   D] All

65. 'Earthing' in potato should be done when plant reaches a height of-
   A] 8-10 cm   B] 15-22 cm
   C] 25-30 cm   D] 30-35 cm

66. ' Powdery mildew' is a serious disease of-
    A] Peach   B] Guava
    C] Mango   D] Papaya

67. The resistance power of crop can be increased by
    application of
    A] Potash   B] Phosphorous
    C] Nitrogen   D] Sulphur

68. The classic example of practical use of colchicine in crop
    improvement is-
    A] F_1 hybrid seed   B] Seedless watermelon
       production
    C] Parthenocarpic fruit   D] Resistant variety
       production   development

69. Pruning is an essential practice for fruiting of-
    A] Jack fruit   B] Grape
    C] Citrus   D] Guava

70. Wintering is practiced
    A] To expose plant to lower temperature
    B] As an alternative to root pruning rose
    C] To protect plant against lower temperature
    D] None

71. Object of precooling is
    A] To fresh the product   B] To improve the quality
    C] Both a and b   D] To remove field heat

72. Kalipak is prepared from
    A] Green nuts   B] Mature nuts
    C] Mature green nuts D] Pipe nuts

73. The pigment present in watermelon is
    A] Carotenoid   B] Anthocynin
    C] Lycopene   D] Both b & c

74. Oxalates are present more in
    A] Leafy vegetables   B] Root vegetables
    C] Bulb crops   D] None

75. For the preservation of fruit juices, concentration of sodium
    benzoate' should be-
    A] 0.1-0.5%   B] 0.06-0.10%
    C] 0.5-1.0%   D] 1.1-2.0%

Scaling
Technique used to
propagate
foundation stock,
wherein the scale
of a bulb are
removed and
planted. This
produces numerous
scale bulb lets form
a single mother
scale.

Scooping
A method of
propagation of
hyacinths and other
bulbs by the
scooping out the
basal plate, the
main goal is to
eradicate apical
dominance so bulb
lets may be
formed.

Megaspore
A spore having the
property of giving
rise to a
gametophyte
(embryo sac)
bearing only a
female gamete.

Microspore
A spore having the property of giving rise to a gametophyte bearing only male gametes.

Park
An open public place protected by low fence and several gates and beautified with different plant materials, statues, fountains, etc.

Gooseneck
Term describing the proper stage of flower development to cut daffodils.

Protectant
A substance that protects an organism against infection by an pathogen.

Backbulb
An old orchid pseudobulb from which the leaves have abscised.

76. Low temperature preservation prevent growth of micro organism due to
A] Reduction in growth of micro organism
B] Increase in growth of micro organism
C] Killing of micro organism
D] All

77. Wine is preserved at
A] 10% alcohol   B] 14% alcohol
C] 8% alcohol   D] 5% alcohol

78. The first president of 'Horticultural Society of India' was-
A] G.S. Cheeema   B] G.S. Randhawa
C] K.L. Chadha   D] M.S. Randhawa

79. The main problem in breeding of chausa & Langra mango is
A] Alternate bearing   B] Spongy tissue
C] Self incompatability D] None

80. Wilt is a common problematic disease of-
A] Chilli   B] Pepper
C] Cucurbits   D] Rose

81. The most processed product from arecanut is
A] Kottapak   B] Kalipa
C] Nuli   D] All

82. Which crop produce maximum foodgrain in India
A] Wheat   B] Rice
C] Sugarcane   D] None

83. Late varieties of cauliflower produce their seeds-
A] Only on hills   B] Only on plains
C] Not produce seeds   D] Arid zones

84. Which is the largest producer of cardamom-
A] Brazil   B] India
C] Argentina   D] Mexico

85. Which is the largest producer of cashew-
A] Andhra Pradesh   B] Karnataka
C] Tamil Nadu   D] Kerala

86. The largest producer of pineapple is-
A] Sri Lanka   B] Thailand
C] Hawaii   D] Malaysia

87. Which is the largest producer of tea in the world-
    A] Brazil   B] China
    C] India   D] Mexico

88. Maximum guava producing state is
    A] UP   B] MP
    C] AP   D] Maharashtra

89. Highest vegetable producing state is
    A] UP   B] Orissa
    C] Bihar   D] MP

90. In which product TSS is 30 brix
    A] Cordial   B] Squash
    C] Jam   D] Jelly

91. Total foodgrain production (MT) in India in year 1998-99 was-
    A] 200   B] 190
    C] 180   D] 170

92. For breeder seed production in cabbage, which method is applied-
    A] Seed to seed method   B] Head to seed method
    C] Seed to head method   D] None of these

93. In India cabbage seed production is limited to-
    A] Plains   B] Nilgiri Hills
    C] Kashmir Valley D] N-E regions

94. Food grain production of India in 1998-99 was
    A] 198 mt   B] 200 mt
    C] 205 mt   D] 208 mt

Matching

95. Match the following :-
    A] Thompson Seedless,   i]   Raisin grapes
       Pusa Seedless
    B] Beauty Seedless,   ii] Table grapes
       Champion
    C] Sultania, Kishmish   iii] Canning grapes
       Chorni
    D] Red Prince, Bhokri   iv] Wine grapes
    E] Perlette, Delight   v]   Juice grapes

Fern
A big group of Pteridophytes abundantly grown in cool, shady and moist places both in the hills and in the plains.

Pubescent
fine hairiness of a leaf, petiole, pedicel, peduncle, or stem

Pure line
Progeny of a single homozygous self-pollinated plant.

Rosette
A crown of leaves radiating from a stem, and at or close to the surface of the ground.

Chloroplast
A small, dense protoplasmic cell inclusion containing chlorophyll, sometimes accompanied by other pigments.

96. Match the following :-
   A] Thrips tolerant chilli   i] Pusa Jwala
   B] T.M.V. resistant   ii] K-2, G-5. Andhra Jyoti
      bell-pepper
   C] Nematode resistant   iii] Pant C-I, Puri Red
      chilli
   D] 'Leafcurl' resistant chilli iv] MDU-1, X-235
   E] Mutant of chilli   v] Yolo Wonder, Bharat

97. Match the following :-
   A] Tomato for processing i] Sel-120
   B] 'Root Knot' tolerant   ii] Pusa Sheetal
      tomato
   C] 'Wilt' tolerant tomato iii] Arka Shreshtha
   D] Hot set tomato   iv] Roma, Punjab Chhuhara
      (upto 35°C)
   E] Cold set tomato   v] H.S.-102. Pusa Hb-1
      (at 8°C)

98. Match the following :-
   A] Tomato hybrids   i] Pusa-120, Roma,
                              H.S.-101
   B] Tomato selections   ii] Yolo Wonder, Callifornia
                              Wonder
   C] Brinjal hybrids   iii] Arka Alok, Arka Vishal,
                              Arka Abha
   D] Capsicum selections iv] Pusa Jwala, Pant C-1, K-2
   E] Chilli hybrids   v] Pusa Anmol, Pusa Anupam,
                              Pusa Kranti

99. Match the following :-
   A] Tomato   i] Pusa Gaurav, Arka Saurav
   B] Carrot   ii] Pusa Parvati, Contender
   C] Ashgourd   iii] Pusa Meghali, Pusa Kesar
   D] French bean iv] Pusa Kasuri
   E] Fenugreek   v] Pusa Shakti, CO-1

100. Match the following :-
   A] Toughening of pea i] Due to insufficient exhausting
   B] Springer cans   ii] Due to defective seaming

C] Corrosion of cans
D] Leakage in cans
E] Bursting of cans

iii] Due to excess pressure
iv] Due to calcium salt in cans
v] Due to acidic product

101. Match the following :-

A]	Triploidy	i]	Cabbage
B]	Stenospermocarpy	ii]	Guava
C]	Cauliflorous bearing	iii]	Seedless Grape
D]	Stumping	iv]	Rose
E]	Irregular bearing	v]	Pomegranate
F]	Root pruning	vi]	Mango
G]	Slipping stage	vii]	Cocoa
H]	Wintering	viii]	Tomato
I]	Bagging	ix]	Seedless Guava
J]	Emasculation	x]	Gladiolus

102. Match the following :-

A]	Turfing and Bricking	i]	Chain Cacti
B]	Cactanae group	ii]	Japanese Garden
C]	Cereanae group	iii]	White Kachnar
D]	Rhipsalidanae group	iv]	Melon Cactus
E]	Coryphanthanae group	v]	Cypress
F]	*Nymphaea ordorata*	vi]	Tourch Cacti
G]	Symbolic of life	vii]	Water Lily
H]	Symbolic of youth & life	viii]	Lawn
I]	Symbolic of immortality	ix]	Water
J]	Pagoda tree	x]	Pin-cushion Cacti

103. Match the following :-

A]	Two/More kinds of fruits	i]	68% T.S.S. at 105°C
B]	One kind of fruits	ii]	Jelly
C]	Two kinds of citrus fruits	iii]	65% T.S.S. at 105°C
D]	End point of jelly	iv]	Marmalade
E]	End point of jam	v]	Jam

Pseudodominance
Apparent expression of the recessive allele in a heterozygous individual when the dominant allele is absent.

H-budding
A double rectangular cutting, resembling the letter H, made in the bark of the stock. The flaps are lifted in the centre for insertion of the bud.

Perennation
A lasting stage, referring particularly to the persistence of fruit long after its usual season of maturity.

Inbreeding
Mating of closely related individuals than ancestry than would be expected under random mating.

Heritability

A measure of genetic relationship between the parent and progeny. In broad sense, it is the ratio of genotypic variance to the phenotypic variance and in narrow sense, it is the ratio of additive genetic variance to phenotypic *i.e.,* total variance.

Slow release fertilizer

Fertilizers that release their nutrients slowly and evenly, over a long period of time

Morphine

An alkaloid present in raw opium (latex), which is released from cuts made in the unriped capsules (fruits) of poppy (*Papaver somniferum*). It is a powerful analgesic and nareotic used as medicine.

104. Match the following :-

A] UPCAR	i]	Kent (England)
B] EMRS	ii]	Kirkee (Pune)
C] VPKAS	iii]	Washington (USA)
D] RRII	iv]	Trichy
E] IWGSC	v]	Almora
F] TBGRI	vi]	Mysore
G] CFTRI	vii]	Kottayam
H] NRCB	viii]	Nagpur
I] NRCC	ix]	Lucknow
J] GKBG	x]	Palode (Kerala)

105. Match the following :-

A] Vector of 'tristeza virus'   i] Mechanical (Knife) transmission

B] Vector of 'Xyloporosis'   ii] Citrus leaf minor (*Phyllocnistis citrela*)

C] Vector of 'Exocortis/ Scaly butt'   iii] Oriental citrus aphid (*Toxoptera citricida*)

D] Vector of 'Greening MLOs'   iv] Citrus psylla (*Diaphorina citri*)

E] Vector of 'Citrus Canker'   v] Bud/Scion wood

106. Match the following :-

A] Vit-A	i]	Cashewnut
B] Vit-B$_1$	ii]	Aonla
C] Vit-B$_2$	ii]	Bael
D] Vit-C	iv]	Jamun
E] Insulin	v]	Mango

107. Match the following :-

A] Vitamin -A	i]	Aonla, Guava
B] Thiamine	ii]	Karonda, Date
C] Riboflavin	iii]	Almond, Pea
D] Ascorbic acid	iv]	Litchi, Wood apple
E] Carbohydrate	v]	Mango, Papaya
F] Protein	vi]	Litchi, Colocasia
G] Fat	vii]	Walnut, Almond

H] Iron
I] Calcium
J] Phosphorous

viii] Cashew nut, Walnut
ix] Dry Apricot, Date
x] Bael, Litchi

108. Match the following :-

A] Vitamin -A
B] Thiamin
C] Iron
D] Vitamin-C
E] Carbohydrate

i] Cashewnut
ii] Banana
iii] Papaya
iv] Datepalm
v] Barbados cherry

109. Match the following :-

A] Vrindavan Garden
B] Garden City of India
C] Heart of a Garden
D] Home of Spices
E] Sikander Bagh

i] Bangalore
ii] Lawn
iii] India
iv] Mysore
v] Lucknow

110. Match the following :-

A] Walnut
B] Wood apple
C] Bread fruit
D] Litchi
E] Loquat
F] Datepalm
G] Pomegranate
H] Banana
I] Papaya
J] Ber

i] Moraceae
ii] Rhamnaceae
iii] Juglandaceae
iv] Arecaceae
v] Caricaceae
vi] Rutaceae
vii] Sapindaceae
viii] Musaceae
ix] Punicaceae
x] Rosaceae

111. Match the following :-

A] Walnut
B] Mango
C] Cashewnut
D] Dry Karonda
E] Bael

i] Richest source of Fat
ii] Richest source of Riboflavin
iii] Richest source of Vit-A
iv] Richest source of Iron
v] Richest source of Protein

112. Match the following :-

A] Water Core
B] Quick Decline
C] Brown Rot
D] Rancidity
E] Fruit Cracking

i] Pear
ii] Pomegranate
iii] Walnut
iv] Apricot
v] Apple

Blinds Shoots
Shoots that remain vegetative under conditions which normally stimulate formation of inflorescence.

Peat moss
The partially decomposed remains of various mosses. This is a good, water retentive addition to the soil, but tends to add the acidity of the soil pH.

Evaporator
An equipment used for removal of moisture from fruit and vegetable juices by evaporation.

Dehydrator
Any device which artificially removes moisture from a product for preservation or for reducing its bulk and weight.

Inlaying
A graft consisting of removing the area of stock and preparing a scion which will fit it exactly.

Dichogamy
Male and female reproductive organs of a hermaphrodite flower mature at different times.

Oleoresin
A natural combination of resinous substances and essential oils present in the fruits of certain crop plants like chilli.

Tolerance
A mechanism of host resistance due to the ability of the plant to withstand the effect of adverse growing condition (drought, pest and disease attack, etc.)

113. Match the following :-
A] Western   i] Self -fertile almond
B] Moreland, Stuart   ii] Self-fertile pecan
C] Desirable, Cheynne iii] Protandrous pecan
D] Franquette, Pedro iv] Pollinizers of walnut
E] Genco, Mollisona   v] Protogynous pecan

114. Match the following :-
A] White   i] Dominant colour in Summer
B] Red   ii] Dominant colour in Winter
C] Yellow iii] Dominant colour in Night Garden
D] Blue   iv] Evergreen colour in a Garden
E] Green   v] Dominant colour in Water Garden

115. Match the following :-
A] Wine/Table grape   i] *V. aestivalis*
B] Summer grape   ii] *V. rupestris*
C] Winter grape   iii] *V. berlandieri*
D] Fox grape   iv] *V. vinifera*
E] Sand grape   v] *V. labrusca*

116. Match the following :-
A] *Withania sominifera*   i] Fabaceae
B] *Asparagus officinalis* ii] Solanaceae
C] *Catharanthus roseus*   (iiii) Apocynaceae
D] *Acorus calamus*   iv] Liliaceae
E] *Glycyrrhiza glabra*   v] Araceae

|||||||||||||||||||||||||||||||||||| **True / False** ||||||||||||||||||||||||||||||||||||

117. Indian Gooseberry is rich source of Ascorbic acid.   T/F

118. Postachionut requires long hot & dry summer.   T/F

119. Seed rate of tomato is lower than brinjal seed rate.   T/F

120. Stone grafting is practised in jack fruit.   T/F

121. All the self-pollinated crops are self-fruitful.   T/F

122. Apricot is seriously affected by leaf curl.   T/F

123. *Butea monosperma* is a flowering shrub.   T/F

124. Edible banana are commonly sterile. T/F
125. India is the largest producer of rubber. T/F
126. Root wilt disease of coconut is caused by MLOs. T/F
127. Seeds in berry are found in meso & endocarp. T/F
128. Sodium benzoate is used for fruit ripening. T/F
129. 'Ring neck' is a disorder of avocado. T/F
130. Apple fruit is developed from superior ovary. T/F
131. Cashew is a tropical nut crop. T/F
132. Fruit necrosis occurs in Anola. T/F
133. Loquat and Apple belong to same family. T/F
134. Palm oil is extracted from fleshy mesocarp of fruit. T/F
135. Plantains belong to AAB group of banana. T/F
136. Seed rate of early cauliflower is more than late ones. T/F
137. 'Brown blast' in rubber is a physiological disorder. T/F
138. Almond is known as Butter fruit also. T/F
139. Almost all varieties of almond are self-unfruitful. T/F
140. Apple is propagated by seeds. T/F
141. Banana is commonly propagated by water suckers. T/F
142. Cavendish group of banana has high female sterility. T/F
143. Effect of pollens on the fruit is known as metaxenia. T/F
144. Lettuce & Artichoke belong to family Compositae. T/F
145. Papain is an enzyme having proteolytic properties. T/F
146. Tea mosquito bug is a major insect-pest of cashew. T/F
147. Banana is a heavy feeder crop. T/F
148. Beverage made from coffee is called as benign stimulant. T/F
149. Budding in ber should be done in August. T/F
150. Grafted jack fruit plant can produce fruits within
     three years. T/F
151. Litchi is widely cultivated in U.P. T/F
152. Mango bears flower on mature shoots terminally. T/F
153. Neelam is a variety of dahlia. T/F

Prehensile
Clasping, coiling in response to touch.

Dormant
A state of rest and reduced metabolic activity in which plant tissues remain alive but do not grow.

Compost
An organic soil amendment resulting from the decomposition of organic matter.

Evaporation
Process by which water returns to the air. Higher temperatures speed the process of evaporation.

Lined
Lightly ridged or ribbed.

Panning
Potting or transplanted root cuttings or bulbs.

154. Phomopsis blight of brinjal causes damage to fruits.   T/F
155. Red rust of mango is caused by a fungus.   T/F
156. Taj Garden was built by Jahangir.   T/F
157. 'Bunchy top' virus is transmitted through mites.   T/F
158. 'Pentagonium' was first variety of cocoa introduced to India from Guatemala.   T/F
159. Among non-alcoholic beverages, coffee ranks first in world.   T/F
160. Grape is pruned regularly for better fruiting.   T/F
161. High C/N ratio results in a low S/R ratio.   T/F
162. Kurrukan is a salt tolerant root stock of mango.   T/F
163. Master Peace is a variety of canna.   T/F
164. Phalsa and Fig are deciduous fruit plants.   T/F
165. Pistachionut and Cashewnut both belong to same family.   T/F
166. Waxing is effective to prolong shelf life of fruits.   T/F

Fill in the Blanks

167. Seedless grapes are used for _____ making.
168. _____ district of Maharashtra is known as 'vine yard of India'.
169. Ganesh is soft seeded variety of _____.
170. _____ is major onion producing state.
171. _____ is major spices producer in the world.
172. Mallika and Amrapalli mangoes are released by _____.
173. Epicotyle grafting is practiced in _____.
174. Amrapalli variety of mango accommodates about _____ plants/ha.
175. CFTRI is situated at _____.
176. Arka Jyoti is a variety of _____.
177. Pusa Sanyog is a hybrid variety of _____.
178. About _____ humidity is recommended for storage of most products.

179. _____ is the largest producer & exporter of cashewnut.

180. _____ is suitable time for beheading in cashewnut.

181. The ratio of length & width of an orchard should be _____.

182. In muskmelon _____ fertilizers increase sweetness.

183. Cucurbits can be grown off season by _____ forcing.

184. *Manihot esculenta* belongs to _____ family.

185. *Citrullus lanatus* is the botanical name of _____.

186. Ratna mango is a cross of _____.

187. Triploid varieties are mostly cultivated in _____ crop.

188. Pineapple belongs to _____ family.

189. *Eriobotrya japonica* is the botanical name of _____.

190. Litchi belongs to family _____.

191. Rose Scented is a _____ season variety of litchi.

192. Grape is commercially propagated by _____ cuttings.

193. Black tip disorder occurs on _____ of mango.

194. Improved Golden Yellow is variety of _____.

195. Pineapple has its origin from _____.

196. Loquat fruits are harvested during _____ months.

197. Apple grows well on _____ soils.

198. _____ varieties are usually self-fruitful.

199. Winter Banana is a cultivar of _____.

200. Fig is _____ in growth habit.

Transverse ridge
A ridge that runs across the stem from one leaf scar to its pair on t he opposite side of the twig.

Annular
Ring-like; said of leaf scars which encircle the bud, or bundle scars which are circular with an opening in the center.

Pterrygosperonon
A compound obtained from sajna (*Moringa pterygosperma*) which acts actively against growing moulds and fungi.

Girdle
Encircling of plant roots, stems, trunks or branches resulting in a constriction of the plant part, or a reduction of water and nutrient flow through the girdled plant part.

An ancient Indian sage was teaching his disciples the art of archery. He put a wooden bird as the target and asked them to aim at the eye of the bird. The first disciple was asked to describe what he saw. He said, "I see the trees, the branches, the leaves, the sky, the bird and its eye."

The sage asked this disciple to wait. Then he asked the second disciple the same question and he replied, "I only see the eye of the bird." the sage said, "Very good, then shoot." The arrow went straight and hit the eye of the bird.

Unless we focus, we cannot achieve our goal. It is hard to focus and concentrate, but it is a skill that can be learned.

Script your success story

Keep

your **E**yes upon

the **G**oal.

Step-Sixteen
Losers say what they choose;
Winners choose what they say.

Multiple Choice

1. Etawah pilot project was started in the year
    A] 1948   B] 1938
    C] 1944   D] 1954

2. Which is useful to prolong shelf life of fresh-perishable commodities-
    A] Ethephon   B] Gibberellin
    C] Brominated carbon D] Florigen

3. Self incompatibility promotes
    A] Autogamy   B] Allogamy
    C] Homogamy   D] None

4. 'Spongy Tissue' prone variety of mango is-
    A] Dashehari   B] Mallika
    C] Amrapali   D] Alphonso

5. Fruit crop propagated by tissue culture on large scale
    A] Papaya   B] Banana
    C] Grape   D] None

6. Tuberose is propagated by
    A] Bulb   B] Corm
    C] Seed   D] None

7. Bougainvillia is propagated by
    A] Layering   B] Cutting
    C] Seed   D] Grafting

8. Chrysanthemum is propagated by-
    A] Leaf cutting   B] Root cutting
    C] Herbaceous cutting D] Budding

9. Jack fruit is commercially propagated by-
    A] Seed   B] Air-Layering
    C] Herbaceous cutting D] Shield budding

10. Lemon is commercially propagated by-
    A] Seed   B] Cutting
    C] Air layering   D] Grafting

Bare root
Plants offered for sale which have had all of the soil removed from their roots.

SEM
The abbreviation for scanning electron microscopy.

Savannah or savanna
Grassland with scattered trees, fairly common in the central plains of the U.S. before development. Bret Rappaport writes, "...it is theorized that humans are genetically predisposed to favor open grass-type landscapes as an artifact of our species' development on the savannas and grasslands of East Africa."

Olifactoric
Attraction of
pollinators by
scent.

Stalked
Said of buds with
scales elongated
perceptibly below
the lowest scales
(alder).

Fringed
Ciliate with glands
or scales rather
than fine hairs.

Subopposite
Said of paired leaf
scars that are
close, but not at
exactly the same
height on the twig
(staggered).

Mucilage
A complex
glutinous
carbohydrate
secreted by certain
plants (Okra,
Basella, Isabgol,
etc.), which
absorbs water
freely.

11. Jack fruit is commercially propagated by-
    A] Seed   B] Inarchng
    C] Budding   D] Layering

12. Gladiolus is propagated through-
    A] Seed   B] Corm
    C] Tuber   D] Cutting

13. Strawberry is propagated through
    A] Suckers   B] Cuttings
    C] Runners   D] None

14. Pineapple is propapated best by-
    A] Crown   B] Slip
    C] Sucker   D] Rhizome

15. The richest sources of protein is-
    A] Pea   B] Fenugreek
    C] Pointed gourd   D] Potato

16. Rose are pruned in month of
    A] August   B] December
    C] October   D] February

17. For getting optimum yield pruning is necessary in-
    A] Phalsa   B] Jack fruit
    C] Pineapple   D] Wood apple

18. The best time for pruning of roses in North India is-
    A] Ist week of October   B] IInd week of October
    C] IIIrd week of October D] Ist week of November

19. The term pruning refer to
    A] Providing shape to   B] Providing space for growth
       plant
    C] Way for better fruiting D] All

20. From the date of pruning to flowering, H.T. rose takes-
    A] 35 days   B] 42 days
    C] 50 days   D] 60 days

21. ' Pruning' in some trees is essential practice-
    A] To get proper size   B] To get strong stature
       & shape
    C] To get heavy bearing   D] To control disease & pest

22. *Fusarium oxsporum f. psidii* is a causal organism of-
     A] Sugarcane wilt   B] Citrus dieback
     C] Guava wilt   D] All of these

23. 'Arka- Puneet', 'Arka Aruna' and 'Arka Neekiran' are new
     varieties of-
     A] Aonla   B] Grape
     C] Tomato   D] Mango

24. ' Punjab Padmini' is a variety of-
     A] Chilli   B] Okra
     C] Onion   D] Litchi

25. Blue/ Purple coloured variety of gladiolus is-
     A] Lucky Star   B] Eurovision
     C] Hermajesty   D] American Beauty

26. Santaram and Pusa Bhadur is variety of
     A] Rose   B] Gladiolus
     C] Chrysanthemum   D] Dahlia

27. ' Pusa Delicious' papaya is a _____ in sex forn.
     A] Gynodioecious   B] Dioecious
     C] Hermaphrodite   D] Andromonoecious

28. ' Pusa Early Dwarf is a variety of-
     A] Pear   B] Walnut
     C] Almond   D] Strawberry

29. ' Pusa Kiran' chaulai is developed by-
     A] Dr. D.P. Singh   B] Dr. P.S. Sirohi
     C] Dr. S.P.S. Raghava   D] Dr. S.K. Dutta

30. ' Pusa Kirti' chaulai developed by-
     A] Dr. N. Sivakami   B] Dr. D.P. Singh
     C] Dr. G. Kalloo   D] Dr. P.S.Sirohi

31. ' Pusa Seedless' is a clone of-
     A] Sonaka   B] Cheema Sahebi
     C] Thompson Seedless   D] Anab-e-Shahi

32. For good quality of Gladiolus flowers, corm diameter should
     be about
     A] 2 cm   B] 3 cm
     C] 4 cm   D] 5 cm

Sympodium
A stem whose successive sections are strictly branches each springing from the one preceding as in the vine.

Thermodormancy
A type of physiological seed dormancy where germination of the freshly harvested seeds is sensitive to temperature *e.g.,* seeds of some cultivars of lettuce and celery do not germinate at temperature above 25°C.

Ligase
An enzyme, which seals the nick or cuts in a DNA strand.

Bur
Any rough or prickly seed envelope.

Interplant
1. To plant seeds of one species between those of another or to plant one crop in rows and another in between rows. 2. To set out young trees among other young trees planted earlier or naturally existing.

Pithy
Sometimes used in the sense of having a large pith and little wood.

Pricking
A method of raising secondary nursery for the crops having very small seeds, in the case of high density sowing, it helps to develop a thinner stand in the nursery. In cole crops and lettuce, pricking is done when first pair of true leaves develop.

33. The best quality tea grown in world is at
A] Nilgiri (India)   B] China
C] Darjeeling (India)   D] Nainital (India)

34. The following quantity of rhizome is enough for planting 1 hac. of ginger-
A] 1200-1400 kg   B] 10 kg
C] 500 kg   D] 2500 kg

35. Two major races of rice are
A] Temperate & tropical   B] Asiatic & japonica
C] European & African   D] Indica & japonica

36. Variety of raddish round in shape-
A] Japanese white   B] Chinese pink
C] Pusa himani   D] Rapid red white tip

37. Most suitable radiation for disinfection of food is
A] U-V rays   B] Gamma rays
C] Alpha rays   D] All

38. ' Radio' and 'White Queen' are tall varieties of-
A] Hollyhock   B] Canna
C] Hibiscus   D] Aster

39. Pungency in radish is due to-
A] Isothiocyanate   B] Allyl propyl di-sulphite
C] Allicin   D] Thiobutenyl

40. ' Rajat Rekha' and 'Swarna Rekha' are commercial varieties of-
A] China Rose   B] Tuberose
C] Rose   D] Amaryllis

41. Following seed rate is recommended for sowing one hactare watermelon
A] 1-1.5 kg   B] 2-2.5 kg
C] 4.5-5.0 kg   D] 5.5 -6.5 kg

42. Sufficient seed rate of cabbage to plant in one hectare-
A] 50-100 g   B] 350-500 g
C] 1000 g   D] 2000 g

43. Suitable seed rate of okra for summer sowing is-
A] 18-22 kg/ha   B] 10-12 kg/ha
C] 30-40 kg/ha   D] 40-50 kg/ha

44. The most economic sex -ratio in cucurbits is-
    A] 25:1   B] 30:1
    C] 20:1   D] 15:1

45. Net capital ratio is given by
    A] Total assets/Total   B] Working assets/Working
         liabilities   liabilities
    C] Current assets/   D] Deferred liabilities/Net
         Current liabilities   worth

46. The preservative recommended by FPO are
    A] SO_2 & Na Benzoate   B] KMS & KNO_3
    C] Acetic acid   D] None

47. Agricultural Scientists Recruitment Board (ASRB) was established during-
    A] 1965   B] 1970
    C] 1973   D] 1975

48. In tomato red colour is due to
    A] Anthocynin   B] Lycopene
    C] Xanthophylls   D] None

49. Forced perspective refers to
    A] Technique used to practice water conservation
    B] Technique of creating illusion of more space in garden
    C] Technique of forcing flowers to come out in off season
    D] Technique of forcing follows plants for early flowering

50. 'Trophy' refers to
    A] The systematic training of shrubs
    B] Growing plants on terraces
    C] Method of growing fruit, vegetables & flower in water based solution
    D] Arrangement of annual, potted plants or flower in water base solution

51. Capacity of refrigeration is measured by
    A] Gram   B] Ton
    C] Pascal   D] Lux

52. NPK are regarded in plant nutrition as-
    A] Micro nutrient   B] Macro nutrient
    C] Trace element   D] Non essential

Fruticose
Shrubby, in the sense of the stems being woody.

Sex-modification
Altering the sequence of flowering and thereby sex ratio in monoecious cucurbits which can be accomplished by exogeneous application of plant growth regulators and some micronutrients.

Corona
An appendage or a series of appendages on the inner side of the corolla of passion flower, oleander, etc. Also called cup, trumpet, coronet.

Crenate -serrate
Having a mixture of blunt and sharp teeth.

Excurrent habit
the cone-shaped form typical of many conifers and a few angiosperms with a clearly identified central leader and growing point at the top of the plant

Ligule
1. A strap shaped organ. 2. A minute projection from the top of the leaf sheath in grasses. 3. The strap shaped corolla in the ray flower of Composites.

Cuneate
Wedge-shaped with an essentially straight side, the structure attached at the narrow end.

Dehiscent
Opening to discharge the seeds or pollen.

Orbiculate
Circular or disc-shaped.

53. Which growth regulator is commercially used is Grape for berry elongation
    A] Etheral   B] Auxin
    C] Cytokinin   D] GA_3

54. Which growth regulator is commonly used as weedicide
    A] ABA   B] IBA
    C] NAA   D] 2, 4

55. Which growth regulator is substitute of low chilling treatment
    A] Auxins   B] Cytokinin
    C] Gibrillin   D] AH_3

56. The growth regulator related with abscission of leaves is
    A] ABA   B] Ethylene
    C] GA   D] Cytokinin

57. First book related to horticulture was on-
    A] Date palm   B] Apple
    C] Litchi   D] Citrus

58. Avenue tree related to Lord Krishna is-
    A] *Anthocephalus cadamba*   B] *Delonix regia*
    C] *Ficus religiosa*   D] *Saraca indica*

59. Term 'bahar' is related to the cultivation of-
    A] *Musa prardisiaca*   B] *Litchi chinensis*
    C] *Psidium guajava*   D] *Malus pumila*

60. Pulsing is related to
    A] Preservation   B] Storage
    C] Cut flower   D] None

61. Calcium carbide releases-
    A] Acetylene gas   B] Benzene gas
    C] Methane gas   D] Ethylene gas

62. An equipment used for removing moisture from fruit & vegetable juices by evaporation is-
    A] Evaporator   B] Transpirator
    C] Autoclave   D] Dehydrator

63. Minimum TSS required for Jam is
    A] 68%   B] 70%
    C] 85%   D] 80%

64. Seed rate of garlic required for one hectare-
    A] 8-10 q cloves   B] 13-22 q cloves
    C] 25-28 q cloves   D] 30-35 q cloves

65. Zinc is required for synthesis of-
    A] Proteins   B] Sugars
    C] Trytophan   D] Fats

66. Light intensity requirement for orchid is
    A] 1000-1500FC   B] 2000-6000FC
    C] 7000-10000FC   D] 10000-12000FC

67. Night temperature requirement for tomato is
    A] 5-10°C   B] 15-20°C
    C] 25-30°C   D] 10-15°C

68. Which method of grafting requires the scion and root-stock
    of same thickness-
    A] Wedge grafting   B] *In-situ* grafting
    C] Inarching   D] Veneer grafting

69. Good storage require
    A] High temperature & high humidity
    B] Low temperature & high humidity
    C] High temperature & low humidity
    D] None

70. Indian Council of Medical Research (ICMR) is situated at-
    A] New Delhi   B] Lucknow
    C] Bangalore   D] Pune

71. Indian Institute of Horticultural Research (IIHR) was
    established in-
    A] 1965   B] 1968
    C] 1976   D] 1980

72. Central Potato Research Institute (CPRI, 1949) was
    shifted from Patna (Bihar) to Shimla (H.P.) in-
    A] 1950   B] 1952
    C] 1960   D] 1956

73. Central Coffee Research Institute is located at
    A] New Delhi   B] Balashur
    C] Tricy   D] Eluru

Thorn
one of various sharp and pointed outgrowths of plants that can arise from stems, leaves, or fruits; botanically, a modified twig which has tiny leaf scars and buds and is usually pointed and sharp; other sharp structures are prickles and spines which are technically different from thorns and each other, but which have the same overall effect

Hamper
Container used for shipping gladious.

Pinching
Removal of the shoot apex to overcome apical dominance and promote lateral shoot development.

Glossary (sidebar)

By passing shoot
Vegetative shoot develops immediately below the flower bud, often occurring azaleas.

Twig
A small branch or shoot of a woody plant.

Off-shoot
A lateral shoot or branch which arises from one of the main stems of a plant.

Leader
The primary or terminal shoot; the trunk of a tree.

Water Sprouts
These are shoots growing form latent adventitious buds on stems or branches.

Die back
Death of shoots, originating at the shoot tips.

---

74. Indian Insititue of Horticulture Research is located at-
    A] Ghethali
    B] Hessarghatta
    C] Godhra
    D] Abohar

75. First Tea Research Station in India was established in-
    A] Jorhat (1904)
    B] Sibsagar (1891)
    C] Shillong (1903)
    D] Darjeeling (1898)

76. The cheapest method of reservation is-
    A] Canning
    B] Freezing
    C] Drying
    D] Preserving

77. 'Woodly aphid' resistant root stock of apple is-
    A] M-27
    B] Crab-apple
    C] M-16
    D] Northern Spy

78. Banana variety resistant to panama wilt disease
    A] Basrai
    B] Poovan
    C] Harichal
    D] Safed velchi

79. Late blight resistant variety of potato is
    A] Kufri badshah
    B] Kufri sindhuri
    C] Kufri megha
    D] None

80. Late blight resistant variety of Potato
    A] Kufri Jyoti
    B] Kufri Giriraj
    C] Kufri Chipsona
    D] All

81. Bacterial wilt resistant variety of Tomato is
    A] Arka Abha
    B] Arka Vardan
    C] Arka Gaurav
    D] All

82. Biological control of ' rhinoceros beetle' of coconut is-
    A] Birds
    B] Baculo virus
    C] Coca mosaic
    D] Malathion

83. Vitamin A rich source fruit is
    A] Ber
    B] Pomegranate
    C] Indian gooseberry
    D] Mango

84. Potato is rich source of
    A] Fat
    B] Starch
    C] Vitamin
    D] Mineral

85. *Capsicum annuum* is a rich source of-
    A] Vit-A   B] Vit-B
    C] Vit-C   D] Vit-E

86. ' Ring budding' is a commercial method of propagation for-
    A] Grape   B] Mango
    C] Ber   D] Aonla

87. Some fruits can be ripen well after their harvesting as well as on the tree. These fruits are known as-
    A] Climacteric   B] Non-Climacteric
    C] False fruits   D] Delicious

88. Rains during ripening of grapes cause-
    A] Fruit drop   B] Berry cracking
    C] Low TSS   D] Low acidity

89. The earliest ripening variety of grape is-
    A] Bharat Early   B] Arka Vati
    C] Perlette   D] Pearl of Cassaba

90. The earliest ripening variety of peach is-
    A] J.H. Hale   B] Safeda
    C] Florda Sun   D] Sharbati

91. ' Roopa' and 'Karan' are improved varieties of-
    A] Walnut   B] Almond
    C] Pecan   D] Datepalm

92. ' Root Knot' disease in vegetables is caused by-
    A] Fungus   B] Virus
    C] Nematode   D] MLO's

93. The best root stock of rose for Northern plains is-
    A] *Rosa indica*   B] *R. bourboniana*
    C] *R. multiflora*   D] *R. wordifolia*

94. Ultra dwarfing rootstock (M-27) of apple was developed by-
    A] Mark Tydeman (1929)   B] Chalmers (1920)
    C] Dunn & Stolp (1970)   D] Vanden Eude (1912)

GLOSSARY

Grafting
The uniting of a short length of stem of one plant onto the root stock of a different plant. This is often done to produce a hardier or more disease resistant plant.

Flaking
Shredding, but with short segments.

Holding
A pre-shipment short-term treatment to cut flowers in which the flowers are placed in solution of sucrose (concentration 20% and above) for a period of 12-24 hours at 100 lux light intensity and at 20-27°C temperature, which prolongs vaselife, promotes opening and improves the colour and size of petals.

Inter Cropping

The programme should be fixed in such a manner that the main fruit crop does not suffer in any way on account of inter-cropping.

Mutant

An organisms or cell showing mutant phenotype due to mutant allele of a gene.

Leggy

condition where a shrub or multi-stemmed small tree does not have significant foliage in its lower portion

Half- or sub - shrub

A plant with stems that are woody at the base, usually dying back to the woody stems or even back to the ground after severe winters. Suffrutescent.

|||||||||||||||||||||||||||| **Matching** ||||||||||||||||||||||||||||

95. Match the following :-

A] x = 12 Chromosomes   i]   Mango
B] x = 17 Chromosomes   ii] Guava
C] x=9 Chromosomes   iii] Loquat
D] x=20 Chromosomes   iv] Ber
E] x = II Chromosomes   v]   Papaya

96. Match the following :-

A] Xenia   i]   Loquat
B] Meta-xenia   ii]   Amaranthus
C] Terminal fruiting   iii] Pistachionut
D] Axillary fruiting   iv]   Litchi
E] C-4 plant   v]   Banana
F] Orthodox seeds   vi]   Guava
G] Recalcitrant seeds   vii] Apple
H] Monocarpism   viii] Datepalm
I] Epigynous fruit   ix]   Citrus
J] Hybrid vigour   x]   Bottle gourd

97. Match the following :-

A] Yellow jasmine   i]   *Trachelospermum jasminoides*
B] Star jasmine   ii] *Petrea volubilis*
C] Railway creeper   iii] *Jasminum humile*
D] Rangoon creeper   iv] *Quisqualis indica*
E] Purple wreath   v]   *Ipomea palmata*

98. Match the following :.-

A] Triploid Bougainvillea   i]   Dr. B.P. Pal (4x =68)
B] Aneuploid Bougainvillea   ii] Sweta
C] Tetraploid Bougainvillea iii] Begum Sikander
D] Bud sport of Trinidad   iv] Surekha
E] Bud sport of Scarlet Queen v] Perfection (3x=51)

99. Match the following
 A] Cascade
 B] Vit A
 C] Hybrid tea
 D] Ipomea batatas
 E] Larkspur
 i] Carrot
 ii] Rose
 iii] Chrysanthemum
 iv] Sweet potato
 v] Annual flower

100. Match the following-
 A] 2n = 28 chromosomes
 B] 2n = 22 chromosomes
 C] 2n = 26 chromosomes
 D] 2n = 24 chromosomes
 E] 2n = 40 chromosomes
 i] *Benincasa hispida*
 ii] *Momordica dioica*
 iii] *Luffa cylindrica*
 iv] *Cucurbita moschata*
 v] *Momordica charantia*

101. Match the following-
 A] Abohar (Punjab)
 B] Pune (M.S.)
 C] Saharanpur (U.P.)
 D] Mashobra (H.P.)
 E] Kahikuchi (Assam)
 i] Citrus Research
 ii] Peach Research
 iii] Apple Research
 v] Mango Research
 v] Grape Research

102. Match the following-
 A] Acalypha
 B] Aurocaria
 C] Sweet pea
 D] Petunia
 E] Zinnia
 i] Indoor plant
 ii] Summer season
 iii] Hanging basket
 iv] Foliage bush
 v] Annual climber

103. Match the following-
 A] *Acroclinium roseum*
 B] *Althea rosea*
 C] *Celosia plumosa*
 D] *Coleus blumei*
 E] *Delphinium ajacis*
 i] Amaranthaceae
 ii] Ranunculaceae
 iii] Malvaceae
 iv] Lamiaceae
 v] Asteraceae

Side arm
Dead arm of grape.

Parallel
Running side by side from the base to tip; especially of veins.

Deltoid
Triangular, with equal sides.

Silicula
A variation of the siliqua which contains fewer seeds and in much shorter (as broad as long) and flattened, seen in *Capsella bursa-pastoris*.

Brushwood
A growth of low shrubs and bushes, name for fallen or cut branches and twigs of trees.

Fructose
Also fruit sugar. Simple form of carbohydrate; $C_6H_2O_6$.

Bulb finger
a human skin irritation resulting form sensitive person handling cut Alstroemeria, Narcissus and tulips and tulip bulbs.

Dissected
Divided in narrow, slender segments.

Attenuate
Showing a long gradual slender taper; usually applied to apices, but equally appropriate for bases of leaves, petals, etc.

Peeper
A bud or a small round swelling during emergence of a sucker from the ground as found in case of banana. The peepers when develop lanceolated leaves are called 'Sword suckers'.

104. Match the following-
A] Alkaline soil   i] Guava canker
B] Copper deficiency   ii] Vector
C] Fungus   iii] Guava wilt
D] Zinc deficiency   iv] Little leaf
E] White fly   v] Gummosis

105. Match the following-
A] Allelopathic effect   i] Pecan nut
B] Heterodi lchogamy   ii] Strawberry
C] Deciduous vine fruit   iii] Cherry
   crop
D] Self-incompatibility   iv] Walnut
E] Mulching   v] Kiwi fruit

106. Match the following
A] Ascorbic acid   i] Yeast
B] Riboflavin   ii] Sunlight
C] Thiamine   iii] Carrot
D] Vitamin-A   iv] Bael fruit
E] Vitamin-D   v] Aonla

107. Match the following-
A] Autogamy   i] Dichogamy
B] Allogamy   ii] Out breeding
C] Apomixis   iii] Homogamy
D] Self-pollination   iv] Inbreeding
E] Cross-pollination   v] Asexual reproduction

108. Match the following-
A] Auxins   i] Genetical dwarfism
B] Gibberellins   ii] Fruit ripening
C] Cytokinin   iii] Cell division
D] Ethylene   iv] Stomatal closing
E] Abscissic acid   v] Apical dominance

109. Match the following-
A] Banana   i] 7.70 lakh ha. (area)
B] Mango   ii] 4.83 lakh ha. (area)
C] Papaya   iii] 0.70 lakh ha. (area)
D] Pomegranate   iv] 15.2 lakh ha. (area)
E] Cashewnut   v] 0.38 lakh ha. (area)

110. Match the following-
    - A] Bougainvillia
    - B] Rose
    - C] Jasmine
    - D] Ber
    - E] Grape
    - i] Pruning in December
    - ii] Pruning time June
    - iii] Pruning in January
    - iv] Pruning in October
    - v] Pruning in May

111. Match the following
    - A] Cauliflower
    - B] Pea
    - C] Turnip
    - D] Tomato
    - E] Okra
    - i] 2-3 kg
    - ii] 400-500 gram
    - iii] 70-80 kg
    - iv] 200-300 gram
    - v] 8-10 kg

112. Match the following-
    - A] China shoe
    - B] Rukmini
    - C] Tree of sadness
    - D] Chitra
    - E] Malti
    - i] *Plumbago auriculata*
    - ii] *Hibiscus rosasinensis*
    - iii] *Ervatomia divericata*
    - iv] *Ixora coccinea*
    - v] *Nyctanthes arbor-tristis*

113. Match the following
    - A] CSIR
    - B] NRCAMP
    - C] CAZRI
    - D] CTCRI
    - E] CIMAP
    - i] Gujarat
    - ii] Lucknow
    - iii] Trivandrum
    - iv] New Delhi
    - v] Jodhpur

114. Match the following-
    - A] Cucumber & Watermelon
    - B] Muskmelon & Pumpkin
    - C] Ridge & Sponge gourd
    - D] Chow-Chow
    - E] Ashgourd (*B. hispida*)
    - i] Edible portion peri & mesocarp
    - ii] Edible portion endocarp
    - iii] Mono type genus
    - iv] Edible portion placentae
    - v] Single seeded & viviparous

115. Match the following-
    - A] Dry flower
    - B] Gajra/Veni
    - i] Marigold
    - ii] Gladiolus

Malodorous
emitting a foul smell or odor

Precuring
A treatment prior to soft wood grafting where the scion attached to the mother plant is kept in leafless condition for 7-10 days for better graft success.

Irrigation
Application of water to soil for the purpose of supplying moisture essential for plant growth.

Osmocote
Slow-release encapsulated fertlizer.

Glandular-ciliate
Fringed with small glands.

Hardpan
The impervious layer of soil or clay lying beneath the topsoil.

Aerate
Loosening or
puncturing the soil
to increase water
penetration.

Solitary
Single, one.

Semi-evergreen
having some leaves
deciduous and
some leaves
persistent
throughout the
winter, with the
degree of
persistent leaves
determined by the
harshness of the
winter season

Starry
A condition in
some Narcissus
species induced by
hot water treatment
prior to stage Pc.

Cordial
This is a sparkling,
clear, sweetened
juice from which all
pulp and other
suspend materials
have been
completely
eliminated.

C] Garland   iii] Helichrysum
D] Bouquet   iv] Rose
E] Button hole   v] Jasmine

116. Match the following-
A] Gene   i] Reduction division
B] Jumping Gene   ii] Barbara McClintock (1950)
C] One gene one enzyme theory   iii] Homotypic division
D] Mitosis   iv] Beadle & Tatum (1941)
E] Meiosis   v] Johannsen (1909)

True / False

117. 'Leaf blight' is a serious disease of tea.   T/F
118. *Aegle mormelos* is monogenic fruit.   T/F
119. Coffee is native of Tropical Africa.   T/F
120. Dahlia is propagated by tuber & cutting.   T/F
121. India is the centre of origin of Jamun.   T/F
122. Paclobutrazol responses regularly in irregular bearing mangoes.   T/F
123. *Petrea vollubilis* flower colour is blue-purple.   T/F
124. Poovan & Basrai banana are resistant to panama wilt.   T/F
125. Vrindavan Garden is situated at Mathura.   T/F
126. 'Katte' disease is serious problem in cardomom.   T/F
127. Bael and wood apple are same fruit.   T/F
128. *Duranta plumerei* is suitable for hanging basket.   T/F
129. *Eloeis guineensis* is botanical name of oil palm.   T/F
130. Geranium & cucumber are day neutral plants.   T/F
131. Gibberellic Acid is very effective for quality improvement of grape.   T/F
132. High Gate is a dwarf bud sport of Gross Michel.   T/F
133. Kufri Sindoori is a medium to late variety of potato.   T/F
134. Mango malformation is caused due to irregular bearing.   T/F

135. Phylloxera is a disease of apple.   T/F
136. Apple & Pecan are deciduous plants.   T/F
137. Flower colour of Ganga & All Gold roses is yellow.   T/F
138. Gooseberry fruits are rich source of Vit-C.   T/F
139. House fly is main pollinator of mango.   T/F
140. India is the 'Home of spices'.   T/F
141. Palm oil is rich source of b-carotene.   T/F
142. Pusa Seedless grape is developed by IIHR.   T/F
143. Smyrna type of figs are commonly grown in India.   T/F
144. Thiourea is effective for rooting in cuttings.   T/F
145. 2, 4-D is useful to control fruit drop in citrus.   T/F
146. Cooking apple Bramley's Seedling has white flesh.   T/F
147. Francis is a variety of phalsa.   T/F
148. Generally abscission layer is formed in roots.   T/F
149. In N. India, grape & roses are pruned same time.   T/F
150. India is leading country in palm oil production.   T/F
151. Kufri Bahar is a white skinned variety of potato.   T/F
152. Whiptail in cauliflower is caused due to excess Mn.   T/F
153. Wood apple is scientifically known as *Malus baccata*.   T/F
154. 'One bud two leaf is the standard picking method
     of tea.   T/F
155. Arecanut is commercially propagated by slips.   T/F
156. Friendship is a variety of rose.   T/F
157. Lily & Ginger stems are botanically rhizome.   T/F
158. *Malus floribunda* is an ornamental red flowering apple.   T/F
159. *Poinciana regia* is a flowering annual.   T/F
160. Rose is pruned in October.   T/F
161. Strawberry is known as 'Christ's thorn' also.   T/F
162. Sweet potato is commercially propagated by cutting.   T/F
163. 'Freedom' apple is highly resistant to apple scabe.   T/F
164. Bael is commonly propagated by seed.   T/F

Bulbil
Also called bulbel. Special mutlicellular body essentially meant for the reproduction of the plant, maybe modification of vegetative bud or floral bud. It develops into plant without formation of seeds. *e.g.* garlic, onion, etc.

Interveinal
Between the veins, specially of a leaf.

Site
A location for a specific purpose; the combination of biotic, climatic and soil conditions of an area. The situation regarding elevation, topography, soil, nearness to water source in considering fruit-planting.

165. Brinjal plant shows perennial growth.   T/F
166. Colour in tea is due to theoflavin.   T/F

Fill in the Blanks

167. Francis is a variety of _____.
168. Ber plant goes into rest during _____ season.
169. Fig goes into dormancy during _____ season.
170. *Annona reticlata* is the botanical name of _____.
171. Jamun belongs to family _____.
172. Bael and Woodapple belongs to _____ family.
173. Caprification is necessary practice in _____.
174. Pungency in onion is due to _____.
175. Pithiness in radish is developed due to _____ deficiency.
176. Yolo Wonder is a variety of _____.
177. Cracking in cabbage head is caused due to _____.
178. Arkel is an early variety of _____.
179. _____ is precursor of Vit-A.
180. Pusa Meghdoot is a variety of _____.
181. Pusa Kesar is a variety of _____.
182. Anemone is a flower group of _____.
183. *Lagerstroemia indica* flowers during _____ season.
184. Seed rate of direct sown onion is _____ per ha.
185. Lettuce belongs to _____ family.
186. Butter Cup and Master Peace are varieties of _____.
187. _____ is used for judging pectin content in extract.
188. Grapes are rich source of _____ acid.
189. 'Curd' is edible part of _____.
190. _____ is responsible for bitterness in chilli.
191. Annual flowers are commonly propagated by _____.
192. _____ is suitable for dried flowers.

193. Candy tuft is a _____ season annual.
194. Fountains & Bridges are important features of _____.
195. *Dracaena* is an indoor _____ plant.
196. Flower colour of *Thevetia nerifolia* is _____.
197. *Jasminum sambac* flowers during _____ season.
198. Polarity is essential consideration in _____.
199. Aonla fruit is botanically a _____.
200. Drupe fruits are developed from _____ ovaries.

Neuter
Asexual, having neither stamen nor pistil.

Bolt
A premature seed-stalk produced by certain plants like onion suddenly under certain conditions, breaking their normal life-cycle.

Exserted
Projecting beyond, as stamens beyond a corolla.

Staminate
An imperfect flower with stamens or pollen producing structures, but with no pistil, or seed producing structure.

Andromonoecious
A sex form where staminate and hermaphrodite flowers are separately produced in the same plant.

A farmer had a dog who used to sit by the roadside waiting for vehicles to come around. As soon as one came he would run down the road, barking and trying to overtake it. One day a neighbor asked the farmer "Do you think your dog is ever going to catch a cat?" The farmer replied, "That is not what bothers me. What bothers me is what he would do if he ever caught one."

Many people in life behave like that dog who is pursuing meaningless goals.

[Goals lead to Purpose in Life.]

Step-Seventeen

Losers use soft arguments but hard words;
Winners use hard arguments but soft words.

Multiple Choice

1. Most suitable rootstock for rose in north India is
   A] *Rosa indica var odorat* B] *Rosa multiflora*
   C] *Rosa bourboriana*   D] All

2. The best rootstock for rose propagation in north India is
   A] *R. canina*   B] *R. indica var odorata*
   C] *R. muttiflora*   D] *R. borboniana*

3. Which citrus rootstock is tolerant to tristiza citrus
   A] Rangpur lime   B] Sweet orange
   C] Karnakhatta   D] All

4. Ultra dwarf rootstock of Apple is
   A] M-111   B] M-104
   C] M-27   D] M-13

5. Salt tolerant rootstock of Grape is
   A] Dogridge   B] Salt creek
   C] Both   D] None

6. Common disorder found in rose cut flower in vase
   decoration is-
   A] Calyx splitting   B] Sleepiness
   C] Bent neck   D] None of these

7. In Southern India, rose is pruned twice a year while in North
   India it is pruned-
   A] One time (Once)   B] Two times (Twice)
   C] Three times (Thrice)   D] Not pruned

8. Which root stock of rose is used commercially in India-
   A] *Rosa indica*   B] *Rosa foetida*
   C] Edward Rose   D] *Rosa damascena*

9. The best time of rose pruning is-
   A] January   B] November
   C] October   D] December

Quiescent
Being in a state of rest, non-active visible bud.

Can
A cylindrical metal (steel coated or aluminum), plastic, paper, or combination of three materials with a hermetically sealed lid ; used for a wide assortment of products.

Canes
A long woody pliable stem rising from the ground. The long shoots of blackberry, grape, etc.

Thatch
The layer of dead stems that builds up under many lawn grasses. Thatch should be removed periodically to promote better water and nutrient penetration into the soil.

Hirsute
Pubescent with coarse or stiff hairs.

Dry Pack storage
The stogare of cut flowers in vapor-poof containers with the stems not in water usually at 0° C.

Dethatch
Process of removing dead stems that build up beneath lawn grasses.

Gummosis
A general disorder of stone fruits, in which exudation and deposit of gum occurs.

Dehydro-freezing
An efficient storage preocedure of the perishables where half the weight of the produce is reduced by warm air drying prior to their freezing.

10. First Floribunda rose variety developed by Dr. B.P. Pal was-
A] Delhi Princess   B] Arunima
C] Delhi Brightness   D] Banjaran

11. In South India, roses are pruned twice a year in the months of-
A] October & January   B] November & March
C] December & July   D] September & February

12. The natural rubber is mostly used in which industry
A] Tyre   B] Sheet
C] Box   D] None

13. Tree of sadness is
A] *Nyctenthis arbotristis* B] *Jacoranda acutifolia*
C] *Hibiscus rosa chinensis* D] *plumeria alba*

14. Jamun has same family as
A] Pomegranate   B] Guava
C] Papaya   D] Sapota

15. San Jose scale is introduced from which country
A] England   B] Spain
C] France   D] Holland

16. Board of Scientific & Industrial Research (BSIR-1940) was renamed as CSIR in-
A] 1942   B] 1945
C] 1950   D] 1952

17. *Phoenix dactylifera* is the scientific name of-
A] Coconut   B] Date palm
C] Chinese palm   D] Areca palm

18. *Cucumis sativus* is the scientific name of-
A] Cucumber   B] Ridge gourd
C] Musk melon   D] Summer squash

19. *Coriandrum sativum* is the scientific name of-
A] Cumin   B] Coriander
C] Fennel   D] Menthi

20. India is second largest producer of vegetables after
A] razil   B] China
C] Africa   D] Japan

21. Storing of seed at freezing temperature
    A] Orthodox   B] Recalcitrant
    C] *Ex-situ* conservation   D] *In situ* conservation

22. Three varieties with two seed rates are replicated thrice, the degree of freedom will be-
    A] (3x2x3)-1=17   B] (3x2x3)-2=16
    C] (3x2x3)-3=15   D] (3x2x3) = 18

23. In Banana seedlessness is due to
    A] Vegetative parthenocarpy   B] Stimulative parthenocarpy
    C] Stenospermocarpy   D] None

24. The age of onion seedlings for rabi season planting should be-
    A] 2-4 weeks   B] 5-6 weeks
    C] 8-10 weeks   D] 12-15 weeks

25. The viability of onion seeds is-
    A] One year   B] Two years
    C] Three years   D] Five years

26. 'Self-incompatibility' in carrot is due to-
    A] Protogyny   B] Protandry
    C] Heterostyly   D] Unfertilization

27. Gladiolus is sensitive to-
    A] CO_2 pollution   B] Nitrogen deficiency
    C] Dust pollution   D] Fluoride pollution

28. Powdery Mildew is a serious problem in-
    A] Pointed gourd   B] Pumpkin
    C] Cucumber   D] Ridge gourd

29. Which is diluted before serving-
    A] Jam   B] Jelly
    C] R.T.S   D] Squash

30. IARI was shifted from Pusa (Bihar) to Pusa (New Delhi) in-
    A] 1905   B] 1929
    C] 1936   D] 1925

Dehydro-canning
An efficient storage procedure of the perishables when the products are dried to half their original weight prior to their canning.

Waxing
A short-term storage technique of fresh fruits and vegetables under ambient conditions by applying wax emulsion containing paraffin wax, triethanol and aleic acid which provides a thin, discontinuous layer on the fruit surface and thus curtails the respiration and transpiration resulting increase in shelf-life. It also helps to keep away the microbes when fungicides like Benlate 50 are used in wax emulsion.

31. 'Zero Energy Cool Chamber' storage was developed by-
    A] S.K. Mukherjee   B] R.N. Singh
    C] S.K. Roy   D] Sant Ram

32. Green manure should be incorporated at its
    A] Seedling stage   B] Fully natured stage
    C] Flowering stage   D] Drying stage

33. Essentially jelly should contain
    A] Pectin, sugar, acid   B] Ripe plum, water, KMS,
       water   sugar
    C] Fully matured guava,   D] Aonla, water, KMS, sugar
       acid, sugar, water

34. The crop showing greatest inbreeding depression
    A] Onion   B] Lettuce
    C] Tomato   D] Lima bean

35. The Garden shows a continuous effect due to
    A] Rythum   B] Shape
    C] Harmony   D] Simplicity

36. ' Sigatoka' disease of banana was first observed in-
    A] Mexico (1911)   B] Fiji (1913)
    C] Malaysia (1918)   D] Philippines (1922)

37. ' Singapore' and 'Ceylon' are varieties of-
    A] Pine apple   B] Oil palm
    C] Papaya   D] Arecanut

38. CIPHET is situated in
    A] Amritsar   B] Nagpur
    C] Hyderabad   D] Ludhiana

39. The largest size fruit grown in India is-
    A] Custard Apple   B] Jack fruit
    C] Pine apple   D] Bael

40. ' Skiffing' is practiced in-
    A] Rubber   B] Coffee
    C] Coconut   D] Tea

41. In a small sample test which method is used
    A] T test   B] F test
    C] Chi square test   D] None

42. To reduce soil erosion what is preferably done
    A] Planting of crops with fibrous roots
    B] Planting tall trees with long spacing
    C] Dividing farm in smaller areas
    D] Keeping gentle slope to land

43. Light textured soil is
    A] Clay
    B] Silt
    C] Sandy
    D] loamy sand

44. In alkali soil which amendment is used
    A] Gypsum
    B] Lime
    C] Sodium carbonate
    D] All

45. An isopernoid alcohol 'Solanesol' is extracted from-
    A] Tobacco
    B] Tomato
    C] Potato
    D] Brinjal

46. Rich source of 'Solanesol' is-
    A] Tomato
    B] Chilli
    C] Brinjal
    D] Tobacco

47. 'Solo' series of papaya was developed by-
    A] N.E. Lee
    B] H.P. Olmo
    C] R.S. Pillai
    D] J.E. Higgins

48. The cheap and richest source of pectin is-
    A] Apple
    B] Bael
    C] Guava
    D] Mango

49. Which is the richest source of Vit-C-
    A] Wood apple
    B] Bael
    C] Citrus
    D] Aonla

50. Grape is a rich source of-
    A] Ascorbic acid
    B] Tartaric acid
    C] Mallic acid
    D] Citric acid

51. Vegetables are the rich source of-
    A] Vitamins
    B] Minerals
    C] Carbohydrate
    D] Fat

52. Tomato is a rich source of-
    A] Vit-A
    B] Riboflavin
    C] Vit-E
    D] Vit-C

Bower
A hut-like structure with four open sides and beautiful ornamental climbers are trained on its roof.

Bud scales
Leaf like structures that surround some flower buds, such as on azaleas.

Florigen
A hormone-like substance or a precursor of such a substance that is produced in the leaves upon receiving photoperiodic message for flower intiation. Florigen remains only a name as the substance has not yet been isolated. However, its transport from leaves to the growing point has been demonstrated up and down stems, across graft unions, and from one plant to another.

Spice
Aromatic or pungent vegetable substance used for flavouring foods, such as cloves, pepper, cardamom, mace, etc.

Collateral
Said of extra or supernumerary buds that are inserted on either side of a normal axillary bud. Said of buds that grow side by side.

Concave
Curved like the inner surface of a sphere.

Gamma-garden
An area subjected to gamma-irradiation for the purpose of irradiating the whole plants in different stages of growth and for varying durations.

53. Suitable time of okra sowing is-
    A] June-July
    B] August-September
    C] October-November
    D] January-February

54. In plant spacing 2x2, how much plant can be accommodated
    A] 2000
    B] 2500
    C] 3000
    D] 3500

55. Which citrus species. are 'non-polyembryonic'-
    A] *C. medica & C. maxima*
    B] *C. reticutata & C. limon*
    C] *C. aurantifolia & C. grandis*
    D] All of them

56. Which citrus species. is 'monoembryonic'-
    A] Pummelo
    B] Mandarin
    C] Sweet orange
    D] Lime

57. Indian Central Spices & Cashewnut Committee was established during-
    A] 1968
    B] 1961
    C] 1955
    D] 1980

58. National Research Centre for Spices (NRCS) was created during-
    A] 1965
    B] 1975
    C] 1980
    D] 1986

59. ' Spices Cess Act' came into existence during-
    A] 1966
    B] 1970
    C] 1986
    D] 1992

60. Directorate of Arecanut & Spices Development (Kalicut) was established in-
    A] 1960
    B] 1964
    C] 1966
    D] 1968

61. All India Spices Development Council was established in-
    A] 1963
    B] 1966
    C] 1971
    D] 1990

62. The leading producer of spices in the world is-
    A] India
    B] Brazil
    C] Mexico
    D] Indonesia

63. Fenugreek & spinach are recommended in which climate-
    A] Cool   B] Warm
    C] Temperate   D] None

64. 'Spongy tissue' disorder in mango was first reported by-
    A] R.N. Singh (1930)   B] Cheema & Dhani (1934)
    C] Swingle (1929)   D] None of these

65. 'Spongy tissue' is a prevalent problem in-
    A] Dashehari   B] Langra
    C] Alphonso   D] Chausa

66. The coffee spp which is widely cultivated
    A] Coffia arabica   B] Coffia robusta
    C] Both   D] None

67. Development of sprouts from the seed tuber mean seed is called
    A] Chitting   B] Sprouting
    C] Pricking   D] Moulding

68. Farmers have staggered harvesting of following as it does not have set harvesting time-
    A] Cassava   B] Onion
    C] Cauliflower   D] Cabbage

69. Plant bearing pistillate and staminate flowers separately on same plant are known as-
    A] Bisexual   B] Monoecious
    C] Dioecious   D] Hermaphrodite

70. In which state first time extension programme was started
    A] UP   B] MP
    C] AP   D] Punjab

71. First Regional Fruit Research Station (RFRS) in U.P. was established at-
    A] Basti   B] Lucknow
    C] Kanpur   D] Saharanpur

72. 'Steckling' roots are found in-
    A] Carrot, Beet   B] Radish, Potato
    C] Cassava, Sweet   D] All of them
       potato

Arbour
A bower; small structure of lattice-work to support vines and provide a shady retreat.

Convex
Curved like the outer surface of a sphere.

Ruffling
An unevenness of the surface of the leaf-blade caused by ridges that develop or become more pronounced from the mid-rib to the lateral margins.

Undulate
A wavy margin or surface.

Liposome
A closed lipid vesicle surrounding an aqueous interior; may be used to encapsulate exogenous materials for ultimate delivery of these into cells by fusion with the cell.

Hanging basket
Container usually suspended form supports in greenhouse and used to suspend plants in the home or garden.

Agro-forestry
A self -sustaining land management system which combines production of agricultural crops with that of tree crops as also with that of livestock simultaneously on the same unit of land.

Tuber
A special kind of swollen modified stem structure that functions as an underground storage organ as in potato, Jerusalem artichoke, etc.

Node
The more or less swollen portion of the twig which bears the leaf or leaves.

73. ' Stocky' or 'Spindle' shaped root development in carrot occurs due to-
A] Mo deficiency
B] High temperature
C] Ca deficiency
D] Mn deficiency

74. ' Stomata Closing' occurs due to-
A] IBA
B] ABA
C] Kinetin
D] GA

75. Which of the ' stone fruits' can be grown successfully in plains of India-
A] Peach & Pear
B] Plum & Pear
C] Almond & Mango
D] Guava & Apple

76. Onion is stored at a temperature of
A] 4-5 °C
B] 0 °C
C] 6-8 °C
D] 10 °C

77. In Apple stratification is done at temperature of
A] 10 °C
B] 4 °C
C] 15°C
D] 0°C

78. In which style of gardening, roads cut to each other at 90° angle-
A] Wild style
B] Formal style
C] Informal style
D] All the above

79. In fruit preservation, sugar acts as preservative by-
A] Osmosis
B] Hydrolysis
C] Fermentation
D] Imbibition

80. In high sugar containing variety of musk melon may contain TSS of-
A] 80%
B] 2%
C] 17%
D] 4%

81. How much sugar is recommended for making jam
A] Above 30%
B] Above 40%
C] Above 70%
D] Above 60%

82. ' Suguna' and 'Sudarsana' are varieties of-
A] Ginger
B] Turmeric
C] Black Pepper
D] Sweet Potato

83. The most suitable 'Filler crop' for litchi orchard is-
    A] Phalsa   B] Papaya
    C] Pineapple   D] Strawberry

84. Which is the most suitable as a filler crop-
    A] Papaya   B] Litchi
    C] Guava   D] Arecanut

85. The fruit suitable for arid zone is
    A] Ber   B] Pomegranate
    C] Fig   D] None

86. Which is the most suitable for hill farming-
    A] Banana   B] Cashewnut
    C] Apple   D] Pineapple

87. The fruit suitable for making jelly is
    A] Guava   B] Lime
    C] Papaya   D] Mango

88. Which tree is not suitable for roadside plantation due to its brittleness-
    A] *Eucalyptus* sp.   B] *Tamariandus* sp.
    C] Amaltas   D] *Azadirachta* sp.

89. The most suitable varieties of peach for growing in U.P. plains.
    A] Elberta & Triumph B] Sharbati & Florida Sun
    C] Noblesse & J .H.   D] Alexander & Early Reverse
       Hale

90. Annual flower suited for planting in shade is
    A] Antirrhinum   B] Stock
    C] Cinaria   D] Naustúrtium

91. The ammonium sulphite used in groundnut mainly supply which micro-nutrient
    A] Mg   B] S
    C] B   D] Mn

92. ' Surinam Cherry' is a suitable root stock for-
    A] Sweet cherry   B] Strawberry
    C] Jamun   D] Aonla

Photosynthesis
Biochemical reactions leading to synthesis of organic compounds, specially carbohydrates from the O_2 of water and carbon of CO_2 in presence of light as the energy source in chlorophyll containing tissues of plants.

Division
Separation of roots system of parent's plant into several units, a method of asexual propagation.

Metsubre
A nutritional disorder of taro (*Colocasia esculenta*) due to calcium deficiency where the defective corm have smooth or concave top, slighty brownish in colour and are of varying size.

Fusiform
Spindle-shaped; tapering to each end from a smaller mid-section.

Malformation
Also called excrescences, symptom found in plants due to disease whereby plant forms show hypertrophy or atrophy but when more localized, show wart, gall, blister, tumour, leaf-curl, hairy root, witches broom (abnormal, tufted appearance of apical part of stem resembling broom), canker (localized lesion with raised margin), and rosette (tufted appearance of leaves). Assists in the identification of pathogen causing disease.

93. Which is the most susceptible to cold injury-
    A] Lime   B] Lemon
    C] Orange   D] Mosambi

94. ' Suvasini' is a hybrid variety of-
    A] *Polianthes tuberosa*   B] *Hibiscus rosa-sinensis*
    C] *Jasminum sambac*   D] None of these

Matching

95. Match the following-
    A] I.I.S.R   i]   Hessarghatta (Kanataka)
    B] I.I.V.R.   ii] Lucknow (U.P.)
    C] I.I.H.R.   iii] Calicut (Kerala)
    D] I.I.T.A   iv] Nigeria (Africa)
    E] F.P.C.I.   v]   Varanasi (U.P.)

96. Match the following-
    A] Japanese Plum   i]   Native to Peru
    B] Kagzi lime   ii] Native to China
    C] Passion fruit   iii] Native to Southern America
    D] Pecan nut   iv] Native to India
    E] Guava   v]   Native to Brazil

97. Match the following-
    A] Jelly Thermometer   i]   To know sugar per cent
    B] Salometer   ii] To know pectin content
    C] Baume Hydrometer   iii] To judge the end point
    D] Jelmeter   iv] To judge T. S. S. per cent
    E] Refractometer   v]   To know salt per cent

98. Match the following-
    A] Kalanchoe   i]   Budding
    B] *Cestrum nocturnum*   ii] Bulb
    C] Rose   iii] Layering
    D] Hippeastrum   iv] Stem cutting
    E] Jasmine   v]   Leaf cutting

99. Match the following-
    A] Kalimpong   i]   Home of Spices
    B] India   ii] King of Annuals
    C] Kerala   iii] Heaven for Orchids

D] Chrysanthemum　　iv] Queen of the East

E] Pansy　　　　　　v] Spices bowl of India

100. Match the following-

A] Kinetin　　　　　i] First crystalline cytokinin

B] Gibberellins　　　ii] First synthetic growth inhibitor

C] Zeatin　　　　　iii] $C_{10} H_9 N_5 O$

D] Maleic hydrazide　iv] Natural growth inhibitor

E] Abscissic Acid　　v] E. Kurosava (1926)

101. Match the following-

A] Lucknow (U.P.)　　i] National Institute of Nutrition

B] Mysore (Karnataka)　ii] Food Technology Institute

C] Hyderabad (A.P.)　iii] Fruit Preservation & Canning Institute

D] Mumbai (M.S.)　　iv] Indian Packaging Institute

E] Nagpur (M.S.)　　v] Central Food & Technological Research Institute

102. Match the following-

A] Maiden-hair fern　i] *Asplenium nidus*

B] Bird's nest fern　　ii] *Polypodium aureum*

C] Sword fern　　　iii] *Platycerium bifurcatum*

D] Staghorn fern　　iv] *Adiantum pedatum*

E] Hare's foot fern　v] *Nephrolepis exaltata*

103. Match the following-

A] Mango　　　　　i] 1.05 m. tonnes (Production)

B] Grape　　　　　ii] 1.76 m tonnes (Production)

C] Banana　　　　iii] 10.20 m. tonnes (Production)

D] Papaya　　　　iv] 2.71 m. tonnes (Production)

E] Pea　　　　　　v] 16.16 m. tonnes (Production)

Plant vase

A non -technical term that describes the rosette of leaves that ascent forming a cup at the base of the plant especially in Bromieiacease family.

Glands

Secreting organs. Leaf teeth and stipules often end in minute glands.

Controlled temperature forcing (CTF)

Procedure for treating Easter lilies in which non precooled bulbs are potted, placed in a controlled temperatures area at approximately 16° C for 2-4 weeks and subsequently, cooled at 2°C-7° for 6-7 weeks prior to being placed in the greenhouse.

Freezing

Means reduction of temperature of the product to such an extent that the activities of micro organism stops.

Bulb programming phase

All temperature treatments given to the bulbs form the time they are harvested until they are placed under greenhouse conditions.

Tendron

A young bud or tender shoot of a plant.

Scandent

Climbing, usually without tendrils.

Cyme

An inflorescence whose single terminal flower opens first, followed by those of secondary and tertiary axes (as in apple).

104. Match the following-
    A] Mango   i] 2020 IU/100g (Vit-A)
    B] Papaya   ii] 1000mg/100g (Vit-C)
    C] Cashew nut   iii] 1191 mg/l00g (Vit-B$_2$)
    D] Bael   iv] 630 mg/100g (Vit-B$_1$)
    E] Barbados cherry v] 4800 IU/100g (Vit-A)~

105. Match the following-
    A] Mango   i] Clubroot resistant Turnip
    B] IHR-1-1   ii] Y.V.M. resistant Okra
    C] Early Badger iii] White rust resistant R. adish
    D] Pusa Sawani iv] Cracking resistant Tomato
    E] Sioux   v] Wilt resistant Pea

106. Match the following-
    A] *Momordica dioica*   i] Gynoecious lines
    B] *Lagenaria siceraria* ii] Dioecious
    C] *Luffa acutangula var. Satputia*   iii] Andromonoecious
    D] *Cucumis sutivus*   iv] Monoecious
    E] *Cucumis melo*   v] Hermaphrodite

107. Match the following-
    A] Pericarp & Placentae i] Strawberry
    B] Endosperm   ii] Pomegranate
    C] Succulent Thalamus   iii] Mesocarp
    D] Juicy seed coat (Aril) iv] Tomato
    E] Muskmelon   v] Coconut

108. Match the following-
    A] Pickling cucumber   i] *Cucumis hardwickii*
    B] Table cucumber   ii] *Cucumis anguria*
    C] Bitter cucumber   iii] *Cucumis ficifolius*
    D] Andromonoecious   iv] *Cucumis sativus*
    E] Tetraploid
       (2n = 48)   v] *Cucumis melo*

109. Match the following-
    A] Pineapple   i] Parthenocarpy
    B] Smyrna fig   ii] Multiple fruit
    C] Almond   iii] Cross protection

D] Acid lime   ·   iv] Caprification

E] Common fig   v] Gametophytic incompatibility

110. Match the following-
- A] Recurrent apomixis   i] Garlic
- B] Nucellar embryony   ii] Chow-Chow
- C] Vegetative apomixis   iii] Citrus
- D] Vivipary   iv] Mangosteen
- E] Parthenogenesis   v] Apple

111. Match the following-
- A] Rusty Red, Valencia   i] Gladiolus
- B] Happy End, Sylvia   ii] Chrysanthemum
- C] Mercedes, Sonia   iii] Hollyhock
- D] Gold Dust, Baggi   iv] French marigold
- E] Giant Double; Powderpuff v] Rose

112. Match the following-
- A] Seedlessness in Corianth grape   i] Parthenogenesis
- B] Seedlessness in Thompson seedless grape   ii] Stimulative parthenocarpy
- C] Seedlessness in banana   iii] Aneuploidy
- D] Seed set without pollination   iv] Stenospermocarpy
- E] Seedlessness in guava   v] Vegetative parthenocarpy

113. Match the following-
- A] Self-unfruitful   i] Dogridge
- B] Salt resistant grape   ii] Arka Mridula
- C] Seedless mango   iii] J. H. Hale
- D] Hybrid of passion fruit iv] Kaveri
- E] Guava hybrid   v] Sindhu

114. Match the following-
- A] Shiitake   i] *Auricularia polytricha*
- B] Wood ear mushroom   ii] *Volvariella volvacea*
- C] White button mushroom iii] *Pleurotus florida*
- D] Oyster mushroom   iv] *Lentinula edodes*
- E] Paddy straw mushroom v] *Agaricus bisporus*

GLOSSARY

Extra-axillary
Above rather than in the axil. Same as supra-axillary.

Allelomorph
A series of more than two alternative forms of a gene.

Rooting room
A controlled temperature facility used to root and satisfy the cold requirement of bulbs.

Polyploid
An organism with more than two set of chromosomes (2x) or genome. It includes triploid (3x), tetraploid (4x), pentaploid (5x), hexaploid (6x), heptaploid (7x), octaploid (8x).

Doubly serrate
Serrations bearing minute teeth on the margins.

Crenate
Scalloped; with rounded teeth.

Superposed
Said of extra buds that appear above the true axillary buds; usually flower buds.

Chasmophytes
A group of lithophytes that are capable of growing on cracks, fissures and crevices of rock, *e.g.,* *Equisetum* sp, wild daisy, many orchids, etc.

Suffrutescent
A plant with stems that are woody at the base, usually dying back to the woody stems or even back to the ground after severe winters. Half- or sub-shrub.

Nematodes
Worm like organisms that can affect roots, stem and foliage.

115. Match the following-
A]	Short-day plant	i]	Carnation
B]	Long-day plant	ii]	Tomato
C]	Low-volume & high-value crops	iii]	Spices
D]	Day-neutral plant	iv]	Coconut
E]	Plantation crop	v]	Chrysanthemum

116. Match the following-
A]	Squash	i]	Fruit juice not < 100%
B]	R. T. S.	ii]	Fruit juice not < 25%
C]	Nectar (mango)	iii]	Fruit juice not <20%
D]	Nectar (pineapple)	iv]	Fruit juice not < 5%
E]	Pasteurized juice	v]	Fruit juice not < 40%

True / False

117. Ganesh cultivar of pomegranate was bred by Dr. G.S. Randhawa. T/F

118. Gerbera is an important cut flower. T/F

119. Mary Palmer is a variety of H.T. Rose. T/F

120. Multitier planting is practiced in arecanut plantation. T/F

121. Railway creeper is a twiner plant. T/F

122. Wintering is a foundation pruning in roses. T/F

123. Bael goes into rest during summer. T/F

124. Cashewnut is native of Sri Lanka. T/F

125. Ganesh is a selection from Alandi. T/F

126. Kangra valley and Mandi districts of H.P. are suitable for tea cultivation. T/F

127. Knolkhol is a root vegetable. T/F

128. Mahara is a double flowered variety of Bougainvillea. T/F

129. Mango bears fruits on past season's shoots. T/F

130. S.R. Cans are used for canning of non-acidic vegetables. T/F

131. Shamrock is a green variety of apple. T/F

132. Strawberry is propagated by budding. T/F

133. 'Tea' is the 'national drink' of China. T/F

134. Cashew apple is rich source of Vit-C. T/F
135. Datepalm does not bear male & female flowers on same plant. T/F
136. Fig goes into dormancy during rainy season. T/F
137. Granulation is a physiological disorder of guava. T/F
138. Jelmeter is used for judging end point of jelly. T/F
139. Pre-curing of scion is necessary for rose cuttings. T/F
140. Root pruning is practised in guava. T/F
141. Rymer is a early ripening variety of apple. T/F
142. Vegetative parthenocarpy is found in grapes. T/F
143. 'Goa' is a variety of cashew nut. T/F
144. Anar butterfly can be controlled by bagging. T/F
145. Apple & Litchi, both require pollinizers. T/F
146. Cashewnut shell liquid (CNSL) is extracted from kernel. T/F
147. For the preservation of mango chutney, sodium benzoate is used. T/F
148. Hand pollination is practised in datepalm. T/F
149. *Hevea brassiliensis* (rubber) has short gestation. T/F
150. Jamun and Guava belong to same family. T/F
151. *Malus sargenti* is apomictic species of apple. T/F
152. T-budding is commercial method of sweet orange propagation. T/F
153. A typical Mughal Garden is circular in design. T/F
154. Alphonso variety of mango thrives best in temperate climate. T/F
155. Caffeine is a product of cocoa. T/F
156. Hexagonal system is the best for land near a city. T/F
157. In aonla, proper budding time is July. T/F
158. Jasmine is very important flower for perfume industry. T/F
159. Mango inflorescence bears mainly male & perfect flowers. T/F

Feathering
Condition in lilies that can occur form late spring to harvest in which all or part of the inner scales developing around daughter axis elongate faster then they are filled.

Erwinia
Genus of bacterium that causes disease of succulent tissue, often called soft rot.

Floral bud
Immature flower that consists of petals, stamens, and pistil.

Seed
A fertilized ripened ovule that contains an embryo.

Long day plant
Plant that flowers when the day length is longer than the critical.

160. Peach requires coolest climate among all the temperate fruits. T/F
161. Strawberry produces seedless fruits. T/F
162. *Annona reticulata* is called as 'Sharifa'. T/F
163. Aphid is chief pollinator of apple. T/F
164. Bael is rich source of Vitamin-B_{12} T/F
165. Fruit fly is a serious insect-pest of banana. T/F
166. H. T. Roses are propagated by budding. T/F

Fill in the Blanks

167. _____ layer must be destroyed during air layering.
168. Fusiform roots are found in _____.
169. Napiform roots are found in _____.
170. Sugarbeet is a _____ type of tap root.
171. Conical form of tap root is found in _____.
172. Late blight of potato is caused by _____ fungus.
173. *Alternaria solani* is causal organism of _____ blight in potato.
174. Sun Scald is a problem of _____.
175. Coconut is _____ drupe.
176. Cashew apple is _____ of fruit.
177. Cocoa belongs to family _____.
178. Cauliferous fruiting is found in _____.
179. Coffee leaf rust is caused by _____.
180. Central Coffee Research Institute (CCRI) is situated at _____.
181. Pecan belongs to family _____.
182. Kiwi fruit seeds require _____ for early germination.
183. Walnut is native to _____.
184. _____ is male sterile cultivar of peach.
185. Saharanpur Prabhat is a hybrid variety of _____.
186. Pear decline is caused by _____.

187. Core breakdown is a storage disorder of _____.
188. _____ is the largest producer of pear in the world.
189. Hachiya is an astringent variety of _____.
190. Persimmon belongs to _____ family.
191. Persimmon has _____ sex form.
192. 'Gamboge' is a physiological disorder of _____.
193. *Tamarindus indica* belongs to family _____.
194. Arka Sahan is an interspecific hybrid of _____.
195. Avocado exhibits _____ type of dichogamy.
196. Dry neck is a physiological disorder of _____.
197. Loquat is commonly propagated by _____.
198. *Manilkara hexandra* (Khirni) is a root stock of for _____.
199. _____ is the richest source of Riboflavin.
200. Jyoti is a soft seeded selection from _____ Anar.

Ascending
Said of a bud that is between spreading and appressed. Curving indirectly or obliquely upward.

Biennial
Ordinarily applied to plants that live only two seasons; during the first season only leaves and stems are produced above ground, while the flowers and seeds are borne the second summer. Here used in a special sense in separating the biennial canes of the raspberries and blackberries from the stems of other woody plants. In these species, the canes themselves are biennial from underground perennial stems.

A man bought a racehorse and put him in a barn with a big sign, "The fastest horse in the world." the owner didn't exercise the horse nor train it to keep it in good shape. He entered the horse in a race and it came last. The owner quickly changed the sign to "The fastest world for the horse." By inaction or not doing what should be done, people fail and they blame luck.

Some people guarantee failure because they get into a project with no dedication or determination. They lack courage, commitment and confidence. They are starting with complacence and call themselves unlucky.

Shallow people believe in **L**uck. People with **S**trength and **d**etermination believe in **C**ause and **E**ffect.

Step-Eighteen

Multiple Choice

1. ' Suvasini' variety of tuberose is a-
   A] 'Single' var.   B] 'Double' var.
   C] 'Coloured' var.   D] Non-scented var.

2. M.S. Swaminathan Research Foundation is situated at-
   A] Chennai   B] Mumbai
   C] New Delhi   D] Mysore

3. ' Sweet orange' is commercially propagated by-
   A] Seed   B] Air layering
   C] Inarching   D] Shield budding

4. Family of sweet potato is
   A] Euphorbiacae   B] Aizoiaceae
   C] Agavaceae   D] Convolvulaceae

5. Family of sweet potato is
   A] Solanaceae   B] Convolvulaceae
   C] Malvaceae   D] Compositae

6. Botanical name of ' Sweet sop' is-
   A] *Annona squamosa*   B] *Annona reticulata*
   C] *Annona muricata*   D] *Annona cherimola*

7. RNA is synthesized in
   A] Cytoplasm   B] Nucleus
   C] Ribosome   D] All

8. Which training system is followed in Pear
   A] Central leader   B] Central modified leader
   C] Open central   D] All

9. Modified leader system of training of fruit trees is-
   A] Now discarded   B] Most acceptable
   C] Unscientific   D] Good for plains only

10. 'Central Leader System' is common method of training in -
    A] Vines   B] Shrubs
    C] Annuals   D] Trees

Habitant
Any plant or organism that is permanent resident of a place or section.

Perennial plant
A plant that lives for more than 2 years, often living for many years. Almost all woody plants and many herbaceous plants are perennials.

Valvate
Applied to bud scales that meet along a definite, usually longitudinal, line without overlapping; the reverse of imbricate.

Therophytes
Annual grasses and herbs that pass through the unfavourable season in the seed stage and complete their life-cycle in one growing season.

11. In 'R. T.S' beverage, total soluble solids should be-
A] < 5%   B] < 10%
C] < 20%   D] <25%

12. Soil productivity takes into accounts
A] Soil fertility only   B] Soil moisture
C] Soil structure   D] Both soil fertility & soil structure

13. Crossing over takes place in
A] *Zygotene*   B] *Pachytene*
C] *Diplotene*   D] *Leptotene*

14. ' Tatura trellis' method of training is developed by-
A] Chalmers & Vanden Eude   B] Dunn & Stolp
C] Post & Fisher   D] Kofranek & Halevy

15. Where does TCA cycle is found
A] C3 plants   B] C4 plants
C] day neutral plant D] long day plant

16. First Hybrid Tea (H.T.) rose was developed by-
A] B.P. Pal   B] Guillot
C] Poulsen   D] Pernet Outcher

17. Most important temperate vegetable is-
A] Indian cauliflower   B] Cabbage
C] Onion   D] Raddish

18. Optimum night temperature for fruit set in tomato
A] 8°C   B] 13°C
C] 17°C   D] 21°C

19. Best storage temperature for Gladiolus spike is
A] 2-4 °C   B] 10-12°C
C] 20 °C   D] 25 °C

20. Time & temperature for vegetable rocessing in canning depend upon
A] Amount of microbes present in food   B] Texture of product
C] Nutrients   D] None

21. Ideal storage temperature of Brinjal is
    A] 10-11 °C   B] 14-15 °C
    C] 8-9 °C   D] 17-18 °C

22. Ideal storage temperature of Papaya
    A] 8-9 °C   B] 9-11 °C
    C] 11-13 °C   D] 5-6 °C

23. Vegetables require a sterilization temperature of-
    A] 100 °C   B] 105 °C
    C] 116 °C   D] 121°C

24. Vegetables requires a sterilization temperature of-
    A] 100°C   B] 105 °C
    C] 116 °C   D] 120 °C

25. During sealing of cans temperature should not fall below-
    A] 60 °C   B] 74 °C
    C] 100 °C   D] 116 °C

26. At which temperature, there is sharp fall in tuberization in potato-
    A] 21°C   B] 31°C
    C] 40°C   D] none

27. In the term Horticulture, the word HORTUS is
    A] Greek   B] Latin
    C] English   D] Italian

28. Which is the 'off season' variety of mango-
    A] Niranjan   B] Banganpalli
    C] Olour   D] Alphonso

29. Which is the 20th century vegetable
    A] Cluster bean   B] Amaranthus
    C] Brinjal   D] Winged bean

30. Which is the anti gibberellin factor
    A] Cycocel   B] BA
    C] Cultar   D] IBA

31. Which is the best concentrate organic manure
    A] FYM   B] Bone meal
    C] Ground nut cake   D] None

Relative humidity
The measurement of the amount of moisture in the atmosphere.

Formal Gardening
It is the application of garden method and material geometrically balanced based on the bilateral symmetry.

Exocortis
A shelling off of the bark of trees.

Sagittate
Arrow-shaped with the basal lobes directed downwards. *e.g.,* leaves of some aroids.

Stipule
an appendage at the base of the petiole, often in pairs and sometimes called auricles (ears)

32. Which is the chief pollinator in passion fruit-
    A] Honey bee   B] House fly
    C] Carpenter bee   D] Birds

33. Which is the climacteric fruit
    A] Pineapple   B] Bael
    C] Citrus   D] Litchi

34. Which is the climber
    A] Hibiscus spp   B] Quisqualis
    C] Ixora spp   D] All

35. Which is useful for the control of mango malformation-
    A] T.I.B.A.   B] I. A. A.
    C] N.A.A.   D] Kenetin

36. 'Ampilography' is the cultivation of-
    A] Cole crops   B] Vine crops
    C] Bulb crops   D] Tuber crops

37. Which is the drought resistant crop
    A] Sweet potato   B] Okra
    C] Cauliflower   D] Potato

38. Which is the drought tolerant crop
    A] Maize   B] Bajara
    C] Sugarcane   D] Sugarbeet

39. Which is the dwarfing rootstock of Guava
    A] French Guava   B] Chinese Guava
    C] American Guava   D] Seedling Guava

40. Which is the early-smooth seeded variety of pea-
    A] Arkel   B] Bonneville
    C] Early Badger   D] Asuji

41. Who is the eminent breeder of grape-
    A] N.E. Lee   B] H.P. Olmo
    C] G.S. Cheema   D] J.E. Higgins

42. Which is the example of Aggregate fruit
    A] Pineapple   B] Strawberry
    C] Bread fruit   D] Bael

43. Which is the fastest method of lawn making
    A] Dibbling   B] Turfing
    C] Seed sowing   D] None

44. Which is the finest variety of Mandarin
   A] Khasi Mandarin   B] Nagpur Mandarin
   C] Kamla Mandarin   D] All

45. Who was the first director of ICAR
   A] B. P. Pal   B] M. S. Randhwa
   C] R. S. Parora   D] M. S. Swaminathan

46. Who was the first to correlate the occurence of 'black tip' of mango with smoke of brick kilns-
   A] Robinson (1908)   B] Woodhouse (1909)
   C] Mukherjee (1925)   D] R.N.Singh (1975)

47. Which of the following colour dominates throughout the year in a garden-
   A] Yellow   B] Red
   C] Green   D] Blue

48. Which of the following crop is suffering from physiological disorder
   A] Sesamum   B] Banana
   C] Mango   D] Litchi

49. Which of the following fruit crop has single seeded nut
   A] Litchi   B] Sapota
   C] Mango   D] Coconut

50. Which of the following fruit is non-climacteric
   A] Persimon   B] Sapota
   C] Avocado   D] Citrus

51. Which of the following fruit ripens first
   A] Banana   B] Apple
   C] Citrus   D] Mango

52. Which of the following hormone increase senescence
   A] Ethylene   B] NAA
   C] IAA   D] ABA

53. Which of the following hormone is most effective in induction of bolting in rosette plant
   A] Auxin   B] GA$_3$
   C] Cytokinin   D] Ethylene

Style
An elongated part of the carpel between the ovary and stigma.

Dysploid
A species in which the chromosome number is more or less than the expected diploid number.

Bleaching
Any process which lightens the colour in a plant by the destruction of the chlorophyll or anthocyanin pigments from preventing the green colour from developing by covering from sunlight.

Colorimeter
An instrument for measuring the colour substances by colour components and standards.

54. Which of the following insect is non poly phagous
    A] Lady bird beetle   B] White grub
    C] Termite   D] Grass hopper

55. Which of the following is acidic igneous rock
    A] Granite   B] Basalt
    C] Limestone   D] Marble

56. Which of the following is class one preservative
    A] Silver nitrate   B] KMS
    C] Sugar   D] Vinegar

57. Which of the following is dioecious fruit crop
    A] Ber   B] Date palm
    C] Mango   D] Guava

58. Which of the following is four angled bean
    A] Winged Bean   B] Cluster Bean
    C] Lima Bean   D] Broad Bean

59. Which of the following is non climacteric fruit
    A] Mango   B] Pineapple
    C] Banana   D] Guava

60. Which of the following is not the hybrid variety of tomato
    A] Karnataka   B] Naveen
    C] Rashmi   D] Pusa Gaurav

61. Which of the following is percursor for auxin synthesis
    A] Methionine   B] Tryptophoan
    C] Adanine   D] None

62. Which of the following is protoandrous
    A] Onion   B] Tomato
    C] Cauliflower   D] French Bean

63. Which of the following is responsible for controlling of fruit drops
    A] Auxin   B] Gibberellins
    C] Cytokinin   D] Ethylene

64. Which of the following is single seeded vegetable
    A] Cucumber   B] Leek
    C] Chow-chow   D] Tomato

65. Which of the following is the flower preservative
    A] KMnO₃   B] KNO₃
    C] HgCl₂   D] none

66. Which of the following is the narcotic drug
    A] Poppy   B] Tea
    C] Coffee   D] All

67. Which of the following is variety of Apricot
    A] Chubattia Anupam B] Chubattia Princess
    C] Chubattia madhu   D] Sunheri

68. Which of the following variety of mango is resistant to
    powdery mildew
    A] Totapuri   B] Rataul
    C] Swarna Jahangir   D] Lal Sundari

69. Which of the following vegetable belong to family
    compositae
    A] Spinach   B] Sorrel
    C] Endive   D] Chive

70. Who was the founder of Indian Botanic Garden (Calcutta}
    A] Robert Kyd (1787) B] George King (1878)
    C] William Llyod   D] None of these
       (1817)

71. Which of the fruit crop require tropical climate
    A] Pineapple   B] Citrus
    C] Banana   D] Litchi

72. Most of the Grape variety is seedless due to
    A] Vegetative   B] Stimulative
       parthenocarpy
    C] Stenospermocarpy D] All

73. Seed production cost is the highest in case of-
    A] Hybrid seeds   B] Foundation seeds
    C] Certified seeds   D] Registered seeds

74. The important factors controlling the keeping quality of 'Cut-
    flowers' are-
    A] Respiration & Tanspiration B] Temperature & Humidity
    C] Shape & Colour of flower   D] Cultural practices

Contour
An imaginary line on the earth surface connecting points of the same elevation.

Cotyledon
The primary leaves of the embryo, present in the seed.

Stomata
A minute pore in the epidermis, especially in the lower surface on the leaf. The "breathing pores" of a leaf.

Umbellifer
Any plant belonging to the family Umbelliferae as carrot, parsley.

Tetratype
A tetrad in which the four meiotic products are different. *e.g.*, AB, aB, Ab and ab, crossing over has occurred in such a tetrad.

Herb
A plant dying to the ground at the end of the season; one whose aerial stems are soft and succulent without appreciable parenchymatous xylem tissue, a plant not woody in texture.

Creeping
Prostrate and spreading over the ground.

Hypsometer
An instrument for determining the height of standing tree from some distance.

Strophiole
A small opening near the hilum.

Feni
A fermented product of the juice from cashew apple.

Bulblet
Small bulbs arising in the leaf axils.

75. Which is the leader exporter of black pepper-
   A] India   B] Indonesia
   C] Brazil   D] Malaysia

76. Which is the leading crop of India-
   A] Mango   B] Banana
   C] Citrus   D] Guava

77. .... is the major pest of Brinjal
   A] Fruit fly   B] Thrip
   C] Fruit & shoot borer   D] Mealy bug

78. Which is the male sterile variety of Peach
   A] Red Heaven   B] Stark Early Glow
   C] J.H. Hale   D] Pratap

79. Which is the metamorphic rock
   A] Marble   B] Basalt
   C] Granite   D] All

80. Which is the monocot crop
   A] Pineapple   B] Date palm
   C] Banana   D] All

81. Which is the most economic cross combination for producing gynodioecious lines in papaya-
   A] $M_2m \times M_2m$ B] $mm \times M_1m$
   C] $M_2m \times mm$ D] $mm \times mm$

82. Which is the most essential for better crop of litchi-
   A] CaO B] $NaCl$
   C] $NaOH$ D] $CaSO_4$

83. Which is the most salt toletant crop
   A] Mango   B] Pomegranate
   C] Banana   D] Amla

84. Which is the most suitable annual for 'edging purpose'-
   A] Holly hock   B] Candytuft
   C] Cosmos   D] Antirrhinum

85. Which is the mutant variety of papaya-
   A] Pusa Nanha   B] Washington
   C] CO-1   D] Solo

86. Carnation is the national flower of
   A] Japan　　　　　B] China
   C] Spain　　　　　D] Holland

87. Which is the non climacteric fruit
   A] Orange　　　　B] Mango
   C] Fig　　　　　　D] Banana

88. Which is the oldest system of training followed in Peach
   A] Multiple stem　　B] Open centre
   C] Central leader　　D] Modified leader

89. 'Cobalt' is the part of-
   A] Vit-A　　　　　B] Vit-B
   C] Vit-C　　　　　D] Vit-D

90. L-49 is the popular variety of-
   A] Pear　　　　　B] Peach
   C] Litchi　　　　D] Guava

91. In jelly marmalade, the ratio of sweet orange and khatta should be-
   A] 2:1　　　　　B] 1:2
   C] 2:3　　　　　D] 3:2

92. Which is the rich source of vitamin C
   A] Cabbage　　　B] Tomato
   C] Chilli　　　　D] All

93. Which root crop is the richest source of a certain vitamin-
   A] Sweet potato　B] Carrot
   C] Radish　　　　D] Dioscorea

94. Which is the richest source of Vit-B$_6$ among all the nuts-
   A] Pecan　　　　B] Walnut
   C] Hazelnut　　　D] Pistachionut

Matching

95. Match the following-
   A] Stone fruit formation　i] Ber
   B] Yellow spot disorder　ii] Annona
   C] Wooly aphid　　　　iii] Apple
   D] Pollen sterility　　　iv] Grape
   E] Millerandage disorder　v] Citrus

Allele
Alternative form of gene that occurs at a given locus in a chromosome.

Serrate
having marginal teeth that lean toward the tip of the leaf; margins can be singly or doubly serrated

Cordate
a leaf base that extends in a curving, dual lobe-like fashion below the top of the petiole; heart-shaped

Pigment
A molecule coloured by the light it absorbs. Most of the natural objects are coloured by their differential absorption of different wavelenghths of the incident white light.

Androecium
Collective term for the male parts of a flower.

Lyophilization
A method of preserving the microorganisms where the organisms are subjected to extreme dehydration in the frozen state and then sealed in a vacuum. Desiccated (lyophilized) cultures of micro-organisms remain viable for many years.

Cleft
Divided to or about the middle into divisions.

Gynandrium
The structure in the orchid flower, which results from the fusion of the male and female portions of the flower.

96. Match the following-

A]	Strawberry	i]	Hexaploid (2n = 90)
B]	Banana	ii]	Octaploid (2n = 56)
C]	Pistachionut	iii]	Tetraploid (2n = 100)
D]	Pseudoananas	iv]	Diploid (2n = 30)
E]	Persimmon	v]	Triploid (2n = 33)

97. Match the following-

A]	Sweet William	i]	Papaveraceae
B]	Californian poppy	ii]	Brassicaceae
C]	Gaillardia	iii]	Asteraceae
D]	Gomphrena	iv]	Caryophyllaceae
E]	Candytuft	v]	Amaranthaceae

98. Match the following-

A]	*Thevetia peruviana*	i]	White Scented
B]	*Daedalcanthus nervosus*	ii]	Red flower
C]	*Ervatomia divericata*	iii]	Yellow flower
D]	*Hamelia patens*	iv]	White flower
E]	*Jasminum sambac*	v]	Blue flower

99. Match the following-

A]	Transgenic plants	i]	A mass of unorganized cells
B]	Micropropagation	ii]	Plant part used for regeneration
C]	Callus	iii]	Foreign D. N. A.
D]	Explant	iv]	Capacity of plant cell to develop into whole plant
E]	Totipotency	v]	Micro cloning

100. Match the following-

A]	Wood apple	i]	*Averrhoa carambola*
B]	Lasoda	ii]	*Feronia limonia*
C]	Jamun	iii]	*Emblica officinalis*
D]	Five corner fruit	iv]	*Cordia mixa*
E]	Aonla	v]	*Syzygium cuminii*

101. Match the following

A]	2,4-D 1	i]	Vitamin
B]	Auxin	ii]	Fungicide
C]	Malathion	iii]	PGR

D] Thiram
E] Riboflavin

iv] Weedicide
v] Pesticide

102. Match the following
A] Beet root
B] Carrot
C] French bean
D] Okra
E] Sweet potato

i] Malvaceae
ii] Convolvulaceae
iii] Chenopodiaceae
iv] Umbelliferae
v] Papilionaceae

103. Match the following
A] ICPH-8
B] APRH-I
C] Pusa Kranti
D] Pusa Meghdoot
E] PHB-10

i] Brinjal
ii] Bajra
iii] Bottle gourd
iv] Rice
v] Pigeonpea

104. Match the following
A] Insect
B] Lapidoptera
C] Acarina
D] B.E.D.
E] Carbofuron

i] Fumigent
ii] Spider mite
iii] Earwig
iv] Moth
v] Rice yellow stem borer

105. Match the following
A] Kallo
B] Tulip
C] Langra
D] Haramadhu
E] Kufri jyoti

i] Himalaya
ii] Potato
iii] Tomato
iv] Melon
v] Mango

106. Match the following
A] Mango
B] Grape
C] Citrus
D] Banana
E] Fig

i] 18
ii] 22
iii] 40
iv] 26
v] 38

107. Match the following
A] Poinsett
B] Arka Harit
C] Ambilli

i] Pumpkin
ii] Bottle gourd
iii] Cucumber

D] pusa vikas   iv] Sponge gourd
E] Pusa chikni   v] Bitter gourd

108. Match the following
A] Pumpkin   i] Asia
B] Water melon   ii] India
C] Snake gourd   iii] Central America
D] Bitter gourd   iv] Africa
E] Ridge gourd   v] Indo-Burma

109. Match the following
A] Pusa basmati   i] Fruit
B] *Staphylo coccus*   ii] Rose
C] Dr. B. P. Pal   iii] Food processing
D] Yellow vein mosaic   iv] Okra
E] Nagpur mandarin   v] Rice

110. Match the following
A] Pusa Jyoti   i] Pumpkin
B] Pusa Deepali   ii] Raddish
C] Pusa Chetki   iii] Bitter Gourd
D] Pusa Naveen   iv] Palak
E] Pusa Vikas   v] Cauliflower

111. Match the following
A] Radio   i] Democratic decentralization
B] Gram adhikari   ii] Agent
C] Willage panchyat   iii] Mass media
D] Hogers   iv] Adoption process
E] Dessimination of information   v] Diffusion

112. Match the following
A] Refractrometer   i] Apple
B] Jelly   ii] Papaya
C] Jam   iii] TSS
D] Cider   iv] Guava
E] Pickle   v] Lactic acid

113. Match the following
A] Rhubarb   i] Spear
B] cauliflower   ii] Head

C] Pea   iii] Leaf stalks
D] Asparagus   iv] Seed
E] Cabbage   v] Curd

114. Match the following
A] Turmeric   i] Pea
B] Dioecious   ii] Scooping
C] Self pollinated   iii] Pant Pitamah
D] Cauliflower   iv] Pointed ourd
E] Coriander   v] Pant Haritama

115. Match the following
A] Malic acid   i] Grape
B] Citric acid   ii] Vinegar
C] Tartaric acid   iii] Pickle
D] Acetic acid   iv] Apple
E] Lactic acid   v] Grape fruit

116. Match the following.:-
A] Marmelosin   i] Mandarin
B] Bromelin   ii] Pineapple
C] Naringin   iii] Orange
D] Hesperidin   iv] Bael
E] Tangererin   v] Grape

True / False

117. Hard endosperm is edible part of arecanut.   T/F

118. In *Hevea brassiliensis* genetic base is very narrow.   T/F

119. Lime fruits are rich in tartaric acid.   T/F

120. Malihabad is the origin place of Dashehari mango.   T/F

121. Thompson Seedless is a variety of grapefruit.   T/F

122. Walnut bears fruits on spurs.   T/F

123. Baby Masquerede is a variety of dahlia.   T/F

124. Bartlett pear requires high chilling.   T/F

125. Coffee is native of Tropical Africa.   T/F

126. Edging plants are suitable for city avenues.   T/F

127. Jack fruit seeds are viviparous.   T/F

Refrigerant
A gas that refrigerates the products placed in a refrigerator by absorbing latent heat from the surrounding. Temperature for evaporation of the gas in liquid form is below the freezing point of water and the latent heat of evaporation is also relatively large. The first refrigerant used in a practical system is ammonia gas. Freons are now used which are non-toxic, non-combustible, non-explosive and leakage in refrigerator usually cause no damage to foods. CO_2 and SO_2 may also be used as refrigerants.

Anthesis
Full flower, the period of pollination.

Hygrometer
An instrument for measuring the relative humidity of the atmosphere.

Gigantism
Also called giantism. The production of luxuriant vegetative growth, usually accompanied by a delay in flowering or fruiting; the appearance of giant forms of a plant.

Moringin
An alkaloid present in the root bark of Sajna (*Moringa* sp.) which raises blood pressure, accelarates heart beat and constricts blood vessels.

Matrix
A place on the root stock that are prepared for joining the scion or bud.

128. Polybags are better than muslin cloth for fruit bagging. T/F
129. Pumpkin & Bottlegourd suits well for transplanting. T/F
130. Red colour in tomato is due to 'Tomatin' T/F
131. Saksham is a variety of mentha. T/F
132. 'Leaf rust' is serious disease of coffee. T/F
133. 'Madhurima' is a variety of mango developed by IARI. T/F
134. Acid content is very important in marmalade preparation. T/F
135. Fire blight is a serious disease of potato. T/F
136. Magness is male-sterile variety of pear. T/F
137. Panama wilt is a disease of grape. T/F
138. Root pruning of guava is effective for *bahar* treatment. T/F
139. *Saraca indica* is grown for foliage beauty. T/F
140. Thiourea is used for breaking dormancy in potato. T/F
141. Turnip & Parsley are rich in Vit-C. T/F
142. 'Solanine' in potato develops due to absence of sun light. T/F
143. Alphonso mango grows best in humid coastal region. T/F
144. Arka Mridula is a variety of guava. T/F
145. Caprification is related to *Carica papaya*. T/F
146. Coorg Honey, variety of papaya is selection from Madhu Bindu. T/F
147. Leaf curl is serious disease of tomato. T/F
148. Muscat is a variety of grape. T/F
149. Oil palm is rich source of beta-carotene. T/F
150. *Prunus avium* is the botanical name of sweet cherry. T/F
151. Quince is vigorous root stock of pear. T/F
152. Ber is commonly propagated by bud-grafting. T/F
153. Bullock's heart and Ramphal are same things. T/F
154. Flemish Beauty is self-fertile pear. T/F
155. Hathijhul is a variety of coconut. T/F

156. Papaya is classified as 'arid zone fruit'. T/F
157. Pineapple can be propagated by layering. T/F
158. Sex determination before flowering is a major problem in papaya. T/F
159. The most of Japanese plum cultivars are self-unfruitful. T/F
160. To protect *Anar* fruits from butter fly, caging technique is used. T/F
161. 'Kinnow' was developed by W.B.Hayes. T/F
162. Alternate bearing is a problem of mango. T/F
163. Butt of gun is manufactured from walnut wood. T/F
164. *Calendula* is a rainy season annual. T/F
165. Citrus peels are rich source of pectin. T/F
166. Double working is commonly followed in litchi. T/F

Fill in the Blanks

167. Phalsa is native of _____.
168. *Ficus glomerata* is a root stock for _____.
169. Seedlessness in litchi is due to _____ parthenocarpy.
170. Dehradun is a variety of _____.
171. _____ is the largest producer of litchi in the world.
172. _____ fruits are the second rich source of Vit-A after mango.
173. *Vitis rotundifolia* is _____ pollinated grape.
174. Illaichi is a variety of _____.
175. Guava canker is a _____ disease.
176. _____ is the largest producer of guava in India.
177. Pineapple guava belongs to family _____.
178. NA-9 variety of anola is renamed as _____.
179. Krishna (NA-5) aonla is a chance seedling of _____.
180. In aonla, fruit bud differentiation takes place during _____.
181. _____ is a monoembryonic citrus species.

EC
Electric conductivity of the saturated soil paste extracts measure din mill mhos per centimeter.

Chipping
For propagation, the sectioning of a bulb in such a way that scale pieces remain attached to the basal plate and a new bulb let forms between the scales.

Plumule
A primary bud in the seed which develops into the primary stem on germination.

Cultivate
1. To till the soil.
2. to loosen the soil and remove weeds among growing plants. 3. to plant, tend, harvest and improve by plant breeding.

Leaf-mould
1. The spongy humus developed under cover of forest vegetation, especially of deciduous trees, in a temperate, humid climate. 2. Decomposed leaf compost used in preparing pot mixture for raising seedlings in pots.

Adsorb
To hold on the surface.

Resupination
A process by which the tip of orchid flower, the uppermost petal, becomes the lower most one through a remarkable twisting of the pedicellate ovary.

Scabrous
Rough or gritty to the touch; rough-pubescent.

182. _____ is the largest producer of lime in the world.
183. Mandarins are commercially propagated by _____.
184. Seedless banana follows _____ growth curve.
185. Most of the cultivated bananas are _____.
186. _____ is the only mutant cultivar of mango.
187. Internal necrosis in mango occurs due to _____ deficiency.
188. _____ is a seedless variety of mango.
189. Night blindness is caused due to deficiency of _____.
190. *Venturia inaequalis* is the causal organism of _____.
191. Black tip of mango occurs due to toxicity of _____.
192. _____ is the botanical name of *Sita Ashok*.
193. Jasminum belongs to family _____.
194. Mango cultivar _____ is more susceptible to spongy tissue.
195. Edible part of turmeric is _____.
196. Cabbage is classified as _____ season crop.
197. _____ variety of mango is tolerant to mango malformation.
198. Walnut is the richest source of _____.
199. The best pruning time for rose in N. India is _____ month.
200. Gladiolus is propagated by _____.

A company wanted to set up a pension plan. In order for the plan to be installed, it needed 100% participation. Everyone signed up except John. The plan made sense and was in the best interest of everyone. John not signing was the only obstacle. John's supervisor and other co-workers had tried to persuade him without success.

The owner of the company called John into his office and said, "John, here is a pen and these are the papers for you to sign to enroll into the pension plan. If you don't enroll, you are fired this minute." John signed right away. The owner asked John why he hadn't signed earlier. John replied, "No one explained the plan quite as clearly as you did."

Motivation comes from outside, such as money, societal approval, fame or fear.

[
Your

Motivation is

your **D**rive and

Attitude.
]

Step-Nineteen

The Losers are always static;
The winners are always in motion.

Multiple Choice

1. Which is the ripening hormone
   A] Auxin   B] NAA
   C] IBA   D] Ethylene

2. Hormone promoting the root tuber growth is
   A] Auxin   B] GA
   C] Cytokinin   D] Ethylene

3. Which is the serious disease of rose
   A] Die back   B] Powdery mildew
   C] Leaf spot   D] Rust

4. Which of the state produce maximum quantity of potato
   A] UP   B] MP
   C] Bihar   D] Rajasthan

5. Which is the summer annual
   A] China aster   B] Zinnia
   C] Coleus   D] Acroclmum

6. Vegetables are dehydrated at the temperature of -
   A] 60-66°C   B] 80-85 °C
   C] 40-50 °C   D] 90-95 °C

7. Lasso is the trade name of
   A] Isoproturon   B] Alachor
   C] Butachlor   D] Atranzine

8. Umran is the var of
   A] Grape   B] Aonla
   C] Ber   D] Apple

9. To check the variability in root stocks one should opt for-
   A] Seedling rootstock   B] Clonal rootstock
   C] Seasoned rootstock   D] Any of above

10. Contender is the variety of french bean which is
    A] Bush type   B] Pole type
    C] Semipole type D] None

Aril
A fleshy appendage of the seed, usually a fleshy seedcoat.

Forskohlin
A diterpeniod present in the tuber of Chinese patato (*Coleus forskohlii*) which is useful in certain pharmaceutical preparations prescribed for and low blood pressure.

Clingstone
Any fruit, especially the varieties of the peach, whose flesh adheres strongly to the seed or stone, even at maturity.

Intersexualism
A term referring to the varying degrees of development of the two sex organs in the same plant, relative maleness and femaleness of . the plant.

11. Nantes is the variety of
    A] Radish   B] Carrot
    C] Onion   D] Turnip

12. Krishna is the variety of
    A] Tomato   B] Brinjal
    C] Chilli   D] Potato

13. ' Thermodormancy' is found in-
    A] Cabbage   B] Lettuce
    C] Cauliflower   D] Leek

14. India is third largest producer of coconut next to-
    A] China & Brazil   B] Japan & Korea
    C] Philippines & Indonesia D] None

15. In which fruit crop three flushes occur in a year-
    A] Jamun   B] Litchi
    C] Loquat   D] Sapota

16. A ' three-way cross hybrid' of carrot is-
    A] Spartan Bonus   B] Nantes
    C] Imperator   D] Pusa Meghali

17. Mango necrosis (Black Tip) is much prominent in-
    A] North India   B] South India
    C] Deccan Plateau   D] Temperate Zone

18. ' Tipburn' in lettuce occurs due to-
    A] 'Ca' deficiency   B] 'B' deficiency
    C] High temperature D] All of these

19. Browning due to boron deficiency is common in-
    A] Radish   B] Cauliflower
    C] Onion   D] Chilli

20. Which instrument is used to check the Brix^0 in jelly-
    A] Jelmeter   B] Refractometer
    C] Brixometer   D] Jelly Thermometer

21. Anaemia is caused due to deficiency of-
    A] Citric acid   B] Ascrobic acid
    C] Folic acid   D] Iron

22. An area laid out to determine the effects of treatment is-
    A] Experimental Plot B] Experimental Design
    C] Experimental Data D] Replication

23. Which fruit crop belongs to family Tiliaceae-
    A] Pineapple   B] Mangosteen
    C] Phalsa   D] All of above

24. Carrot belong to family
    A] Compositae   B] Umbeliferae
    C] Legumnioceae   D] None

25. Loquat belongs to family
    A] Lauraceae   B] Rosaceae
    C] Sapindaceae   D] None

26. Clove belong to family
    A] Leguminoceae   B] Myrtaceae
    C] Stirculiaceae   D] Rosaceae

27. Bael belong to family
    A] Sapindaceae   B] Rutaceae
    C] Rosaceae   D] Myrtaceae

28. Tapioca belongs to family-
    A] Solanaceae   B] Convolvulaceae
    C] Araceae   D] Euphorbiaceae

29. Celery belongs to family-
    A] Umbelliferae   B] Liliaceae
    C] Graminae   D] Aizoaceae

30. Most resistant to nematodes among flowers is
    A] Rose   B] Gladiolus
    C] Merigold   D] Jasmine ,

31. India has to provide food for its population that is now
    A] 1 million   B] 1 billion
    C] 600 million   D] 2 billion

32. Phalsa belong to the family
    A] Bromeliaceae   B] Tiliaceae
    C] Rosaceae   D] Rutaceae

33. Chrysanthemum belongs to the family-
    A] Compositae   B] Malvaceae
    C] Caryophyllaceae D] Umbelliferae

34. Larkspur belong to the family
    A] Scrophulariaceae B] Rarunculaceae
    C] Labiatae   D] Cruciferae

Whorled
An arrangement of three or more structures arising from a single node.

Trifoliate
A compound leaf with three separate leaflets as in French bean, cowpea, etc.

Xeriscaping
Conservation of water through creative landscaping.

Nucleotide
A molecule formed by the union of phosphoric acid, a pentose sugar (ribose of deoxyribose) and an organic base derived from purine or pyrimidine, (adenine, guanine, cytosine, thymine or uracil). Nucleotide is the unit of DNA and RNA organisation

Larva

An immature stage through which some types of insects must pass before developing into adults. Caterpillars are the larvae of moths and butterflies, and grubs are the larvae of beetles. Larvae are typically wormlike in appearance.

Tree age

The length of time from seed germination or from initial budding or vegetative propagation.

Polycarpic

Flowering and fruiting many times.

Mites

A group of tiny animals related to spiders, many of which feed on plants.

35. Crop tolerant to water salinity
   A] Barley   B] Rice
   C] Wheat   D] Potato

36. Gerkhins belong to which genus
   A] Cucurbita   B] Cucumis
   C] Citrulus   D] None

37. NARS refers to which
   A] National Ayurvedic Research System
   B] National Agriculture Research System
   C] National Agriculture Research Scheme
   D] National Agriculture Review System

38. Which is a salt tolerant crop-
   A] Litchi   B] Sugar beet
   C] Muskmelon   D] Walnut

39. Syngamy refer to
   A] Fusion of sperm with nucleus
   B] Fusion of one sperm with egg
   C] Fusion of sperm with synergids
   D] None

40. For sugar translocation in Grape which of the following element is necessary
   A] Bo   B] Ca
   C] Mg   D] Phosphorus

41. Which plants can be transplanted successfully without a ball of earth i.e. bare rooted-
   A] Mango, Guava   B] Jamun, Aonla
   C] Apple, Pear   D] Litchi, Papaya

42. What is transplastomics
   A] Transfer of   B] Transfer of chloroplast
      mitochondria
   C] Transfer of cytoplasm D] All

43. Emision of ethylene during transportation of cut flower cause a disorder which is called as-
   A] Bent neck   B] Bud opening
   C] Sleepiness   D] Calyx splitting.

44. The height of a tree is measured by-
    A] Wood Clipper   B] Sunnto Clonometer
    C] Anemometer   D] Windwane

45. Due to pruning of trees, usually fruiting results in-
    A] Increase in total yield   B] Reduction in total yield
    C] Increase in nitrogen   D] Decrease in carbohydrate
    level

46. ' Tristeza virus' in citrus is transmitted by-
    A] *Myzus persicae*   B] *Pentalonia nigronervosa*
    C] *Daucus dorsalis*   D] *Toxoptera citricida*

47. Family of tuberose is
    A] Malvaceae   B] Rosaceae
    C] Compositae   D] Amarylliadaceae

48. ' Tunicated bulbs' are found in-
    A] Lily   B] Onion
    C] Amaryllis   D] All of them

49. "Brown turkey" is
    A] Rootstock of Fig   B] Rootstock of Sapota
    C] Rootstock of Jackfruit D] Variety of Avocado

50. The cultivated turmeric is
    A] Tetraploid   B] Triploid
    C] Diploid   D] None

51. In one turn of genetic model of DNA, how much distance is
    covered
    A] 3.4A   B] 34A
    C] 36 A   D] 38 A

52. Most expanding type of clay mineral is
    A] Montmorillonite   B] Illite
    C] Vermiculite   D] Kaolinite

53. The fruit type of loquat is
    A] Berry   B] Pome
    C] Drupe   D] Nut

54. Maximum acreage under garlic is in-
    A] Gujarat   B] Rajasthan
    C] Orissa   D] Bihar

55. Approximate area under irrigation in India is
    A] 30 MH   B] 70 MH
    C] 50 MH   D] 90 MH

56. Golden shower under north Indian condition flower in-
    A] Summer   B] Rainy
    C] Spring   D] Winter

57. Maximum acreage under rubber plantation is in-
    A] Kanataka   B] Kerala
    C] Tamil Nadu   D] Andhra Pradesh

58. Brix degree is an unit for measurement of-
    A] Salt content in juice   B] Sucrose content in juice
    C] Total solids in juice   D] Juice content of fruit

59. First 'State Agricultural University' of India is situated at-
    A] Kanpur   B] Pantnagar
    C] Solan   D] Bangalore

60. From sowing of nursery upto harvesting of bulbs, onion takes about-
    A] 125 days   B] 200 days
    C] 250 days   D] 300 days

61. Which is used as a 'source of resistance' for mosaic/leafcurl in papaya breeding-
    A] *Carica papaya*   B] *Carica microcarpa*
    C] *Carica cauliflora*   D] *Carica pentagona*

62. Which fruit crop is used as an emblem of the United Nations-
    A] Apple   B] Mango
    C] Olive   D] Walnut

63. Which is used as an ethylene absorbant-
    A] KCl   B] $KMNO_4$
    C] K_2SO_4 D] KNO_3

64. 'Chloropicrin' is used as-
    A] Soil sterilant   B] Soil fumigant
    C] Storage fumigant   D] Contact herbicide

65. Which is used for artificial ripening of banana and mango-
    A] $KMnO_4$ B] $K_2S_2O_5$
    C] CaC_2 D] CaO

66. Which is used for deblossoming purpose in mango
    A] NAA   B] GA
    C] Paclobutrazol   D] ABA

67. Which chemical used for dehydration of flowers-
    A] CaO   B] CaC_2
    C] Borax   D] All of them

68. Refractometer is used for measuring-
    A] Salt per cent   B] Acid per cent
    C] Sugar per cent   D] Fibre content

69. Yeast is used for
    A] Marmalade   B] Cider
    C] Jelly   D] Jam

70. Standard unit used in expressing refrigeration capacity is
    A] Calorie   B] Ampere
    C] Ton   D] Joule

71. CCC is used in grape for -
    A] Increase vegetative   B] Increase TSS
       growth
    C] Increase fruitfulness D] None

72. Rose root stock commercially used in North India is-
    A] *Rosa foetida*   B] *Rosa bourboniana*
    C] *Rosa chinensis*   D] All the above

73. Diogenin is used in treatment of
    A] Blood cancer   B] HIV
    C] Sexual disorder   D] All

74. Quincunx method used in -
    A] Fruits   B] Vegetables
    C] Flowers   D] Potato

75. Yeast is used in
    A] Nectar   B] Cordial
    C] Syrup   D] Cidar

76. Spontaneous mutation used mostly for
    A] Ornamentals   B] Gruits
    C] Vegetables   D] Spice

Sweetening
A term used to denote the odor of lilies after storage under anaerobic conditions.

Penetrance
Ability of a gene to express itself in the individual carrying it.

Viability
Capacity of seed to germinate is called viability.

Stage'G'
The term used to indicate that the gynocecium (pistil) has formed in the flower of tulip in the bulb itself.

Pesticide
A chemical used to kill an organism considered a pest.

Divided
Deeply lobed when applied to leaves; separated to the base into divisions.

Solubridge
An instrument used to measure the electrical conductivity of the soil solution.

Leaching
Applications of water to media to reduce level of soluble salts.

Protogynous
Having the stigma receptive to pollen before the pollen is released from the anthers of the same flower.

Ethylene scrubbers
Chemicals used to remove ethylene form the atmosphere.

Harvest
A single deliberate action to separate a crop from its growth medium (a plant part from the plant or the whole plant from soil).

77. Most commonly used oil extraction method from flowers
A] Distillation　　　　B] Meceration
C] Solvent extraction　D] None
　　method

78. Most commonly used training system in grapes is
A] Kniffin　　　　B] Bower
C] Head　　　　　D] Telephone

79. Which is useful to control fruit drop In citrus-
A] NAA　　　　B] GA_3
C] 2, 4-D　　　D] ABA

80. Pure line var is developed in-
A] Self pollinated crop　B] Cross pollinated
C] Beet　　　　　　　　D] Cabbage

81. Pusa mukta var of cabbage is resistant to-
A] Virus　　　　B] Black rot
C] Bolting　　　D] Caterpillar

82. In which var of cauliflower seed production is not possible in north Indian plains
A] Pusa deepali　　B] Pusa subhra
C] Pusa synthetic　D] Pusa snowball -I

83. Pusa delicious var of papaya is
A] Dioecious　　　B] Monoecious
C] Gynodioecious　D] None

84. Which kind var of tomato need staking-
A] Indeterminate　B] Determinate
C] Dwarf　　　　　D] None

85. Sweet organe var. 'Pineapple' is a suitable indicator for-
A] Exocartis　　B] Tristeza
C] Greening　　　D] Leaf curl

86. *Beta vulgaris* var. *benghalensis* is the botanical name of-
A] Sugar beet　　B] Spinach beet
C] Fenugreek　　D] Iris potato

87. ' Varad' is a low chilling variety of
A] Apple　　　B] Pear
C] Peach　　　D] Plum

88. 'Chitra' are varietie of -
    A] Bougainvillea   B] Marigold
    C] China rose   D] Rose

89. Highest hybrid varieties in vegetables is grown in
    A] Tomato   B] Cabbage
    C] Brinjal   D] Carrot

90. Most of varieties of mango grown in south India are-
    A] Polyembryonic   B] Monoembryonic
    C] Seedless   D] Disease free

91. Most peach variety bear flower of following colour-
    A] Yellow   B] Red
    C] White   D] Pink

92. Production of a new variety by crossing of genetically different parent is known as-
    A] Recombination   B] Pedigree selection
    C] Hybridization   D] Variation

93. First rose variety developed, by Dr. B.P. Pal was-
    A] Rajhans   B] Sujata
    C] Gulzar   D] Rose Sherbet

94. Dwarf mango variety is
    A] alphanso   B] amrapalli
    C] ratna   D] deshehari

Matching

95. Match the following:
    A] 2n = 22   i]   Radish
    B] 2n = 14   ii] Watermelon
    C] 2n = 18   iii] Sponge Gourd
    D] 2n = 90   iv] Cucumber
    E] 2n = 26   v]   Sweet potato

96. Match the following:-
    A] Amphimixis   i]   East & Mangelsdrof (1925)
    B] Apomixis   ii] Koelreuter (1763)
    C] Self-incompatibility iii] Asexual reproduction

Radical
Of or pertaining to the root.

Resistance
Ability of a host to withstand the effect of unfavourable conditions which is the inherent attribute of the host.

Sterile
Barren, not able to produce seed.

Landscape
1. The sum total of the characteristics that distinguish a certain area of the earth's surface from other areas, such as soil types, vegetations, hills, valleys, cultivated fields, streams, etc.
2. To beautify a terrain by planting of trees, bushes, flowering herbs and with ornamental features, such as terraces, rock gardens, wall, etc.

D] Gametophytic self-incompatibility — iv] Stout (1917)
E] Male sterility — v] Sexual reproduction

97. Match the following:-
A] Bitter gourd — i] Source of tartaric acid
B] Radish — ii] Source of sugar
C] Beetroot — iii] Rich in ascorbic acid
D] Grape — iv] Useful in diabetes
E] Turnip — v] Useful in urinary complaints

98. Match the following:-
A] C I T H — i] Lucknow (U.P.)
B] C I A H — ii] Srinagar (J & K)
C] C I P H E T — iii] Gurgaon (Haryana)
D] N H B — iv] Bikaner (Rajasthan)
E] C I S H — v] Abohar (Punjab)

99. Match the following:-
A] Cocoa stem borer — i] *Nephoperyse eugraphela*
B] Rhinoceros beetle — ii] *Zeuzara aurantii*
C] Red palm weevil — iii] *Raoiella indica*
D] Arecanut mite — iv] *Oryetes rhinoceros*
E] Chiku moth — v] *Rhichophorus ferrugiensis*

100. Match the following:-
A] Colchicine — i] I A R I (Pusa)
B] Genotype & Phenotype — ii] Blakeslee (1937)
C] Dominant gene — iii] R R
D] Recessive gene — iv] Johannsen (1903)
E] Gamma garden — v] rr

101. Match the following:-
A] Cucumber — i] Fleshy Receptacle
B] Custard apple — ii] Fleshy Pericarp
C] Datepalm — iii] Pericarp
D] Fig — iv] Meso & Endocarp
E] Pineapple — v] Fleshy Thalamus

102. Match the following:-
    A] Diploid (AA) banana   i] Bodles Altafort
    B] Triploid (AAA) banana   ii] Namarai, Kadali
    C] Tetraploid (AAAA) banana   iii] Pusa Seedless, Thompson Seedless
    D] Diploid (AB) banana   iv] Dwarf Cavendish, Gross Michel
    E] Triploid (AAB) banana   v] Kuribontha, Kanchkela
    F] Triploid (ABB) banana   vi] Kunnan, Safed Velchi
    G] Tetraploid (ABBB) banana   vii] Perlette, Beauty Seedless
    H] Culinary banana   viii] Blue Teparod
    I] Spur pruning   ix] Poovan, Rasthali
    J] Cane pruning   x] Monthan, Nendran

103. Match the following:-
    A] Gametophytic incompatibility   i] Bhan *et al* (1969)
    B] Sporophytic incompatibility   ii] Hughes & Babcock (1950)
    C] Kinnow   iii] Lynch (1941)
    D] Veneer grafting   iv] East & Mangelsdorf (1925)
    E] Stone grafting in India   v] H.B. Frost (1935)

104. Match the following:-
    A] Marsh, Ruby, Thompson Seedless   i] Seedless Sweet orange
    B] Washington Naval   ii] Seedless var. of Grape fruit
    C] Satsuma   iii] Seedless Mango
    D] Eureka, Lisbon, Assam Lemon   iv] Seedless Mandarin
    E] Sindhu   v] Seedless var. of Lemon

Acclimatization
Adjustment or increase in tolerance shown by a species in the course of several generations in a changed environment, inuring to a new habitat, medium or set of conditions.

Umbel
Usually a flat-topped flower cluster, with pedicels and peduncles arising from a common point, resembling the supports of an umbrella.

Brix
A measure of the total soluble solids in fruit and vegetable juice, calibrated in terms of pure sucrose *i.e.* 1° Brix = 1% sucrose ; measured with a refractometer.

Curing
A post-harvest treatment of tuber and bulb crops by exposing them to relatively high temperature and high humidity to facilitate the drying of upper skin and suberization of the outer tissue, periderm formation and healing of injured surface, reduces the moisture loss through transpiration, prevents microbial attack, reduces rotting and resulting more shelf-life.

Bonsai
It comprises a tree or shrub planted in a small container for developing as a miniature plant showing the general appearance of that plant species found in nature.

105. Match the following:-
A] *Millingtollia hortensis*   i] Pagoda tree
B] *Magnolia grandiflora*   ii] Coral tree
C] *Spathodea companulata*   iii] Jasmine tree
D] *Plumeria alba*   iv] Fountain tree
E] *Erythrina cristagalli*   v] Lily tree

106. Match the following:-
A] N R C (Cashewnut)   i] Puttur (Karnataka)
B] N RC (Citrus)   ii] Nagpur (M.S.)
C] N R C (Banana)   iii] Trichy
D] N R C (Agro forestry)   iv] Boriavi (Gujarat)
E] N R C (Medicinal Plants)   v] Jhansi (U.P.)

107. Match the following:-
A] NRC (Spices)   i] Solan (H.P.)
B] NRC (Oilpalm)   ii] Elura(A.P.)
C] NRC (Orchids)   iii] Pune (M.S.)
D] NRC (Grape)   iv] Calicut (Kerala)
E] NRC (Mushroom)   v] Gangtok (Sikkim)

108. Match the following:-
A] Onion   i] Pectin
B] Cashewnut   ii] Anacardiaceae
C] Tomato   iii] Tunicated bulb
D] Jelly   iv] Salad
E] Lettuce   v] Fruit vegetable

109. Match the following:-
A] Precursor of IAA   i] Auxin
B] Basipetal movement   ii] Methionine
C] Precursor of Ethylene   iii] 3-Acetyl CoA
D] Precursor of Cytokinin   iv] Old leaves
E] Precursor of Vit-A   v] Beta-Carotene
F] Precursor of Gibberellic acid   vi] Young fruits
G] Site of Synthesis of IAA & GA   vii] GA & Cytokinin

H] Site of Synthesis of
Cytokinin

viii] Tryptophan

I] Site of Synthesis of
ABA

ix] Young leaves & buds

J] Acropetal movement

x] Adenine

110. Match the following:-

A] Protandry

i] Annona, Rose

B] Protogyny

ii] Kadamba, Kachnar

C] Dimorphism

iii] Marigold, Coriander

D] Trimorphism

iv] Primrose

E] Chiropteriphily

v] Oxalis

111. Match the following:-

A] Rich in Iron

i] Guava (6.9%)

B] Rich in Phosphorous

ii] Walnut (64.5%)

C] Rich in Calcium

iii] Cashewnut (0.45%)

D] Rich in Fibre

iv] Litchi (0.21%)

E] Rich in Fat

v] Dry Karonda (39.1%)

112. Match the following:-

A] Rockery

i] Suvashini

B] Japanese garden

ii] Poonam

C] Gladiolus

iii] Formal garden

D] Aster

iv] Pargoda

E] Tuberose

v] Cut flower

113. Match the following:-

A] Spur bearing

i] N.A.A.

B] Stone fruit

ii] Gibberellic acid

C] 'Single' flower

iii] Aonla

D] Bahar treatment

iv] Guava

E] Fruit drop

v] I.B.A.

F] Rooting

vi] Tuberose

G] Cell elongation

vii] Peach

H] Mango malformation

viii] Apple

I] Fruit ripening

ix] ABA

J] Abscission layer

x] Ethylene

Espalier
Process of training
a tree or shrub so
its branches grow
in a flat pattern.

Trainer
An intermediate or
overtopped tree
which by its
shading hastens the
natural pruning of
crop tree.

Bole
The stem of a tree.

Smudging
A practice of
smoking trees like
the mango
commonly
followed in The
Philippines to
induce flowering
and produce an
off-season crop as
smoke contains
acetylene not
successful in India
in producing an
early blossom.

Crenulate
Finely crenate.

Pedicel
Flower stalk.

Dichotomous
Forked in pairs.

Channeled
Grooved
lengthwise.

Indurate
Hardened.

Bearded
Having long hairs.

Dimorphic
Having two forms.

Cucullate
Hooded.

Hippocrepiform
Horseshoe-
shaped.

Ternate
In threes.

Foliage
Leaves.

Linear
Long and narrow.

Lactiferous
Milky.

114. Match the following:-
A] Sweet orange   i]   Susceptible to 'Gummosis'
B] Rangpur lime, Grape   ii]   Susceptible to 'Foot rot'
    fruit, Lemon
C] Acid lime, Lemon,   iii]   Susceptible to 'Cankar'
    Sour orange
D] Rangpur lime,   iv]   Susceptible to Exocortis/
    Citrange, T. orange   Scaly butt
E] Sour orange, Sweet   v]   Susceptible to tristeza/Quick
    lime   decline

115. Match the following:-
A] Tetraploid mango   i]   Combodiana
B] Polyembryonic mango   ii]   Sour orange
C] Monoembryonic citrus   iii]   Pineapple
D] Polyembryonic citrus   iv]   Pummelo
E] Monocot fruit   v]   Vellaicollumban

116. Match the following:-
A] Theobromin   i]   Anti gibberellin action
B] Paclobutrazol   ii]   Citrus
C] Monocot   iii]   Carambola
D] Degreening   iv]   Banana
E] Golden Star (variety)   v]   Cocoa

True / False

117. Green Gauge is a self-sterile variety of plum.   T/F

118. Jelly can be prepared from unripe mango.   T/F

119. Kalyanpur Baramasi is a variety of bitter gourd.   T/F

120. Pear is more toterant to wet soils than apple.   T/F

121. 'Mattocking' is practised in canna & pineapple.   T/F

122. Blanching is common in fruit canning.   T/F

123. Cashew brandy 'Fenni' is prepared from cashew apple.   T/F

124. Curd in cauliflower is a result of hypoplasea.   T/F

125. Degletnoor is a variety of coconut.   T/F

126. Grape originated in Afghanistan.   T/F

127. *Ipomoea palmata* is a flowering climber.   T/F
128. Kinnow is an intervarietal hybrid.   T/F
129. Pear decline is caused by bacteria.   T/F
130. The main crop of *Smyrna* fig requires pollination.   T/F
131. 'Tuti-fruiti' is made from papaya.   T/F
132. Alphonso is a commercial mango variety grown widely in U.P.   T/F
133. Bitterness in chilli is due to capsanthin.   T/F
134. Blanching activates enzymes responsible for tissue discolouration.   T/F
135. Dwarf Cavendish is a mutant of banana.   T/F
136. Flame peeling is suitable for onion.   T/F
137. Montezuma is a popular variety of canna.   T/F
138. Phylloxera is a serious problem in citrus growing.   T/F
139. Pink end (Pink calyx) is physiological disorder of pear.   T/F
140. Sericulture is related to mulberry.   T/F
141. 'Freedom' is scabe resistant variety of peach.   T/F
142. Ambri is a cultivar of peach.   T/F
143. Angoorlata is a variety of grape vine.   T/F
144. Bending is practised in mango.   T/F
145. Jelly can be prepared from ripe mango.   T/F
146. Mango pickle is a 'national pickle'.   T/F
147. Most of peach cultivars are autogamous.   T/F
148. Phyllanthoid branching is found in aonla.   T/F
149. 'Gola' var. of ber is suitable for humid regions.   T/F
150. Amrapalli is suitable for high density orcharding.   T/F
151. *Anthocephalus cadamba* is a fruit tree.   T/F
152. Banana is a rich source of Vit-C.   T/F
153. Ber is propagated by ring budding.   T/F
154. Ber is suitable for sodic soils.   T/F
155. June Elberta is self-sterile variety of peach.   T/F

Serrulate
Minutely serrate.

Glabrate
Nearly glabrous.

Standard
1. A small tree commonly produced by grafting a weeping or dwarf form on a trunk of the desired height. 2. The erect petals of an iris flower, as opposed to the broader and often drooping falls petals. 3. The upper, more or less erect petals of a pea-like flower which consists of three types of petals: standard, wings and keel.

Pinked
Notched.

Secund
One sided.

Exfoliating
Peeling away.

Xylem
A complex tissue of tough, fibrous, lignified elongated cells forming vessels and woody tissue, part of the vascular bundle responsible for upward movement of water and minerals from roots to the above ground parts of the plant.

Aculeate
Prickly.

Stoloniferous
Producing stolons.

Baccate
Pulpy, fleshy.

Rufous
Reddish brown.

Vernal
Related to spring.

Tubercle
A miniature tuber, tuber-like structure or projection.

Hassock

156. Lettuce & Celery are used as salad.   T/F
157. Shamrock is green variety of apple.   T/F
158. Mr. Alexander Coutts (1887) introduced a large number of apple varieties in India.   T/F
159. There are 500-1000 flowers/inflorescence in mango.   T/F
160. Lalit & S-120 are nematode resistant varieties of tomato. T/F
161. Mridula (H-61) is a hybrid variety of pomegranate.   T/F
162. N-53 is an onion variety suitable for Kharif season.   T/F
163. N-53 variety of onion is suitable for kharif season.   T/F
164. L-49 guava is developed by Dr. G.S. Randhawa.   T/F
165. Lucknow-49 is a popular variety of guava.   T/F
166. M-27 is a dwarfing root stock of mango.   T/F

Fill in the Blanks

167. Radish is a _____ day plant.
168. Browning in cauliflower occurs due to _____ deficiency.
169. Application of fertilizers with irrigation water is known as _____.
170. Strawberry grows well in _____ climate.
171. F.P.A. is the abbreviated form of _____.
172. Jelmeler is used for _____ test in fruit extracts.
173. Papain is prepared from _____ fruits.
174. Chilli fruit is botanically _____.
175. Removal of growing point of a shoot along with few leaves is known as _____.
176. _____ is the most common shrub for evergreen hedge.
177. Fruit Preservation and Canning Institute at Lucknow was established in _____.
178. Pusa Bedana is a seedless hybrid of _____.
179. Pusa Bedana is a cross of _____.
180. *Althea rosea* is a _____ is a season annual.

181. The average yield of tomato crop is _____ per hectare.
182. Bael belongs to the family _____.
183. Rumani is the dwarfing rootstock of _____.
184. Ber should be pruned during _____ months.
185. Niranjan is an off season cultivar of _____.
186. Bitter gourd is a _____ plant by sex form.
187. Okra has _____ somatic chromosomes.
188. Banana is commercially propagated by _____ suckers.
189. _____ budding is commercial method of aonla propagation.
190. _____ state is the largest producer of guava.
191. A cross between a hybrid and one of its parent is known as _____.
192. _____ is the origin centre of cowpea.
193. The superiority of F_1 hybrid over both of its parents is known as _____.
194. Botanically ber fruit is a _____.
195. Litchi is commercially propagated by _____.
196. _____ is the botanical name of Amaltas.
197. China Aster is commercially propagated by _____.
198. *Ficus glomerata* is nematode resistant root stock of _____.
199. Phalsa is the native of _____.
200. Pusa Harit is a variety of _____.

There was a man who made a living selling balloons at a fair. He had all colors of balloons, including red, yellow, blue and green. Whenever business was slow, he would release a helium-filled balloon into the air and when the children saw it go up, they all wanted to buy one. They would come up to him, buy a balloon, and his sales would go up again. He continued this process all day. One day, he felt someone tugging at his jacket. He turned around and saw a little boy who asked, "If you release a black balloon, would that also fly?" Moved by the boy's concern, the man replied with empathy, "Son, it is not the colour of the balloon, it is what is inside that makes it go up."

It is what is inside that counts. The thing inside of us that makes us go up is our attitude.

It is **O**ur

Attitude that

make us to **G**o

Up.

Step-Twenty

The Losers are complex;
The winners are simple.

Multiple Choice

1. Which is an early variety of apple-
    A] Winter Banana   B] Rymer
    C] Red Delicious   D] Fairy

2. Dwarf Cavendish variety of Banana has Genome of
    A] ABB   B] A AB
    C] AAA   D] AB

3. First scented variety of gladiolus was-
    A] Pink Pefume   B] Lucky Star
    C] Manohar   D] None of these

4. Bangalore blue variety of grape is having parents
    A] *V. vinefera x*   B] *V. vinefera x v. asiatica*
    V. lubrisca
    C] *Both*   D] None

5. Most of variety of onion grown in India are-
    A] Short day   B] Long day
    C] Day neutral   D] None

6. The best variety of peach for Northern plains is-
    A] Alexander   B] J.H. Hale
    C] Sharbati   D] Chaubatia Red

7. 'Kufri Swarna' variety of potato is resistant to-
    A] Late blight   B] Early blight
    C] Cyst nematode   D] Tuber rotting

8. Kufri Chandramukhi is a variety of-
    A] Early potato   B] Medium potato
    C] Late potato   D] Summer potato

9. Leguminous leafy vegetable is
    A] Cowpea   B] Fenugreek
    C] Spinach   D] Palak

10. Which is the oldest Vegetable Research Centre in U.P.-
    A] Pantnagar   B] Varanasi
    C] Kalyanpur   D] Saharanpur

Spines
Sharp outgrowths of the twig, sometimes but not always paired at the nodes. Similar to "Prickle".

Appressed
Flattened against the twig; not spreading.

Facination
The fusion of two floral stalks to form a single floral stalk.

Diploid
an organism having two genomes.

Decurrent habit
plant form typical of angiosperm trees and shrubs which are more round-headed and spreading, lacking a main leader to the top of the plant

Tender plants
Plants which are unable to endure frost or freezing temperatures.

Bud
A bud is undeveloped and elongated stems composed of a very short axis of meristem cells from which embryonic leaves, lateral buds, flower parts or all tree arises.

Rudiment
The beginning of an undeveloped plant part.

Primordium
A group of undifferentiated cells.

Stress
Any environmental factor potentially unfavourable to the plants.

Plantation
Large scale Agricultural unit devoted to the production of particular fruit crop or any other commodity.

11. Fruit & vegetables are rich in dietary fibres which (protect) from -
    A] Colon cancer   B] Protein deficiency
    C] Fat deficiency   D] Sugar deficiency

12. Which is a monocot vegetable-
    A] Carrot   B] Fenugreek
    C] Pointed gourd   D] Colocasia

13. ' Veneer grafting' in mango was first developed by-
    A] Traub & Auchter   B] Lynch (1941)
    (1934)
    C] R.N. Singh (1974)   D] S.K. Roy (1946)

14. ' Vikram' is an improved variety of-
    A] Lemon   B] Grape fruit
    C] Acid lime   D] Mandarin

15. 'Leaf curl virus' of tomato is transmitted by
    A] White fly   B] Nematode
    C] House fly   D] Butter fly

16. Which are water soluble vitamins-
    A] Vit-A and C   B] Vit-A and D
    C] Vit-B and C   D] Vit-B and E

17. A rich source of Vit-B_1 (Thiamine) is-
    A] Mango   B] Guava
    C] Cashew nut   D] Karonda

18. On the basis of Vit-C content which statement is true-
    A] Aonla > Chilli > Guava > Barbados cherry
    B] Wood apple > Aonla > Chilli > Guava
    C] Barbados cherry > Aonla > Chilli > Guava
    D] Guava > Bael > Citrus > Aonla

19. The term viviparous means
    A] Germination of seed in field
    B] Germination of seed in field
    C] Germination of seed on monoecious plant
    D] Germination of seed on undesirable place

20. Flower colour of *Petrea volubillis* is-
    A] White   B] Red
    C] Purple   D] Yellow

21. Taj garden was built by-
    A] Babar   B] Akbar
    C] Noorjehan   D] Shahjahan

22. Irish famine was caused by
    A] *Phytopthora infestans* B] *Alternaria solani*
    C] *Fusarium oxysporum*   D] None

23. Term genetics was given by
    A] Mendal   B] Bateson
    C] Punnet   D] None

24. Dwarf wheat was introduced from
    A] Brazil   B] Turkey
    C] Mexico   D] USA

25. First KVK was started at
    A] Nilokhari   B] Rajasthan
    C] Pondichery   D] None

26. Growing of vegetables in water is known as-
    A] Water harvesting   B] Hydrology
    C] Hydroponics   D] Floating garden

27. TSS of watermelon is
    A] 10-12° Brix   B] 14-16° Brix
    C] 18-22° Brix   D] 20-22° Brix

28. European carrot varieties seeds well in-
    A] Plains   B] Hills
    C] Anywhere   D] No seed formation

29. Origin of wheat is
    A] Mexico   B] Turkey
    C] Africa   D] India

30. The crop which can be grown in acidic soils
    A] Spinach   B] Cauliflower
    C] Watermelon   D] French Bean

31. Seed of which crop does not have endosperm
    A] Rose   B] Orchid
    C] Carnation   D] Gerbera

32. 'Extract' of which crop is used as insecticide-
    A] Turmeric   B] Garlic
    C] Onion   D] Chilli

Compound leaf
A leaf of two or more leaflets, in some cases (Citrus) the lateral leaflets may have been lost and only the terminal leaflet remain.

Amphidiploid
A species of a type of plant derived from doubling the chromosome in the F_1 hybrid of two species, an allopolyploid. In an amphidiploid, the two species are known, whereas in other allopolyploids they may not be known.

Quadrat
A small field study unit or sample area, usually a square metre or a milliacre size, established for the purpose of detailed observation.

Syncarp
An aggregate fruit with united carpels such as the mulberry and pineapple.

Transplanting
The process of digging up a plant and moving it to another location.

Ovule
The egg-containing unit of an ovary, which after fertilization becomes the seed.

Hotbed
A propagation structure made up of large wood box or frame with a sloping and tight-fitting glass lid and provision of heat generation below the propagating medium. Seedlings can be started and leafy cuttings rooted in the hot beds early in the season when outside temperature is low.

33. Productivity of which fruit crop is maximum in India
A] Mango   B] Pineapple
C] Banana   D] Grape

34. A jelly in which fruit peels are suspended is known as-
A] Jam   B] Marmalade
C] Ketchup   D] Cocktail

35. In India which fruit tops in production
A] Mango   B] Banana
C] Apple   D] Citrus

36. The nutrient which improves fruit quality & giving rigidity to shoots & spur etc.
A] Phosphorous   B] Potassium
C] Nitrogen   D] Calcium

37. In anthurium which is cut flower spp
A] A. odarum   B] A. adreanum
C] Both   D] None

38. In mango which is dwarfing rootstock
A] Deshahari   B] Neelum
C] Alphonso   D] Vellacolumban

39. Flower crop which is grown by corm
A] Gladiolus   B] Iris
C] Canna   D] Amaryllis

40. The crop which is not transplanted
A] Tomato   B] Brinjal
C] Chilli   D] None of the above

41. In India which is the second rank pulse crop
A] Bengal gram   B] Pigeon pea
C] Lentil   D] Moong

42. In air layering, which layer of tissues is removed-
A] Xylem & Phloem   B] Phloem & Cambium
C] Xylem only   D] Pholem only

43. In litchi which part is edible
A] Endocarp   B] Epicarp
C] Mesocarp   D] Rind

44. Name of which plant is associated with an eminent person-
    A] Marigold   B] Dahlia
    C] Balsam   D] China rose

45. In freezing which principle is followed
    A] Condensation   B] Crystallization
    C] Drying   D] All

46. A plant which provides pollen grain for pollination in another plant is known as-
    A] Pollinizer   B] Pollinator
    C] Genotype   D] Donor

47. Fruit in which pulp is outgrowth of seed
    A] Litchi   B] Fig
    C] Loquat   D] Grape

48. In mustard which substance is present
    A] Allyl propyl   B] Diallyl disulphide
       disulphide
    C] Isothiocynate   D] None

49. Pummelo prefer which type of atmosphere-
    A] Dry   B] High humid
    C] Cool-dry   D] Arid-temperate

50. In mustard which type of incompatability exist
    A] Sporophytic   B] Gametophytic
    C] Both   D] None

51. ' Wild' style of gardening was discovered by-
    A] Lamark   B] Pasteur
    C] William Robinson   D] B.P. Pal

52. Which agency is associated with agricultural marketing-
    A] APEDA   B] NAFED
    C] NDDB   D] NABARD

53. Pathogen associated with Bengal Famine was related to
    A] Wheat   B] Rice
    C] Maize   D] Sugarcane

54. 'Transgenic Plants' with complete resistance to insects have been developed in
    A] Tomato   B] Potato
    C] Tobacco   D] All of them

Tenderometer

An instrument by which toughness of the seed coat and firmness of pulp is determined and is mostly used to determine the seed quality of pea where high value of tenderometer indicates low quality.

Vernalization

Cold-moist treatment applied to a seed, plant, or bulb to induce or hasten the development of the capacity for flowering.

Dividing

The process of splitting up plants, roots and all that have began to get bound together. This will make several plants from one plant, and usually should be done to mature perennials every 3 to 4 years.

55. Person related with Green Revolution in India is
    A] Dr. M.S. Swaminathan B] Dr. N.S. Randhawa
    C] Dr. B.P. Pal   D] Dr. R.S. Paroda

56. Nitrogenous fertilizer with highest nitrogen content
    A] Urea   B] Anhydrous ammonia
    C] Ammonium chloride   D] None

57. The shrub with scented flower is
    A] Jasmine   B] Nasturtium
    C] Both   D] None

58. Semi hard wood cutting is mostly used to propagate the following-
    A] Evergreen fruit trees   B] Vegetable crops
    C] Peach   D] Apple

59. Mango hybridization work in India was first started by-
    A] Cheema & Dhani   B] Burns & Prayag (1911)
       (1908)
    C] R.N. Singh (1925)   D] Woodhouse (1898)

60. Banana breeding work in India was first started in-
    A] 1925   B] 1938
    C] 1949   D] 1971

61. Guava improvement work in India was initiated during-
    A] 1902   B] 1907
    C] 1911   D] 1919

62. Apple improvement work in India was initiated during-
    A] 1956   B] 1962
    C] 1974   D] 1980

63. The systematic work on improvement of oil palm in India was started during-
    A] 1968   B] 1976
    C] 1982   D] 1986

64. Vikramarya is world's first tungro virus associated with -
    A] Maize   B] Rice
    C] Wheat   D] Tomato

65. Debt net worth ratio is
 A] Deffered liabilities/ B] Total cost/total benefit
 total assets
 C] Total benefit/total cost D] None

66. By crossing 'Hermaphrodite x Female' parents of papaya, the ratio of hermaphrodite and female will be-
 A] 1:1 B] 1:2
 C] 2:1 D] 5:3:2

67. Gynomonoecious means -
 A] All female flowers B] Female & bisexual flower
 C] Only male flower D] None

68. Litchi fruits ripen during -
 A] August-September B] October-November
 C] February-March D] June-July

69. Which shows alkaline reaction -
 A] Auxin B] Gibberellic acid
 C] Cytokinin D] Ethylene

70. IIVR is located at -
 A] Bangalore B] Hissar
 C] Varanasi D] Bareily

71. Botanically strawberry is a -
 A] Berry fruit B] Aggregate fruit
 C] Sorosis fruit D] Pome fruit

72. Which is correctly matched -
 A] Carrot-Fusiform B] Radish-Conical
 C] Beet-Napiform D] Potato-Corm

73. CFTRI is located at-
 A] Chennai B] Mysore
 C] Lucknow D] Chandigarh

74. Which is not correct -
 A] CITH, Srinagar B] CMRS, Lucknow
 C] ICAR, Lucknow D] CFTRI, Mysore

75. Choose the incorrect matching -
 A] Coriander-Seed Spice B] Saffron-Crosin
 C] Capsanthin-Turmeric D] Penniyur-2-Black pepper

Farm
An area of land used for agriculture either to raise crops or pasture or to maintain livestock.

Phytometer
A plant or plants used for measuring the physical factors of the habitat in terms of physiological activities.

Gypsum
Calcium sulfate, used to alter pH or the medium.

Gibberellic acid
Chemical compound used to break flower bud dormancy or stimulate shoot elongation.

Dark storage
A term used to describe the time foliage plants remains in darkness within shipping containers during transits or storage.

76. Botanically Arvi is a -
    A] Crom   B] Tuber
    C] Bulb   D] Rhizome

77. Betel vine -
    A] Dioecious   B] Monoecious
    C] Hermaphrodite   D] Andromonoecious

78. *Lagerstroemia indica* is a -
    A] Flowering climber B] Foliage tree
    C] Flowering shrub   D] Flowering annual

79. Which is incorrectly matched -
    A] Hollyhock-Annual B] Kachnar-Fruit tree
    C] Marigold-Annual   D] Poinsettia-Bush

80. Brinjal is originated from -
    A] India   B] South Africa
    C] China   D] Europe

81. Apomixis is found in -
    A] Jamun   B] Guava
    C] Loquat   D] Bottle gourd

82. Polyembryony is found in -
    A] Jamun   B] Mango
    C] Citrus   D] All of these

83. Litchi fruits ripen during -
    A] January-February B] February-March
    C] April-May   D] May-June

84. *Poinciana regia* flowers during -
    A] January   B] May
    C] August   D] October

85. 'Onion' contains -
    A] Na   B] S
    C] P   D] Mo

86. Select the incorrect pair -
    A] Oleoresin- Capsicum B] Essential oil-Nutmeg
    C] *Attar*-Rose   D] Pyrethrum-Chrysanthemum

87. Which pair is incorrect -
    A] Pepper-*Capsicum annuum*   B] Mint-Menthol
    C] Ginger-Rhizome   D] Nutmeg-Tree spice

88. Which is matched incorrectly -
    A] Shield budding-Aonla   B] King of fruits-Mango
    C] Polyembryony-Citrus   D] Inarching-Guava

89. Strawberry is propagated by -
    A] Shield budding   B] Rhizome
    C] Stolon   D] Stooling

90. During snowfall, plants -
    A] Show minimum life function   B] Show maximum transpiration
    C] Become more active   D] Do not respire

91. Who discovered the Vitamin -
    A] Summer   B] Funk
    C] Lunin   D] Sangour

92. The costliest seeds are -
    A] Truthful seeds   B] Certified seeds
    C] Foundation seeds   D] Breeder seeds

93. *Colocasia esulants* is a -
    A] Tuber   B] Bulb
    C] Rhizome   D] Crom

94. Auxins translocate through-
    A] Xylem   B] Cambium
    C] Phloem   D] Root

Matching

95. Match the following:
    A] Yeast   i] 88°C
    B] Mould   ii] 66°C
    C] Spore forming Bacteria   iii] 79°C
    D] Blanching   iv] 82°C
    E] Pasteurization of Juice   v] 100°C

Saprophyte
An organism that uses dead organic material for food.

Miticide
Pesticide that is used to control mites.

Vpm
Vapor parts million used to measure gases.

Damping off
A fungus, usually affecting seedlings and causes the stem to rot off at soil level. Sterilized potting soil and careful sanitation practices usually prevent this.

Meadow
A grassland, occurring usually in a moist shady depression.

Class
A group that includes varieties of similar magnitude.

96. Match the following:
    A] Blossom endrot   i] Beet root
    B] Riceyness   ii] Cucumber
    C] Forking   iii] Cauliflower
    D] Bottleneck   iv] Tomato
    E] Brown Heart   v] Carrot

97. Match the following:
    A] Change scion   i] Rose
    B] T budding   ii] Guava
    C] Approach grafting   iii] Top working
    D] Mound layering   iv] Inarching
    E] Repair of   v] Bridge grafting stem
        unproductive

98. Match the following:
    A] Chrysanthemum   i] *Cosmos sulphereus*
    B] Jasmine   ii] *Rosa spp.*
    C] African marigold   iii] *Chsysanthamum indicum*
    D] Cosmos   iv] *Tagetes erecta*
    E] Rose   v] *Jasminum spp.*

99. Match the following:
    A] Clove   i] Cucurmine
    B] Pepper   ii] Cineol
    C] Ginger   iii] Eugenol
    D] Turmeric   iv] Piperin
    E] Cardamom   v] Ginzebarine

100. Match the following:
    A] Diffenbachia   i] Fern
    B] Mamillaria   ii] Climber
    C] Nephrolepsis   iii] Cactus
    D] Clitoria   iv] Succulent
    E] Euphorbia   v] House plant

101. Match the following:
    A] Duranta   i] Preservation
    B] Respiration   ii] Repetition
    C] Pansy   iii] Fruit
    D] Phalsa   iv] Hedge
    E] Pickle   v] Annual

102. Match the following:
    A] Enterprise
    B] Marketing
    C] Regulated market
    D] Demand
    E] Optimal money

    i] Fixed prices
    ii] Elasticity
    iii] Complementary
    iv] Function
    v] Bank cheques

103. Match the following:
    A] GA$_3$
    B] DNA
    C] Metasystox
    D] Glucose
    E] Cellulose

    i] Cell wall
    ii] Nucleic acid
    iii] Sugar
    iv] Insecticide
    v] Growth regulator

104. Match the following:
    A] Guava
    B] Papaya
    C] Pineapple
    D] Mango
    E] Grape

    i] South East Asia
    ii] Caspian region
    iii] Mexico
    iv] Tropical America
    v] Brazil

105. Match the following:
    A] IIHR
    B] IISR
    C] IARI
    D] IVRI
    E] CPCRI

    i] New Delhi
    ii] Kasargod
    iii] Bangalore
    iv] Calicut
    v] Varanasi

106. Match the following:
    A] Indes
    B] Cabbage
    C] Hindustan
    D] New world
    E] *Allium cepa*

    i] Brinjal
    ii] Capsicum
    iii] Cole crops
    iv] Onion
    v] Mountain range

107. Match the following:
    A] Kiwi
    B] Begonia
    C] Brussels sprout
    D] Mexican
    E] Cumin

    i] Spice
    ii] Fruit
    iii] Ornamental
    iv] Wheat
    v] Vegetable

Swamp
A completely or partially vegetated area subject to permanent inundation in stagnant or slowly flowing water.

Asexual propagation
Reproduction by vegetative means, such as cuttings, division.

Vegetative
referring to non-reproductive structures or growth

Fissured
Torn lengthwise, in vertical furrows, as applied to bark, or to pith, for which the more general term spongy is used.

Suspension
A liquid in which very fine solid particles are dispersed but not dissolved.

108. Match the following:
    A] Litchi   i] Slips
    B] Grape   ii] Suckers
    C] Banana   iii] Hardwood cutting
    D] Mango   iv] Air layering
    E] Pineapple   v] Veneer Grafting

109. Match the following:
    A] Mango   i] China
    B] Banana   ii] Europe
    C] Citrus   iii] Malaysia
    D] Pomegranate   iv] Indo-Burma
    E] Pear   v] Iran

110. Match the following:
    A] Marigold   i] Terminal cutting
    B] Tuberose   ii] Corm
    C] Carnation   iii] Bulb
    D] Canna   iv] Seed
    E] Gladiolus   v] Rhizome

111. Match the following:
    A] Millardat   i] Athrecea
    B] Elbert Mayer   ii] Aldicarb
    C] Granules   iii] Etawah project
    D] Silk   iv] B. M.
    E] Earthworm   v] Compost

112. Match the following:
    A] Peach   i] Leafy vegetable
    B] Swiss chord   ii] Stone fruit
    C] Chick pea   iii] Legume
    D] Wheat   iv] Sugar crop
    E] Sugarbeet   v] Cereal

113. Match the following:
    A] Pummelo   i] *Citrus grandis*
    B] Grape fruit   ii] *Citrus paradii*
    C] Sweet lime   iii] *Citrus limitoides*
    D] Orange   iv] *Citrus reticulata*
    E] Sweet orange   v] *Citrus sinensis*

114. Match the following:
     A] Pusa purple long   i] Bio pesticide
     B] Neem   ii] Brinjal
     C] Arka jay   iii] Dolichos bean
     D] Roma   iv] Watermelon
     E] 4D2009   v] Tomato

115. Match the following:
     A] Tikka disease   i] Tomato
     B] Whiptail   ii] Sugarcane
     C] Blossom end rot   iii] Groundnut
     D] Root rot   iv] Mango
     E] Spongy tissue   v] Cauliflower

116. Match the following:
     A] 4n   i] Cell organelle
     B] F_l ii] Pureline
     C] Fixed   iii] Hybrid
     D] Allosome   iv] Tetraploid
     E] Mitocondria   v] Sex chromosome

True / False

117. M-27 is a super dwarf root stock of wood apple.   T/F

118. M-20 root stock of apple is suitable for high density planting.   T/F

119. Snowball-16 cauliflower seeds well in Rajasthan.   T/F

120. F-12/1 rootstock is clonal selection of Mazzard.   T/F

121. In India, economic life of tea bush is 40-50 years.   T/F

122. Punjab NR-7 is a variety of tomato.   T/F

123. For planting, offshoot weight of datepalm should be 10-15kg.   T/F

124. H.T. Rose takes 42-45 days for flowering after pruning.   T/F

125. Arecanut takes 6-8 years for bearing.   T/F

126. Citrus seeds stand viability for 1-3 weeks only.   T/F

127. Seed rate of onion for direct sowing is 8-10 kg/ha.   T/F

128. Okra can be stored best at 8-9°C temperature.   T/F

Puberulent
Minutely hairy, when viewed with a hand lens.

Erosion
The wearing away, washing away, or removal of soil by wind, water or man.

Bulk density of soil
Weight per unit volume, such as grams per cubic centimeter.

Rootbound
A condition which exists when a potted plant has outgrown its container. The roots become entangled and matted together, and the growth of the plant becomes stunted. When repotting, loosen the roots on the outer edges of the root ball, to induce them to once again grow outward.

Volatile

A compound is volatile when it evaporates or vaporizes at ordinary temperatures on exposure to the air.

Slipping stage

The time when the inflorescence emerges for the leaf sheaths.

Determinate

Said of an inflorescence when the terminal flower opens first and the prolongation of the axis is thereby arrested.

Graft union

the region where rootstock and scion come together; there can be slightly deformed growth at the union that is noticeable, but does not effect the function of the tree

129. About 10% fruits are processed in India.   T/F

130. Hexagonal system accommodates 15% more plants over square system.   T/F

131. Hexagonal system accommodates about 25% more plants.   T/F

132. Shelling per cent of pea is about 45%.   T/F

133. In high density orcharding, a mango orchard accommodates about 2000 plants/ha.   T/F

134. 'Head space' in cans should not be less than 2.5 cm.   T/F

135. For better fruiting, C:N ratio should be 3:2.   T/F

136. The ratio of length & width of orchard should be 2:3.   T/F

137. For proper growth & fruiting, C/N ratio should be 2:3.   T/F

138. Bael takes 11 months to ripen on the tree.   T/F

139. Papaya seeds remain viable for 4 years at 10°C.   T/F

140. Ripe banana peels constitute 22-30% of whole fruit.   T/F

141. Tomato develops better colour at temp. upto 29 °C.   T/F

142. The most of the garden rose varieties are tetraploid (4x=28).   T/F

143. 'Duke cherry' is a tetraploid (4x=32) cherry.   T/F

144. High temperature (>38 °C) causes 'blank nuts' in walnut.   T/F

145. Drip irrigation saves water about 40-60%.   T/F

146. Most fruits & vegetables have 40-60% water.   T/F

147. Red colour of tomato becomes dark red at 40°C.   T/F

148. Rancidity in walnut kernels occurs due to very high temp. (>43 °C).   T/F

149. Mango contains about 4800mg Vit-A/100g pulp.   T/F

150. World's 51 per cent sugar is manufactured from sugarbeet.   T/F

151. *Areca catechu* prefers temperature of 15°C to 38°C.   T/F

152. Root vegetables should be washed with 25 to 50 ppm chlorine.   T/F

153. Cayenne is a triploid (3x=75) variety of pineapple.   T/F
154. *In-situ* budding in ber should be done in 90 days old seedling.   T/F
155. Asparagus can be stored best at 10°C.   T/F
156. Spray of 200 ppm NAA in Oct.-Nov. controls mango malformation.   T/F
157. From seed sowing to bulb production, onion takes about 225 days.   T/F
158. Processing temperature for vegetables is 115 to 121 °C.   T/F
159. Hybrid vigour was first time observed in tomato in 1908.   T/F
160. Anab-e-Shahi grape was discovered by R.S. Pillay in 1930.   T/F
161. In India jam, jelly and squash preparation was started in 1930.   T/F
162. R.R.I.I. (Kottayam) was established in 1955.   T/F
163. Tea Research Association was established in 1964.   T/F
164. First hybrid of tomato was available for cultivation in 1973.   T/F
165. There are 4000 plants/ha of banana in 'intensity orcharding'.   T/F
166. Average yield of coconut is 10000 to 14000 nuts/ha/year.   T/F

Fill in the Blanks

167. An uninjured garlic clove contains _____ amino acid.
168. _____ is commercial method of pomegranate propagation.
169. *Eriobotrya japonica* is the native of _____.
170. _____ is the largest producer of pear in the world.
171. Generally fruit crops prefer a pH range of _____.
172. _____ is the largest producer of walnut in India.
173. *Artocarpus heterophyllus* is the native of _____.

Nursery bed
A prepared area where seed is sown or into which transplants or cuttings are raised for sowing or planting in gardens or fields. In other words the nursery industry involves the production and distribution of different kinds of planting materials.

Articulate
Having nodes or joints where separation may naturally occur.

Cymose
A type of branching where the growth of the main stem is definite and the lateral branches grow more vigorously than the terminal ones giving the plant a dome-shaped appearance.

174. Mango has _____ somatic chromosome numbers.
175. Seed rate of cauliflower is _____.
176. Pant C-1 chilli is tolerant to _____ virus.
177. Dwarf banana should be planted at _____ distance.
178. _____ cultivar of mango is suitable for high density planting.
179. _____ species of *Jasminum* is suitable for perfumery.
180. _____ method is the most common system of planting.
181. Potato can be stored at 38°- 40°F F at _____ R.H. for 30-45 days.
182. _____ plants provide viable pollen grains to self-incompatible varieties of a fruit crop.
183. Chilling requirement of apple at 7.2°C is _____ hours.
184. _____ requires the warmest climate among temperate fruits.
185. *Pterospermum acerifolium* is the botanical name of _____.
186. *Duranta plumerei* is a suitable shrub for _____ making.
187. Jamun is the rich source of _____ hormone.
188. *Lagenaria siceraria* is the botanical name of _____.
189. For sugar patients _____ gourd is the best remedy.
190. Headquarters of N.H.B. is located at _____.
191. CAZRI is located at _____.
192. Stone weevil is a serious pest of _____.
193. Grape requires judicious _____ for better fruiting.
194. _____ is the edible part of Fenugreek.
195. Cracking of tomato occurs due to _____ deficiency.
196. Jam contains _____ T.S.S. at end point.
197. *Solanum melongena* is the native of _____.
198. Arka Navneet is a hybrid variety of _____.
199. Subjecting fruits/vegetables to boiling water for few minutes is known as _____.
200. Removal of outer covering of fruits is termed as _____.

Step One

Multiple Choice

1. B	17. D	33. A	49. D	65. B	81. C
2. B	18. B	34. B	50. A	66. B	82. D
3. D	19. B	35. A	51. A	67. B	83. B
4. A	20. C	36. A	52. A	68. D	84. C
5. C	21. A	37. D	53. C	69. A	85. B
6. C	22. A	38. B	54. B	70. D	86. A
7. A	23. C	39. A	55. D	71. B	87. B
8. C	24. B	40. A	56. B	72. A	88. B
9. C	25. B	41. C	57. B	73. B	89. B
10. C	26. B	42. B	58. A	74. C	90. A
11. A	27. D	43. B	59. B	75. A	91. C
12. B	28. D	44. C	60. B	76. A	92. D
13. C	29. C	45. B	61. B	77. D	93. D
14. D	30. D	46. C	62. B	78. A	94. A
15. B	31. A	47. C	63. C	79. C	
16. B	32. D	48. A	64. A	80. B	

Matching

95. A ii B iv C i D v E iii
96. A ii B v C iv D i E iii
97. A ii B v C ii D i E iv
98. A v B iii C i D ii E iv
99. A ii B iii C i D v E iv
100. A iv B i C v D ii E iii
101. A iv B ii C i D v E iii
102. A ii B iv C i D v E iii
103. A v B viii C i D vii E ii F vi G iii H x I iv J ix
104. A ix B v C i D ii E x F iii G iv H vi I vii J viii
105. A v B iv C i D ii E iii
106. A iv B i C v D ii E iii
107. A iv B i C iii D v E ii
108. A iv B i C ii D v E iii

109. A iv B i   C iii D ii E v
110. A ii B iv C i   D v   E iii
111. A ii B i   C iv D iii E v
112. A iv B i   C v   D iii E ii
113. A ii B iv C i   D v   E iii
114. A v   B iii C i   D iv E ii
115. A iii B v   C i   D ii E iv
116. A ii B iv C i   D v   E iii

True / False

117. T	126. T	135. T	144. T	153. T	162. T
118. T	127. T	136. F	145. F	154. T	163. F
119. T	128. T	137. T	146. F	155. F	164. T
120. F	129. F	138. T	147. F	156. T	165. T
121. F	130. F	139. F	148. F	157. F	166. F
122. T	131. T	140. F	149. T	158. T	
123. T	132. F	141. T	150. T	159. F	
124. T	133. T	142. T	151. T	160. T	
125. F	134. F	143. F	152. T	161. F	

Fill in the Blanks

167. Parthenocarpic
168. Banana
169. Pineapple
170. Leaves
171. Root
172. Andromonoecious
173. Gynoecious
174. Rose
175. Carrot
176. Roots
177. Layering
178. Dioecious

179. Pectin
180. Vit-C
181. Vit-B
182. Mango
183. Suckers
184. Papaya
185. Top working
186. Monoecious
187. Vit-A
188. Mango
189. Papaya

190. Preservation
191. Enzyme
192. Papaya
193. Stem cutting
194. Stem
195. Veneer grafting
196. Late blight
197. Aonla
198. Banana
199. 300mg
200. *Vitis*

Step Two

Multiple Choice

1.	A	17.	C	33.	B	49.	C	65.	C	81.	A
2.	D	18.	B	34.	C	50.	C	66.	B	82.	A
3.	C	19.	A	35.	B	51.	C	67.	B	83.	B
4.	D	20.	C	36.	D	52.	C	68.	D	84.	C
5.	A	21.	C	37.	A	53.	A	69.	B	85.	C
6.	B	22.	C	38.	A	54.	C	70.	B	86.	C
7.	B	23.	C	39.	A	55.	B	71.	C	87.	C
8.	C	24.	C	40.	C	56.	A	72.	D	88.	B
9.	A	25.	A	41.	B	57.	C	73.	B	89.	C
10.	B	26.	C	42.	C	58.	A	74.	B	90.	C
11.	A	27.	B	43.	B	59.	A	75.	C	91.	D
12.	D	28.	B	44.	D	60.	B	76.	B	92.	D
13.	A	29.	C	45.	B	61.	B	77.	B	93.	A
14.	B	30.	B	46.	C	62.	B	78.	C	94.	C
15.	A	31.	C	47.	C	63.	B	79.	B		
16.	B	32.	A	48.	D	64.	B	80.	C		

Matching

95. A v B i C iv D ii E iii
96. A iv B i C vi D ii E vii F iii G x H ix i viii J v
97. A ii B v C i D iv E vii F iii G ix H vi i x J viii
98. A iii B v C i D vii E viii F x G ii H ix i vi J iv
99. A vi B viii C i D ii E iv F vii G v H x i iii J ix
100. A iii B v C i D viii E ii F vii G iv H x i ix J vi
101. A iv B i C iii D v E ii
102. A iv B ii C v D iii E i
103. A iv B i C iii D v E ii
104. A viii B v C i D ix E vii F ii G vi H iii i iv J ix
105. A iii B v C i D vii E iv F ix G vi H x i viii J ii
106. A ii B iv C i D v E iii
107. A iii B v C vii D i E viii F iv G vi H x i ii J ix
108. A iii B v C i D iv E ii

109. A v B vii C ix D i E x F viii G ii H iii i iv J vi
110. A v B iii C vii D viii E i F ix G vi H ii i x J iv
111. A i B iii C ii D iv E v
112. A ii B iii C v D iv E i
113. A ii B i C iv D iii E v
114. A ii B v C i D iii E iv
115. A i B ii C v D iii E iv
116. A ii B v C iii D i E iv

True / False

117. T	126. T	135. T	144. T	153. T	162. T
118. T	127. F	136. F	145. F	154. F	163. F
119. T	128. T	137. T	146. T	155. T	164. T
120. T	129. T	138. T	147. T	156. F	165. T
121. F	130. F	139. F	148. T	157. F	166. F
122. T	131. T	140. T	149. F	158. F	
123. T	132. T	141. T	150. T	159. F	
124. F	133. T	142. F	151. T	160. F	
125. F	134. T	143. T	152. F	161. F	

Fill in the Blanks

167. Apple	179. Brinjal	190. Bael
168. Okra	180. Jelly	191. Lime
169. Rose	181. Taiwan	192. Mango
170. Short day	182. Datepalm	193. Carrot
171. Aonla	183. Rose	194. Litchi
172. Cauliflower	184. MLOs	195. Added colour
173. Varanasi	185. Cauliflower	196. Rose
174. Jammu & Kashmir	186. Kinnow	197. Muskmelon
175. Rhizome	187. Marmalade	198. Citrus
176. Assam	188. Tomato	199. CCC:NN
177. Kochia	189. Ber	200. 15%
178. Datepalm		

Step Three

Multiple Choice

1.	A	17.	A	33.	D	49.	C	65.	D	81.	B
2.	D	18.	A	34.	C	50.	C	66.	A	82.	B
3.	A	19.	B	35.	C	51.	B	67.	A	83.	B
4.	A	20.	A	36.	C	52.	C	68.	A	84.	C
5.	D	21.	D	37.	B	53.	A	69.	A	85.	B
6.	C	22.	C	38.	C	54.	C	70.	B	86.	A
7.	D	23.	A	39.	C	55.	C	71.	C	87.	B
8.	D	24.	B	40.	D	56.	C	72.	B	88.	B
9.	C	25.	C	41.	D	57.	B	73.	C	89.	B
10.	C	26.	D	42.	C	58.	C	74.	A	90.	A
11.	C	27.	D	43.	B	59.	B	75.	D	91.	D
12.	A	28.	D	44.	B	60.	B	76.	A	92.	B
13.	C	29.	D	45.	B	61.	C	77.	B	93.	A
14.	C	30.	B	46.	A	62.	A	78.	D	94.	A
15.	D	31.	A	47.	B	63.	C	79.	A		
16.	D	32.	A	48.	A	64.	D	80.	D		

Matching

95. A ii B iv C iii D v E i
96. A iii B v C i D ii E iv
97. A ii B v C i D iii E iv
98. A iv B i C v D ii E iii
99. A iii B i C vi D iv E vii F ii G x H v I viii J ix
100. A iL B iv C i D v E iii
101. A ii B iv C v D i E iii
102. A iii B i C v D ii E iv
103. A v B i C iv D ii E iii
104. B A ii B iv C iii D v E i
105. A iv B i C v D ii E iii
106. A iii B i C v D vii E ii F ix G iv H x I viii J vi
107. A iii B v C iv D ii E i
108. A iv B vi C ix D i E vii F ii G iii H x I v J viii

109. A i   B iiL C ii   D v   E iv
110. A iii B v   C i   D ii   E iv
111. A v   B iii C i   D iv   E ii
112. A iv   B vi C x   D i   E vii F ix   G ii   H v   I viii J iii
113. A ii   B vi C iv   D vii E ix   F x   G viii H v   I i   J iii
114. A iii B i   C v   D vii E ix   F viii G x   H vi   I iv   J ii
115. A iv   B i   C v   D iii E ii
116. A ii   B iv C iii D v   E i

True / False

117. T	126. F	135. F	144. T	153. F	162. T
118. T	127. F	136. T	145. T	154. T	163. F
119. F	128. F	137. F	146. T	155. T	164. T
120. F	129. T	138. T	147. T	156. T	165. T
121. T	130. T	139. T	148. T	157. F	166. F
122. T	131. T	140. F	149. T	158. T	
123. F	132. F	141. F	150. F	159. T	
124. F	133. F	142. F	151. T	160. T	
125. F	134. T	143. T	152. F	161. F	

Fill in the Blanks

167. Mango	179. Guava	190. Coconut
168. Lime	180. Tomato	191. Stem
169. Papaya	181. Pothos	192. Papaya
170. Oct	182. Hexagonal	193. Pumpkin
171. Umran	183. Winter	194. Peach
172. Sucker	184. Bougainvillia	195. Winter
173. M-9	185. GA3	196. Rose
174. Hazelnut	186. Grape	197. Vit-A
175. Snowball	187. Ber	198. Grape
176. PPL	188. Honey suckle	199. Thiourea
177. Muskmelon	189. Vegetative parthenocarpy	200. CCC
178. Mango		

Step Four

Multiple Choice

1.	A	17.	C	33.	C	49.	C	65.	A	81.	C
2.	A	18.	B	34.	D	50.	A	66.	A	82.	C
3.	A	19.	A	35.	D	51.	C	67.	B	83.	B
4.	D	20.	D	36.	A	52.	A	68.	A	84.	B
5.	A	21.	D	37.	B	53.	D	69.	C	85.	A
6.	A	22.	A	38.	C	54.	C	70.	B	86.	C
7.	D	23.	C	39.	A	55.	C	71.	C	87.	C
8.	B	24.	C	40.	B	56.	D	72.	C	88.	B
9.	A	25.	B	41.	A	57.	C	73.	B	89.	A
10.	B	26.	D	42.	A	58.	B	74.	B	90.	C
11.	A	27.	B	43.	D	59.	B	75.	A	91.	B
12.	C	28.	B	44.	C	60.	B	76.	C	92.	A
13.	B	29.	B	45.	B	61.	A	77.	C	93.	C
14.	C	30.	B	46.	A	62.	B	78.	A	94.	A
15.	D	31.	B	47.	A	63.	B	79.	D		
16.	A	32.	B	48.	C	64.	B	80.	B		

Matching

95. A iv B v C i D ii E iii F viii G ix H vi I x J vii
96. A ii B iv C i D iii E v
97. A v B vii C ix D i E vi F iii G viii H ii I x J iv
98. A iv B i C iii D v E ii
99. A ii B iv C i D v E iii
100. A iii B v C i D iv E ii
101. A iii B v C i D ii E iv
102. A iv B i C v D ii E iii
103. A ii B iv C v D i E iii
104. A iv B ii C i D v E iii
105. A v B ii C iv D iii E i
106. A iv B i C iii D v E ii
107. A v B i C Jv D ii E iii
108. A iv B i C vi D ii E iii F ix G v H x I vii J viii

109. A ii B iii C v D iv E i
110. A vi B ix C i D iii E vii F iv G x H v I viii J ii
111. A iii B i C v D ii E iv
112. A i B iii C v D vii E x F ix G iv H vi I viii J ii
113. A ii B iv C iii D i E v
114. A iii B i C iv D v E ii
115. A iii B iv C i D v E ii
116. A x B i C vi D viii E ii F vii G iii H iv I v J ix

True / False

117. T	126. F	135. T	144. T	153. T	162. T
118. T	127. T	136. T	145. F	154. T	163. T
119. F	128. T	137. F	146. T	155. T	164. T
120. F	129. F	138. T	147. T	156. F	165. T
121. T	130. F	139. T	148. T	157. T	166. F
122. T	131. T	140. F	149. T	158. F	
123. T	132. T	141. F	150. T	159. T	
124. F	133. T	142. T	151. T	160. T	
125. T	134. T	143. T	152. T	161. T	

Fill in the Blanks

167. Ethylene
168. Citrus
169. T-budding
170. Queens flower
171. NaOH
172. Watemelon
173. Embryo abortion
174. Cauliflower
175. I.B.A. ─
176. Ethylene
177. Olour
178. K.M.S.

179. Pusa Reshmi
180. Crown grafting
181. Tomato
182. Kamini
183. Amrapalli
184. Day neutral
185. GA
186. Persimmon
187. C
188. Delonix
189. Sword Sucker
190. Date

191. Rosaceae
192. Ethrel
193. Mango
194. Virus
195. Malvaceae
196. ICAR
197. ICAR
198. Citrus
199. Jack fruit
200. Banana

Step Five

Multiple Choice

1.	C	17.	B	33.	C	49.	B	65.	D	81.	A
2.	B	18.	D	34.	B	50.	A	66.	B	82.	B
3.	C	19.	D	35.	A	51.	C	67.	B	83.	C
4.	C	20.	C	36.	C	52.	C	68.	B	84.	B
5.	B	21.	B	37.	B	53.	C	69.	A	85.	B
6.	B	22.	D	38.	D	54.	A	70.	B	86.	C
7.	C	23.	B	39.	B	55.	C	71.	C	87.	C
8.	C	24.	D	40.	B	56.	D	72.	B	88.	C
9.	B	25.	C	41.	C	57.	A	73.	B	89.	B
10.	B	26.	C	42.	A	58.	C	74.	B	90.	C
11.	D	27.	B	43.	B	59.	B	75.	C	91.	D
12.	A	28.	B	44.	A	60.	C	76.	B	92.	A
13.	D	29.	C	45.	D	61.	B	77.	B	93.	C
14.	C	30.	B	46.	A	62.	D	78.	D	94.	A
15.	A	31.	D	47.	A	63.	B	79.	C		
16.	C	32.	D	48.	B	64.	C	80.	B		

Matching

95. A iii B vii C ix D vi E x F i G iv H ii I v J viii
96. A ii B iv C iii D v E i
97. A iv B i C iii D v E ii
98. A iv B i C iii D v E ii
99. A ii B iv C i D v E iii
100. A ii B iv C i D v E iii
101. A iii B v C iv D i E ii
102. A ii B iii C i D iv E v
103. A ii B i C iv D v E iii
104. A iii B vi C i D ii E x F ix G iv H vii I v J viii
105. A iv B iii C i D ix E v F viii G iv H ii I x J vii
106. A v B i C iii D ii E iv
107. A iii B v C iv D ii E i
108. A ii B iii C v D iv E i

109. A i   B ii   C iii D v   E iv
110. A ii   B iv   C i   D v   E iii
111. A ii   B iv   C i   D iii E v
112. A iv   B vi   C viii D i   E iii F x G ii H ix I vii J v
113. A i   B iii C v   D ii   E iv
114. A iv   B Y   C iii D i   E ii
115. A iii B v   C i   D iv   E ii
116. A i   B iv   C ii   D v   E iii

True / False

117. F	126. T	135. T	144. T	153. T	162. T
118. T	127. T	136. T	145. F	154. F	163. T
119. T	128. T	137. F	146. F	155. F	164. F
120. T	129. F	138. T	147. F	156. F	165. T
121. F	130. T	139. T	148. T	157. T	166. T
122. T	131. T	140. T	149. F	158. F	
123. F	132. T	141. T	150. T	159. T	
124. F	133. F	142. F	151. F	160. T	
125. F	134. F	143. F	152. T	161. T	

Fill in the Blanks

167. Apomictic
168. Pineapple
169. Monoecious
170. Rose
171. Citrus
172. Grape
173. Neelam
174. Mango
175. IBA
176. Ethrel
177. Banana
178. Banana

179. Humid coastal
180. Phalsa
181. Guava
182. Banana
183. Slips
184. Cauliflower
185. Apple
186. GA_3
187. Leaf curl
188. Corm
189. Andromonoecious

190. Late
191. Hybrid
192. Pea
193. Bacterial
194. S
195. Banana
196. Beri-beri
197. Algal
198. Primary
199. Hexagonal
200. Inarching

Step Six

Multiple Choice

1.	A	17.	C	33.	D	49.	C	65.	A	81.	A
2.	A	18.	B	34.	D	50.	B	66.	A	82.	A
3.	A	19.	C	35.	D	51.	C	67.	B	83.	B
4.	B	20.	D	36.	D	52.	A	68.	C	84.	C
5.	A	21.	B	37.	D	53.	A	69.	C	85.	D
6.	C	22.	D	38.	B	54.	A	70.	B	86.	A
7.	C	23.	D	39.	A	55.	B	71.	B	87.	B
8.	B	24.	C	40.	C	56.	B	72.	B	88.	C
9.	C	25.	B	41.	A	57.	C	73.	A	89.	C
10.	B	26.	A	42.	A	58.	A	74.	C	90.	C
11.	D	27.	A	43.	D	59.	C	75.	C	91.	C
12.	A	28.	A	44.	D	60.	D	76.	B	92.	B
13.	A	29.	B	45.	D	61.	C	77.	B	93.	A
14.	A	30.	C	46.	A	62.	C	78.	D	94.	D
15.	D	31.	C	47.	C	63.	A	79.	A		
16.	B	32.	C	48.	A	64.	A	80.	B		

Matching

95. A iv B vii C ix D i E x F ii G v H iii I vi J viii
96. A iii B vi C i D vii E iv F ii G ix H x I viii J v
97. A iv B i C v D ii E iii
98. A ii B iv C i D v E iii
99. A iii B v C i D ii E iv
100. A ii B iv C i D v E iii
101. A iii B v C i D ii E iv
102. A iv B ii C v D iii E i
103. A iv B i C v D ii E iii
104. A iii B v C i D iv E ii
105. A v B x C vii D i E viii F ii G iv H ix I vi J iii
106. A iii B v C ii D iv E i
107. A ii B v C iv D iii E i
108. A iii B i C vi D ii E viii F iv G ix H vii I x J v

109. A ii B iv C iii D v E i
110. A ii B iv C i D v E iii
111. A iv B i C v D ii E iii
112. A v B iv C ii D iii E i
113. A ii B iv C iii D v E i
114. A x B v C viii D i E ii F vii G iii H ix I iv J vi
115. A iv B ii C v D i E iii
116. A iv B i C v D ii E iii

True / False

117. T	126. T	135. T	144. T	153. T	162. T
118. T	127. F	136. F	145. T	154. F	163. T
119. T	128. T	137. F	146. T	155. F	164. F
120. F	129. F	138. T	147. F	156. T	165. T
121. T	130. F	139. T	148. F	157. T	166. T
122. F	131. T	140. T	149. T	158. T	
123. T	132. T	141. F	150. T	159. T	
124. T	133. F	142. T	151. F	160. F	
125. T	134. T	143. T	152. F	161. T	

Fill in the Blanks

167. Triploids
168. Ratna
169. Deciduous
170. Temperate
171. Apple
172. Peach
173. Diploid
174. Apple
175. Self-fruitful
176. Peach
177. Cashewnut
178. Self-unfruitful

179. Papaya
180. Mid
181. Brazil
182. Califonia Advance
183. Hardwood
184. Barbados cherry
185. Apple
186. Ber
187. Datepalm
188. Pectin
189. Hybrid

190. Tristeza
191. Apple
192. Dashari x Neelam
193. Pome
194. Polygamous
195. Self-sterile
196. Monoecious
197. Triploid
198. 8-10 kg
199. 600- 750g
200. October

Step Seven

Multiple Choice

1.	D	17.	C	33.	C	49.	D	65.	B	81.	C
2.	C	18.	C	34.	C	50.	B	66.	B	82.	B
3.	D	19.	D	35.	B	51.	A	67.	A	83.	B
4.	D	20.	B	36.	B	52.	D	68.	C	84.	A
5.	C	21.	D	37.	A	53.	C	69.	A	85.	C
6.	C	22.	A	38.	A	54.	C	70.	B	86.	B
7.	D	23.	B	39.	D	55.	B	71.	A	87.	C
8.	B	24.	B	40.	B	56.	D	72.	A	88.	D
9.	A	25.	B	41.	D	57.	D	73.	B	89.	A
10.	A	26.	C	42.	A	58.	B	74.	D	90.	A
11.	C	27.	C	43.	C	59.	B	75.	C	91.	C
12.	C	28.	B	44.	A	60.	A	76.	A	92.	C
13.	C	29.	B	45.	C	61.	B	77.	C	93.	D
14.	B	30.	C	46.	B	62.	A	78.	C	94.	A
15.	D	31.	C	47.	A	63.	C	79.	A		
16.	D	32.	A	48.	A	64.	B	80.	D		

Matching

95. A ii B v C i D vii E iii F x G iv H ix I vi J vii
96. A iii B i C v D ii E iv
97. A iv B i C v D ii E iii
98. A viii B ii C vi D i E x F vii G v H ix I iv J ii
99. A iv B i C iii D v E ii
100. A iii B v C i D iv E ii
101. A v B vi C viii D ii E i F iii G ix H x I vii J iv
102. A ix B iv C vii D i E x F v G ii H vi I iii J viii
103. A ii B iv C v D i E iii
104. A iii B i C v D iv E ii
105. A iii B ii C v D i E iv
106. A v B iii C ii D i E iv
107. A iii B i C v D iv E ii
108. A v B iii C i D ii E iv

109. A iv B viii C ii D x E vi F i G v H iii I ix J vii
110. A ix B vi C i D v E ii F viii G iii H x I vii J iv
111. A iv B i C ii D Y E iii
112. A v B i C iii D ii E viii F iv G ix H x I vi J vii
113. A vi B viii C i D x E iii F v G ix H viii I iv J ii
114. A v B x C i D vii E ii F iii G ix H viii I vi J iv
115. A iii B v C i D vii E x F viii G ii H vi I iv J ix
116. A v B x C vii D i E ii F ix G iii H vi I iv J viii

True / False

117. F	126. T	135. F	144. T	153. T	162. T
118. T	127. F	136. T	145. T	154. F	163. T
119. F	128. T	137. T	146. T	155. T	164. T
120. T	129. T	138. T	147. T	156. T	165. T
121. F	130. F	139. T	148. F	157. F	166. F
122. F	131. T	140. T	149. F	158. T	
123. T	132. F	141. T	150. T	159. F	
124. T	133. T	142. F	151. T	160. F	
125. T	134. F	143. T	152. F	161. T	

Fill in the Blanks

167. December
168. Tuber
169. Tomato
170. Pusa Ruby
171. Cucurbitaceae
172. Alkaline
173. Cutflowers
174. Cassava
175. Soft
176. Fruits
177. *Fusarium*
178. White fly

179. *Rosa bourboniana*
180. H_2O
181. Apple
182. Berry
183. Peduncle
184. Banana
185. Protein
186. 33%
187. Recurrent
188. 600
189. Boron
190. Budding

191. Cane
192. Banana
193. 13°C
194. Polyembryonic seeds
195. Tribute
196. Monty
197. Croton
198. Nendran
199. Sorosis
200. Mulberry

Step Eight

Multiple Choice

1.	D	17.	D	33.	C	49.	A	65.	A	81.	B
2.	B	18.	C	34.	D	50.	B	66.	A	82.	C
3.	A	19.	C	35.	B	51.	B	67.	C	83.	A
4.	B	20.	D	36.	A	52.	A	68.	A	84.	D
5.	A	21.	B	37.	B	53.	C	69.	D	85.	D
6.	B	22.	D	38.	C	54.	B	70.	A	86.	C
7.	A	23.	C	39.	C	55.	D	71.	B	87.	A
8.	B	24.	B	40.	B	56.	A	72.	B	88.	D
9.	B	25.	B	41.	D	57.	A	73.	A	89.	C
10.	C	26.	C	42.	C	58.	B	74.	C	90.	C
11.	B	27.	A	43.	C	59.	C	75.	D	91.	D
12.	B	28.	C	44.	A	60.	A	76.	C	92.	A
13.	A	29.	A	45.	B	61.	A	77.	C	93.	A
14.	D	30.	C	46.	C	62.	B	78.	C	94.	C
15.	C	31.	D	47.	A	63.	B	79.	D		
16.	B	32.	D	48.	A	64.	A	80.	B		

Matching

95. A iii B i C vi D viii E ii F iv G ix H v I x J vii
96. A v B iii C i D iv E ii
97. A iv B i C v D ii E iii
98. A iii B v C i D ii E iv
99. C A iii B i C v D iv E ii
100. A vi B viii C x D i E ii F ix G iv H vii I v J iii
101. A iv B i C iii D v E ii
102. A iii B i C ii D iv E v
103. A x B vii C i D iv E ix F v G ii H vi I viii J iii
104. A iv B i C vi D viii E ii F ix G v H vii I iii J x
105. A vi B ix C i D vii E ii F viii G x H iii I v J iv
106. A v B vii C i D iii E ii F iv G ix H x I viii J vi
107. A iv B i C ii D v E iii
108. A vi B iv C viii D i E vii F iii G x H ii I v J ix

109. A ii B iv C i D v E iii
110. A v B i C iii D ii E iv
111. A v B iii C i D ii E iv
112. A iv B vii C ix D i E viii F ii G v H x I vi J iii
113. A iii B iv C i D v E ii
114. A ix B iv C ii D i E vi F viii G vii H x I v J iii
115. A iii B v C i D vii E x F viii G iv H vi I ix J ii
116. A iii B vi C v D ii E i F iv G vii H ix I viii J x

True / False

117. T	126. T	135. F	144. F	153. T	162. F
118. T	127. F	136. F	145. F	154. T	163. T
119. T	128. F	137. T	146. F	155. T	164. F
120. T	129. T	138. T	147. T	156. F	165. F
121. T	130. F	139. T	148. T	157. T	166. T
122. T	131. T	140. F	149. T	158. T	
123. F	132. T	141. F	150. F	159. T	
124. T	133. F	142. T	151. T	160. T	
125. F	134. F	143. T	152. F	161. T	

Fill in the Blanks

167. Dec-Jan
168. Twice
169. Tea
170. Salt tolerant
171. Water
172. Frost
173. Oil palm
174. First
175. Rangpur lime
176. Housefly
177. Quince-A
178. MLOs
179. Ambri
180. Trifoliate
181. July
182. Vegetative
183. Self-incompatibility
184. Hard
185. Seed
186. 44°C
187. Stimulative parthenocarpy
188. Litchi
189. Litchi
190. Papaya
191. K
192. Wild Rose
193. Acetylene
194. First
195. William Robinson
196. Neutral
197. Primary
198. Shahjehan
199. Mughal
200. Sikander Bagh

Step Nine

Multiple Choice

1. A	17. B	33. D	49. A	65. D	81. B
2. D	18. C	34. A	50. B	66. A	82. A
3. C	19. D	35. B	51. B	67. C	83. B
4. A	20. B	36. C	52. C	68. C	84. A
5. A	21. D	37. C	53. A	69. C	85. C
6. D	22. A	38. C	54. C	70. C	86. C
7. C	23. D	39. C	55. A	71. C	87. A
8. B	24. B	40. B	56. B	72. D	88. A
9. C	25. C	41. B	57. B	73. C	89. C
10. A	26. C	42. A	58. B	74. D	90. C
11. A	27. A	43. D	59. A	75. A	91. A
12. A	28. B	44. C	60. D	76. A	92. C
13. D	29. C	45. A	61. A	77. A	93. C
14. B	30. B	46. B	62. D	78. D	94. A
15. B	31. A	47. C	63. C	79. A	
16. C	32. A	48. B	64. D	80. C	

Matching

95. A i B ii C iii D iv E v
96. A ii B v C i D iv E iii
97. A v B iii C iv D ii E i
98. A iii B v C iv D L E ii
99. A iv B ii C v D iii E i
100. A iv B i C ii D v E iii
101. A iii B v C viii D vi E ix F i G ii H iv I vii J x
102. A ii B iv C v D iii E i
103. A v B iii C i D iv E ii
104. A v B iii C i D ii E iv
105. A iii B i C v D ii E iv
106. A ii B iv C i D v E iii
107. A iv B i C v D iii E ii
108. A v B i C iii D iii E ii

109. A iii B i C v D iv E ii
110. A ii B iv C i D iii E v
111. A ii B iv C v D i E iii
112. A iv B vi C i D vii E ii F viii G v H iii I ix J x
113. A ii B iv C iii D v E i
114. A i B ii C iv D v E iii
115. A iii B v C i D vi E ii F viii G x H iv I vii J ix
116. A iv B vii C i D ix E ii F iii G x H v I vi J viii

True / False

117. T	126. T	135. F	144. T	153. F	162. F
118. T	127. T	136. T	145. T	154. T	163. T
119. F	128. F	137. T	146. F	155. F	164. T
120. F	129. F	138. F	147. T	156. F	165. T
121. F	130. T	139. T	148. T	157. T	166. T
122. F	131. F	140. F	149. T	158. F	
123. T	132. T	141. T	150. T	159. T	
124. F	133. T	142. T	151. F	160. T	
125. T	134. T	143. F	152. T	161. T	

Fill in the Blanks

167. Summer
168. Lal Bag
169. Cool dry
170. *Hiptage*
171. 150mg
172. *Ficus repens*
173. Rose Sherbet
174. Delphinidin
175. Nyctaginaceae
176. Scarlet Queen
177. Sunny
178. Chandrama

179. Kokette
180. India
181. Iridaceae
182. Tuberose
183. Double
184. Surface flowering
185. *Victoria regia*
186. Orchid
187. Caladium
188. Swama Chameli
189. Jasmine

190. Gold Mohar
191. FebMarch
192. Jasmine
193. Pusa Reshmi
194. $AgNO_3$
195. Persian garden
196. Dioecious
197. Monocotyledonous
198. Brinjal
199. Mango
200. Flower arrangement

Step Ten

Multiple Choice

1.	D	17.	C	33.	D	49.	B	65.	D	81.	D
2.	A	18.	B	34.	D	50.	A	66.	A	82.	A
3.	D	19.	D	35.	B	51.	B	67.	D	83.	B
4.	A	20.	A	36.	B	52.	D	68.	C	84.	C
5.	C	21.	C	37.	D	53.	D	69.	B	85.	A
6.	B	22.	A	38.	C	54.	B	70.	A	86.	A
7.	D	23.	B	39.	B	55.	B	71.	B	87.	B
8.	B	24.	B	40.	C	56.	B	72.	B	88.	B
9.	C	25.	C	41.	A	57.	A	73.	A	89.	C
10.	B	26.	C	42.	A	58.	A	74.	A	90.	B
11.	C	27.	C	43.	B	59.	C	75.	B	91.	D
12.	B	28.	D	44.	B	60.	C	76.	B	92.	D
13.	C	29.	D	45.	C	61.	D	77.	A	93.	D
14.	A	30.	A	46.	C	62.	A	78.	A	94.	A
15.	D	31.	B	47.	D	63.	B	79.	B		
16.	B	32.	B	48.	A	64.	C	80.	D		

Matching

96. A iv B vii C i D ix E v F ii G x H vi I viii J iii
97. A ii B iv C v D iii E i
98. A x B vii C ii D vi E ix F i G iv H v I viii J iii
99. A iv B ii C v D iii E i
100. A iii B v C i D vii E ix F iv G ii H vi I iii J x
101. A v B viii C vi D ii E i F iii G iv H vii I x J ix
102. A iii B v C i D ii E iv
103. A ii B v C iv D iii E i
104. A iii B vi C viii D ix E i F x G ii H v I iv J vii
105. A iv B i C v D ii E iii
106. A v B viii C i D vii E ii F iv G x H vi I ix J iii
107. A vi B i C v D i E viii F vii G iv H x I ii J ix
108. A iii B i C v D iv E ii
109. A v B i C iv D iii E ii

110. A ii B viii C iv D x E vi F ix G v H iii I vii J i
111. A iv B i C v D iii E ii
112. A ix B i C v D viii E ii F iii G x H iv I vi J vii
113. A viii B v C i D x E iii F ix G ii H ii I vi J vii
114. A ii B v C iii D i E
115. A v B i C iv D iii E ii
116. A iii B v C i D ii E iv

True / False

117. F	126. T	135. F	144. T	153. T	162. F
118. T	127. T	136. F	145. F	154. T	163. T
119. T	128. T	137. F	146. F	155. T	164. T
120. F	129. T	138. T	147. T	156. F	165. F
121. T	130. T	139. F	148. F	157. T	166. T
122. T	131. F	140. T	149. T	158. F	
123. T	132. F	141. T	150. T	159. F	
124. F	133. T	142. T	151. F	160. T	
125. T	134. F	143. T	152. F	161. T	

Fill in the Blanks

167. Apple
168. Mango
169. Chrysanthemum
170. Herbaceous
171. Guava
172. Ber
173. Ring
174. Kochia
175. Litchi
176. WB Hayes
177. J&K
178. Bonsai
179. Japanese
180. Peach
181. Housefly
182. Onion
183. Radish
184. Mango
185. Aonla
186. Anacardiaceae
187. Rosaceae
188. Twice
189. October
190. Dioecious
191. Tea
192. April-May
193. Gulmohar
194. Foliage
195. Yellow
196. Blue
197. Guava
198. Andromonoecious
199. Cucurbitaceous
200. India

Step Eleven

Multiple Choice

1. C	17. C	33. B	49. A	65. D	81. B
2. C	18. D	34. A	50. C	66. A	82. B
3. A	19. C	35. B	51. A	67. B	83. C
4. A	20. D	36. C	52. C	68. D	84. B
5. B	21. A	37. C	53. A	69. C	85. A
6. B	22. C	38. B	54. B	70. B	86. A
7. B	23. A	39. B	55. A	71. B	87. B
8. A	24. B	40. A	56. B	72. D	88. C
9. B	25. B	41. B	57. C	73. B	89. C
10. C	26. A	42. A	58. B	74. B	90. C
11. D	27. A	43. A	59. B	75. A	91. D
12. B	28. D	44. B	60. B	76. B	92. A
13. C	29. D	45. A	61. D	77. C	93. B
14. A	30. A	46. D	62. A	78. A	94. D
15. D	31. A	47. A	63. B	79. B	
16. D	32. C	48. A	64. A	80. D	

Matching

95. A iii   B i   C iv   D v   E ii
96. A v   B i   C iii   D iv   E ii
97. A ii   B iv   C i   D v   E iii
98. A iv   B i   C v   D iii   E ii
99. A iii   B i   C v   D ii   E iv
100. A iii   B i   C v   D ii   E iv
101. A iv   B i   C iii   D v   E ii
102. A ii   B i   C iv   D v   E i ii
103. A ii   B iii   C v   D i   E iv
104. A ii   B v   C iv   D iii   E i
105. A ii   B iv   C iii   D v   E i
106. A ii   B i   C iv   D v   E iii
107. A i   B ii   C iv   D v   E iii
108. A ii   B iv   C iii   D v   E i

109. A ii B iv C i D iii E v
110. A iii B v C i D iv E ii
111. A ii B iv C v D iii E i
112. A iv B i C v D ii E iii
113. A v B ii C iv D i E iii
114. A viii B iv C i D iii E x F ix G v H vi I vii J ii
115. A iii B i C iv D ii E v
116. A ii B iv C iii D v E i

True / False

117. T	126. T	135. F	144. F	153. T	162. F
118. F	127. F	136. F	145. F	154. T	163. T
119. F	128. F	137. T	146. T	155. F	164. F
120. F	129. F	138. T	147. F	156. F	165. F
121. T	130. F	139. T	148. F	157. T	166. F
122. F	131. T	140. T	149. T	158. T	
123. F	132. F	141. F	150. F	159. T	
124. T	133. T	142. T	151. T	160. F	
125. T	134. T	143. T	152. T	161. F	

Fill in the Blanks

167. Alphonso	179. Shahjehan	190. Pea
168. M-27	180. Youth & Life	191. Tropical
169. Dioecious	181. Mughal Gardens	192. Lotus
170. Ber	182. Wild Rose	193. Cauliflower
171. Pectin	183. Formal style	194. Northern Spy
172. Litchi	184. Annuals	195. Vellaikollumban
173. Three	185. Foliage	196. Totapuri Red
174. Japanese	186. English	197. *Poincirus*
175. Wreath	187. Floribunda	198. Herring bone
176. Mysore	188. Foliage	199. Knolkhol
177. Jodhpur	189. Dolichos bean	200. Acidic
178. Lucknow		

Step Twelve

Multiple Choice

1.	C	17.	D	33.	B	49.	D	65.	D	81.	D
2.	B	18.	A	34.	A	50.	D	66.	A	82.	B
3.	B	19.	C	35.	D	51.	B	67.	D	83.	A
4.	C	20.	B	36.	B	52.	C	68.	C	84.	D
5.	C	21.	D	37.	B	53.	B	69.	D	85.	C
6.	A	22.	C	38.	C	54.	C	70.	C	86.	A
7.	A	23.	A	39.	D	55.	B	71.	D	87.	D
8.	D	24.	B	40.	B	56.	B	72.	C	88.	D
9.	D	25.	D	41.	A	57.	A	73.	D	89.	D
10.	B	26.	B	42.	A	58.	D	74.	D	90.	B
11.	A	27.	B	43.	C	59.	C	75.	C	91.	A
12.	B	28.	A	44.	C	60.	D	76.	B	92.	B
13.	D	29.	D	45.	B	61.	D	77.	B	93.	D
14.	B	30.	D	46.	D	62.	A	78.	D	94.	C
15.	D	31.	D	47.	B	63.	A	79.	C		
16.	D	32.	B	48.	B	64.	A	80.	B		

Matching

95. A iv B vi C i D ii E iii F viii G v H ix I viii J x
96. A iii B v C i D ii E iv
97. A iii B v C i D iv E ii
98. A ii B iv C i D iii E v
99. A i B iv C ii D v E iii
100. A iv B iii C i D v E ii
101. A iv B i C v D ii E iii
102. A v B vii C i D ix E iii F viii G ii H x I iv J vi
103. A iv B iii C i D v E ii
104. A iv B v C i D iii E ii
105. A iii B v C i D iv E ii
106. A viii B i C v D x E iii F vii G ix H vi I iv J ii
107. A iii B i C iv D v E ii
108. A iv B ii{ C)vD iii E i

109. A iv B vi C i D viii E ii F x G vii H ix I iii J v
110. A ii B Iv C i D v E iii
111. A iii B i C iv D ii E v
112. A i B ii C iii D iv E v
113. B A ii B iv C i D iii E v
114. A ii B vi C iv D viii E ix F x G iii H v I vii J i
115. A iv B vi C i D viii E ii F x G v H vii I iii J ix
116. A ix B iii C vii D i E x F iv G v H vi I ii J viii

True / False

117. T	126. F	135. F	144. F	153. T	162. T
118. F	127. T	136. F	145. T	154. T	163. T
119. F	128. T	137. T	146. T	155. T	164. F
120. F	129. T	138. F	147. T	156. F	165. F
121. T	130. T	139. T	148. T	157. T	166. F
122. T	131. T	140. T	149. T	158. T	
123. T	132. T	141. F	150. T	159. T	
124. T	133. F	142. T	151. F	160. F	
125. T	134. F	143. F	152. F	161. T	

Fill in the Blanks

167. Brussel's Sprout
168. Curd
169. Potato
170. Dr Kihara
171. Pointed gourd
172. Leaf hopper
173. Brinjal
174. Capasinthin
175. Broccoli
176. Jelly
177. Pistachionut
178. Mughal

179. *Citrus reticulata*
180. *Citrus grandis*
181. Aphid
182. Banana
183. Pineapple
184. UmranUmran
185. 300 days
186. GolaGola
187. Vit-A
188. Iron
189. April

190. *Momoridica dioica*
191. Araceae
192. Grape
193. Black tip
194. North
195. Etiolation
196. Hyperplasea
197. Chlorosis
198. Boron
199. Manganese
200. Mango

Step Thriteen

Multiple Choice

1.	D	17.	D	33.	C	49.	A	65.	C	81.	B
2.	C	18.	C	34.	C	50.	C	66.	B	82.	C
3.	A	19.	B	35.	B	51.	A	67.	C	83.	A
4.	D	20.	B	36.	B	52.	C	68.	B	84.	A
5.	D	21.	C	37.	A	53.	C	69.	B	85.	B
6.	C	22.	D	38.	C	54.	A	70.	A	86.	C
7.	C	23.	C	39.	A	55.	B	71.	C	87.	C
8.	B	24.	D	40.	A	56.	B	72.	D	88.	B
9.	D	25.	D	41.	B	57.	D	73.	B	89.	D
10.	B	26.	D	42.	B	58.	B	74.	D	90.	D
11.	B	27.	D	43.	D	59.	A	75.	A	91.	D
12.	D	28.	A	44.	C	60.	C	76.	C	92.	A
13.	D	29.	B	45.	A	61.	A	77.	C	93.	B
14.	B	30.	A	46.	D	62.	B	78.	C	94.	B
15.	C	31.	D	47.	B	63.	C	79.	B		
16.	C	32.	B	48.	B	64.	B	80.	B		

Matching

95. A iii B v C i D iv E ii
96. A iv B i C viii D x E ii F ix G v H viii I vi J iii
97. A iv B i C v D ii E iii
98. A iii B v C i D iv E ii
99. A iv B ii C i D v E iii
100. A iii B v C i D vii E ix F viii G vi H iv I ii J x
101. A i B ii C iv D v E iii
102. A iii B viii C vi D i E iv F x G v H ix I ii J vii
103. A iv B i C vi D ix E ii F v G x H vii I iii J viii
104. A iii B v C i D vii E ix F ii G iv H x I viii J vi
105. A v B ii C i D iii E iv
106. A iv B i C iii D ii E v
107. A ii B v C iii E iv
108. A v B i C iv D vii E ix F x G ii H vi I viii J iii

109. A i   B iii   C v   D ii   E iv
110. A v   B viii   C x   D i   E ii   F ix   G iv   H vii   I vi   J iii
111. A iii   B vi   C i   D viii   E ii   F x   G iv   H v   I vii   J ix
112. A iv   B ii   C i   D v   E iii
113. A vi   B i   C iv   D ix   E ii   F viii   G x   H v   I iii   J vii
114. A ii   B iv   C v   D iii   E i
115. A ii   B iv   C v   D iii   E i
116. A iii   B v   C vii   D x   E viii   F i   G ix   H vi   I iv   J ii

True / False

117. T	126. T	135. F	144. T	153. T	162. F
118. T	127. T	136. F	145. T	154. T	163. F
119. T	128. T	137. T	146. T	155. F	164. F
120. T	129. T	138. T	147. F	156. T	165. F
121. T	130. T	139. T	148. T	157. T	166. F
122. T	131. F	140. F	149. T	158. T	
123. T	132. F	141. T	150. F	159. T	
124. T	133. T	142. T	151. F	160. F	
125. F	134. T	143. F	152. F	161. T	

Fill in the Blanks

167. Grapefruit
168. Chenopodiaceae
169. Sardar Guava
170. Rose
171. Sweet orange
172. Pectin
173. Grape
174. Rutaceae
175. Watermelon
176. Tiliaceae
177. Doobgrass
178. Citriculture

179. Yellow Vein Mosaic
180. Brinjal
181. Dioecious
182. *Carissa carandus*
183. Processing
184. Bonsai
185. Climacteric peak
186. Lycopene
187. Brinjal
188. Bougainvillia
189. MLOs

190. Gibberellins
191. Aril
192. Knolkhol
193. 1-2 weeks
194. NaOH
195. Whiptail
196. Tomato
197. Sweet orange
198. Pruning
199. Ethrel
200. Ethylene

Step Fourteen

Multiple Choice

1.	A	17.	C	33.	D	49.	A	65.	D	81.	A
2.	A	18.	C	34.	B	50.	C	66.	A	82.	D
3.	A	19.	C	35.	B	51.	D	67.	A	83.	C
4.	A	20.	A	36.	B	52.	D	68.	C	84.	B
5.	C	21.	A	37.	B	53.	D	69.	C	85.	C
6.	A	22.	C	38.	C	54.	D	70.	C	86.	D
7.	B	23.	D	39.	C	55.	B	71.	B	87.	C
8.	A	24.	B	40.	A	56.	C	72.	B	88.	A
9.	A	25.	D	41.	A	57.	C	73.	D	89.	A
10.	C	26.	B	42.	B	58.	C	74.	A	90.	C
11.	C	27.	D	43.	A	59.	B	75.	D	91.	B
12.	A	28.	B	44.	B	60.	B	76.	B	92.	B
13.	A	29.	A	45.	C	61.	C	77.	A	93.	A
14.	D	30.	B	46.	D	62.	D	78.	D	94.	C
15.	B	31.	B	47.	D	63.	B	79.	D		
16.	D	32.	C	48.	B	64.	C	80.	B		

Matching

95. A ii B iii C v D i E iv
96. A ii B iv C i D v E iii
97. A ix B iv C vi D i E x F ii G viii H v I iii J vii
98. A iv B vi C i D x E vii F ii G ix H v I viii J vii
99. A v B i C vii D ii E ix F iii G viii H vi I x J iv
100. A ii B iv C i D v E iii
101. A v B ii C iv D i E iii
102. A ii B iv C iii D v E i
103. A iv B i C v D ii E iii
104. A ii B i C iv D iii E v
105. A ii B v C iv D iii E i
106. A v B i C iii D v E ii
107. A iv B i C iii D v E ii
108. A iv B ii C v D iii E i

109. A iii B v C i D ii E iv
110. A iv B i C iii D ii E v
111. A v B iii C iv D ii E i
112. A v B i C iii D ii E iv
113. A i B iii C v D ii(E}iv
114. A iv B i C v D iii E ii
115. A iv B vii C i D ii E vi F ix G v H viii I x J iii
116. A iv B i C iii D v E ii

True / False

117. F	126. T	135. T	144. F	153. F	162. F
118. T	127. T	136. F	145. T	154. T	163. T
119. T	128. F	137. F	146. F	155. T	164. T
120. T	129. F	138. T	147. T	156. T	165. T
121. T	130. T	139. T	148. T	157. T	166. T
122. T	131. F	140. T	149. F	158. T	
123. T	132. F	141. F	150. T	159. T	
124. T	133. T	142. T	151. T	160. T	
125. T	134. T	143. F	152. F	161. F	

Fill in the Blanks

167. Alternate
168. Vit-C
169. Mango
170. Malvaceae
171. New Delhi
172. ICAR
173. Mango
174. Panama
175. Litchi
176. Citrus
177. Cuttings
178. Grapes
179. Citrus
180. Okra
181. Pea
182. Rosaceae
183. IBA
184. Pineapple
185. Tomato
186. Bacterial
187. Tomato
188. Vitamins
189. Seeds
190. Amrapalli
191. Carbohydrate
192. Putrefication
193. Lucknow
194. October
195. December
196. Datepalm
197. Evergreen
198. 100
199. 25 x 25
200. Temperate

Step Fifteen

Multiple Choice

1.	A	17.	A	33.	B	49.	C	65.	B	81.	A
2.	C	18.	B	34.	D	50.	C	66.	C	82.	B
3.	B	19.	B	35.	B	51.	C	67.	A	83.	A
4.	C	20.	A	36.	A	52.	A	68.	B	84.	B
5.	D	21.	A	37.	A	53.	B	69.	B	85.	D
6.	A	22.	D	38.	C	54.	C	70.	D	86.	B
7.	A	23.	B	39.	B	55.	B	71.	D	87.	C
8.	C	24.	C	40.	D	56.	D	72.	D	88.	A
9.	A	25.	D	41.	A	57.	A	73.	D	89.	B
10.	D	26.	A	42.	D	58.	C	74.	A	90.	A
11.	B	27.	B	43.	C	59.	D	75.	B	91.	A
12.	B	28.	A	44.	B	60.	D	76.	C	92.	B
13.	B	29.	C	45.	D	61.	B	77.	B	93.	C
14.	C	30.	B	46.	A	62.	A	78.	B	94.	A
15.	C	31.	C	47.	A	63.	D	79.	C		
16.	C	32.	C	48.	A	64.	B	80.	B		

Matching

95. A iii B v C i D iv E ii
96. A ii B v C i D iii E iv
97. A iv B i C iii D v E] ii
98. C A iii B i C v D ii E iv
99. A i B iii C v D ii E iv
100. C A iv B i C v D ii E iii
101. A ix B iii C vii D i E vi F ii G x H iv I v J viii
102. A viii B iv C vi D i E x F vii G ix H iii I v J ii
103. A v B ii C iv D iii E i
104. A ix B i C v D vii E iii F x G vi H iv I viii J ii
105. A iii B v C i D iv E ii
106. A v B i C iii D ii E iv
107. A v B viii C x D i E ix F iii G vii H ii I vi J iv
108. A iii B i C iv D v E ii

109. A iv B i C Ji D iii E v
110. A iii B vi C i D vii E x F iv G ix H viii I v J ii
111. A i B iii C v D iv E ii
112. A v B i C iv D iii E ii
113. A ii B v C iii D iv E i
114. A iii B ii C i D v E iv
115. A iv B i C iii D v E ii
116. A ii B iv C iii D v E i

True / False

117. T	126. T	135. T	144. T	153. F	162. T
118. T	127. T	136. T	145. T	154. T	163. T
119. T	128. F	137. T	146. T	155. F	164. T
120. F	129. T	138. F	147. T	156. F	165. F
121. F	130. F	139. T	148. T	157. F	166. T
122. F	131. T	140. F	149. F	158. T	
123. F	132. T	141. F	150. T	159. T	
124. T	133. T	142. T	151. F	160. T	
125. F	134. T	143. T	152. T	161. T	

Fill in the Blanks

167. Raisin
168. Nasik
169. Pomegranate
170. Maharashtra
171. India
172. IARI
173. Mango
174. 1600
175. Mysore
176. Watermelon
177. Cucumber
178. 85-90%

179. India
180. Dec-Feb
181. 3:2
182. Potassic (K_2O)
183. Warm
184. Euphorbiaceae
185. Watermelon
186. Neelam x Alphonso
187. Banana
188. Bromeliaceae
189. Loquat

190. Sapindaceae
191. Mid
192. Hardwood
193. Fruits
194. Loquat
195. Brazil
196. March-April
197. Heavy
198. Diploid
199. Apple
200. Deciduous

Step Sixteen

Multiple Choice

1.	A	17.	A	33.	C	49.	D	65.	C	81.	A
2.	C	18.	A	34.	A	50.	B	66.	B	82.	B
3.	B	19.	A	35.	D	51.	B	67.	B	83.	D
4.	D	20.	B	36.	D	52.	B	68.	C	84.	B
5.	B	21.	C	37.	B	53.	D	69.	D	85.	C
6.	A	22.	C	38.	B	54.	D	70.	B	86.	C
7.	A	23.	D	39.	A	55.	C	71.	B	87.	A
8.	C	24.	B	40.	B	56.	A	72.	D	88.	B
9.	A	25.	C	41.	B	57.	C	73.	B	89.	D
10.	C	26.	A	42.	B	58.	A	74.	B	90.	C
11.	A	27.	A	43.	A	59.	C	75.	A	91.	A
12.	B	28.	D	44.	D	60.	C	76.	C	92.	C
13.	C	29.	B	45.	C	61.	A	77.	D	93.	B
14.	B	30.	A	46.	A	62.	A	78.	B	94.	A
15.	B	31.	C	47.	C	63.	A	79.	A		
16.	C	32.	D	48.	B	64.	B	80.	A		

Matching

95. A iv B iii C v D i E ii
96. A iii B viii C i D vi E ii F ix G iv H v I vii J x
97. A iii B i C v D iv E ii
98. A v B iii C i D ii E iv
99. A iii B i C ii D iv E v
100. A ii B v C iii D i E iv
101. A ii B v C iv D iii E i
102. A iv B i C v D iii E ii
103. A v B iii C i D iv E ii
104. A iii B v C i D iv E ii
105. A iv B i C v D iii E ii
106. A v B i C iv D iii E ii
107. A iv B ii C v D iii E i
108. A v B i C iii D ii E iv

109.	A ii	B iv	C iii	D v	E i
110.	A ii	B iv	C i	D v	E iii
111.	A ii	B iii	C i	D iv	E v
112.	A ii	B iv	C v	D i	E iii
113.	A iv	B i	C v	D iv	E ii
114.	A iv	B i	C ii	D v	E iii
115.	A iii	B v	C i	D ii	E iv
116.	A v	B ii	C iv	D iii	E i

True / False

117. T	126. T	135. F	144. F	153. F	162. T
118. T	127. F	136. T	145. T	154. T	163. T
119. T	128. F	137. T	146. T	155. F	164. T
120. T	129. T	138. T	147. F	156. F	165. T
121. T	130. T	139. T	148. F	157. F	166. T
122. T	131. T	140. T	149. F	158. T	
123. T	132. T	141. T	150. F	159. T	
124. T	133. T	142. T	151.	160. T	
125. F	134. F	143. F	152. F	161. F	

Fill in the Blanks

167. Aonla
168. Summer
169. Rainy
170. Bullock's Heart
171. Myrtaceae
172. Rutaceae
173. Fig
174. Alylpro'yle disulphide
175. Water
176. Capsicum
177. Late harvesting
178. Pea
179. Carotene
180. Bottlegourd
181. Carrot
182. Dahlia
183. Rainy
184. 25 kg
185. Composite
186. Canna
187. Jelmeter
188. Tartaric
189. Cauliflower
190. Capsaicin
191. Seeds
192. Helichrysum
193. Winter
194. Japanese garden
195. Foliage
196. Yellow
197. Summer
198. Cuttings
199. Capsule
200. Superior

Step Seventeen

Multiple Choice

1. A	17. B	33. C	49. D	65. C	81. D
2. B	18. A	34. A	50. B	66. A	82. B
3. D	19. B	35. C	51. B	67. A	83. B
4. C	20. B	36. B	52. A	68. A	84. A
5. C	21. C	37. C	53. A	69. B	85. A
6. C	22. A	38. D	54. B	70. C	86. A
7. A	23. A	39. B	55. A	71. D	87. A
8. B	24. C	40. D	56. A	72. A	88. A
9. C	25. A	41. A	57. B	73. B	89. B
10. A	26. B	42. C	58. D	74. B	90. C
11. C	27. D	43. D	59. C	75. A	91. B
12. A	28. B	44. A	60. C	76. B	92. C
13. A	29. D	45. A	61. B	77. B	93. A
14. B	30. C	46. D	62. A	78. B	94. A
15. A	31. C	47. D	63. C	79. A	
16. A	32. A	48. C	64. B	80. C	

Matching

95. A iii B v C i D iv E ii
96. A ii B iv C v D iii E i
97. A iii B v C i D ii E iv
98. A v B iii C i D ii E iv
99. A iii B i C v D iv E ii
100. A iii B v C i D iv E ii
101. A iii B v C i D iv E ii
102. A iv B i C v D iii E ii
103. A iii B i C v D ii E iv
104. A v B i C iv D iii E ii
105. A i B iii C v D ii E iv
106. A ii B iv C v D i E iii
107. A iv B v C i D ii E iii
108. A ii B iv C i D v E iii

109. A ii   B iv   C v   D iii   E i
110. A v   B iii   C i   D ii   E iv
111. A iv   B i   C v   D ii   E iii
112. A ii   B iv   C v   D i   E iii
113. A iii   B i   C v   D iv   E ii
114. A iv   B i   C v   D iii   E ii
115. A v   B i   C iii   D ii   E iv
116. A ii   B iv   C v   D iii   E i

True / False

117. F	126. T	135. T	144. T	153. F	162. F
118. T	127. F	136. T	145. T	154. F	163. F
119. F	128. T	137. F	146. F	155. F	164. T
120. T	129. T	138. F	147. T	156. T	165. F
121. T	130. T	139. F	148. T	157. T	166. T
122. F	131. T	140. T	149. F	158. T	
123. T	132. F	141. F	150. T	159. T	
124. F	133. T	142. F	151. T	160. F	
125. T	134. T	143. F	152. T	161. F	

Fill in the Blanks

167. Cambium
168. Radish
169. Turnip
170. Cuttings
171. Napiform
172. *Phytophthora infestans*
173. Early
174. Arecanut
175. One seeded
176. Fleshy peduncle
177. Sterculiaceae

178. Cocoa
179. *Hamelia vestatrix*
180. Balahanur
181. Juglandaceae
182. Stratification
183. Iran
184. JH Hale
185. Peach
186. MLOs
187. Pear
188. Italy
189. Persimmon

190. Ebenaceae
191. Monoecious
192. Mangosteen
193. Leguminoseae
194. Annona
195. Protogynous
196. Avocado
197. Seeds
198. Sapota
199. Bael
200. Bassein Seedless

Step Eighteen

Multiple Choice

1.	B	17.	B	33.	B	49.	A	65.	B	81.	A
2.	A	18.	C	34.	B	50.	D	66.	A	82.	A
3.	D	19.	A	35.	C	51.	C	67.	C	83.	D
4.	D	20.	D	36.	B	52.	D	68.	D	84.	B
5.	B	21.	B	37.	A	53.	B	69.	C	85.	A
6.	A	22.	B	38.	B	54.	A	70.	A	86.	C
7.	C	23.	C	39.	B	55.	A	71.	C	87.	A
8.	B	24.	C	40.	D	56.	C	72.	B	88.	B
9.	B	25.	B	41.	B	57.	B	73.	A	89.	B
10.	D	26.	A	42.	C	58.	A	74.	A	90.	D
11.	B	27.	B	43.	B	59.	B	75.	A	91.	A
12.	C	28.	A	44.	C	60.	D	76.	B	92.	C
13.	B	29.	D	45.	A	61.	B	77.	C	93.	C
14.	A	30.	C	46.	B	62.	A	78.	C	94.	B
15.	B	31.	B	47.	C	63.	A	79.	A		
16.	B	32.	C	48.	C	64.	C	80.	D		

Matching

95. A ii B v C iii D i E iv
96. A ii B v C iv D iii E i
97. A iv B i C iii D v E ii
98. A iii B v C iv D ii E i
99. A iii B v C i D ii E iv
100. A ii B iv C v D i E iii
101. A iv B iii C v D ii E i
102. A iii B iv C v D i E ii
103. A v B iv C i D iii E ii
104. A iii B iv C ii D i E v
105. A iii B i C v D iv E ii
106. A iii B v C i 9 D ii E iv
107. A iii B v C ii D i E iv
108. A iii B iv C ii D v E i

109. A v B iii C ii D iv E i
110. A iv B v C ii D iii E i
111. A iii B iv C i D ii E v
112. A iii B iv C ii D i E v
113. A iii B v C iv D i E ii
114. A iii B iv C i D ii E v
115. A iv B v C i D ii E iii
116. A iv B ii C v D iii E i

True / False

117. T	126. F	135. F	144. T	153. T	162. T
118. T	127. T	136. T	145. F	154. T	163. T
119. F	128. F	137. F	146. T	155. F	164. F
120. T	129. F	138. T	147. T	156. F	165. T
121. F	130. F	139. F	148. T	157. F	166. F
122. T	131. T	140. T	149. T	158. T	
123. F	132. T	141. T	150. T	159. T	
124. T	133. T	142. F	151. F	160. F	
125. T	134. T	143. T	152. T	161. F	

Fill in the Blanks

167. India
168. Fig
169. Stimulative
170. Litchi
171. China
172. Papaya
173. Cross
174. Ber
175. Fungal
176. Uttar Pradesh
177. Myrtaceae
178. Neelam

179. Banarasi
180. March-April
181. Pummelo
182. India
183. T-budding
184. Double sigmoid
185. Triploids
186. Rosica
187. Boron
188. Sindhu
189. Vitamin-A

190. Apple Scab
191. SO_2
192. *Saraca indica*
193. Oleaceae
194. Alphonso
195. Rhizome
196. Cool
197. Bhadauran
198. Fat
199. October
200. Corm

Step Nineteen

Multiple Choice

1. D	17. A	33. A	49. A	65. C	81. B
2. A	18. D	34. B	50. A	66. A	82. D
3. A	19. B	35. A	51. B	67. D	83. C
4. A	20. B	36. B	52. A	68. C	84. A
5. B	21. D	37. D	53. B	69. B	85. C
6. A	22. A	38. B	54. D	70. C	86. B
7. B	23. C	39. D	55. C	71. A	87. A
8. C	24. B	40. A	56. C	72. B	88. A
9. B	25. B	41. C	57. B	73. C	89. A
10. A	26. B	42. D	58. C	74. A	90. A
11. B	27. B	43. C	59. B	75. D	91. D
12. B	28. D	44. B	60. D	76. A	92. C
13. B	29. A	45. B	61. C	77. A	93. D
14. C	30. C	46. D	62. C	78. B	94. B
15. C	31. B	47. D	63. B	79. C	
16. A	32. B	48. B	64. B	80. A	

Matching

95. A ii B iv C i D v E iii
96. A v B iii C iv D i E ii
97. A iv B v C ii D i E iii
98. A ii B iv C v D iii E i
99. A ii B iv C v D iii E i
100. A ii B iv C iii D v E i
101. A iv B ii C iii D i E v
102. A ii B iv C i D vi E ix F v G viii H x I vii J iii
103. A iv B ii C v D iii E i
104. A ii B i C iv D Y E iii
105. A iii B v C iv D i E ii
106. A i B ii C iii D v E iv
107. A iv B ii C v D iii E i
108. A iii B ii C v D i E iv

109. A viii B i C ii D x E v F iii G ix H vi I iv J vii
110. A iii B i C iv D v E ii
111. A v B iii C iv D i E ii
112. A iii B iv C v D ii E i
113. A viii B vii C vi D iv E iii F v G ii H ii I x J ix
114. A i B ii C iii D iv E v
115. A v B i C iv D ii E iii
116. A v B i C iv D ii E iii

True / False

117. T	126. F	135. T	144. F	153. T	162. T
118. F	127. T	136. T	145. F	154. T	163. T
119. T	128. T	137. F	146. T	155. T	164. F
120. T	129. F	138. F	147. T	156. T	165. T
121. T	130. T	139. T	148. T	157. T	166. F
122. F	131. T	140. T	149. F	158. T	
123. T	132. F	141. F	150. T	159. T	
124. F	133. F	142. F	151. F	160. T	
125. F	134. F	143. F	152. F	161. T	

Fill in the Blanks

167. Long
168. Boron
169. Fertigation
170. Temperate
171. Food Product Act
172. Pectin
173. Papaya
174. Berry
175. Pinching
176. *Clerodendron inerme*
177. 1949
178. Watermelon

179. Tetra-2xPusa Rasal
180. Winter
181. 250q
182. Rutaceae
183. Mango
184. May
185. Mango
186. Monoecious
187. 2n=36
188. Sword
189. Shield

190. Uttar Pradesh
191. Back cross
192. Central Africa
193. Heterosis
194. Drupe
195. Air layering
196. *Cassia fistula*
197. Seed
198. Fig
199. India
200. Spinach

Step Twenty

Multiple Choice

1.	D	17.	C	33.	C	49.	B	65.	C	81.	A
2.	C	18.		34.	B	50.	A	66.	C	82.	D
3.	B	19.	C	35.	B	51.	C	67.	D	83.	C
4.	A	20.	C	36.	B	52.	B	68.	D	84.	B
5.	A	21.	D	37.	B	53.	B	69.	C	85.	B
6.	C	22.	C	38.	D	54.	D	70.	C	86.	B
7.	C	23.	B	39.	A	55.	A	71.	B	87.	A
8.	A	24.	C	40.	D	56.	B	72.	C	88.	C
9.	B	25.	C	41.	B	57.	C	73.	B	89.	C
10.	C	26.	C	42.	B	58.	A	74.	C	90.	A
11.	A	27.	A	43.	C	59.	B	75.	C	91.	B
12.	D	28.	B	44.	B	60.	C	76.	A	92.	D
13.	B	29.	B	45.	B	61.	B	77.	A	93.	D
14.	C	30.	C	46.	A	62.	A	78.	C	94.	C
15.	A	31.	B	47.	A	63.	B	79.	B		
16.	C	32.	B	48.	C	64.	B	80.	A		

Matching

95. A ii B iii C i D v E iv
96. A iv B iii C v D ii E i
97. A iii B i C iv D ii E v
98. A iii B v C iv D i E ii
99. A iii B iv C iv D i E ii
100. A v B iii C i D ii E iv
101. A iv B ii C v D iii E i
102. A iii B iv C i D ii E v
103. A v B ii C iv D iii E i
104. A iii B iv C v D i E ii
105. A iii B iv C i D v E ii
106. A v B iii C i D ii E iv
107. A ii B iii C v D iv E i
108. A iv B iii C ii D v E i

109.	A	iv	B	iii	C	i	D	v	E	ii
110.	A	iv	B	iii	C	i	D	v	E	ii
111.	A	iv	B	iii	C	ii	D	i	E	v
112.	A	ii	B	i	C	iii	D	v	E	iv
113.	A	i	B	ii	C	iii	D	iv	E	v
114.	A	ii	B	i	C	iii	D	v	E	iv
115.	A	iii	B	v	C	i	D	ii	E	iv
116.	A	iv	B	iii	C	ii	D	v	E	i

True / False

117. F	126. T	135. T	144. T	153. T	162. T
118. F	127. F	136. F	145. T	154. T	163. T
119. F	128. T	137. F	146. F	155. F	164. T
120. T	129. F	138. T	147. F	156. T	165. T
121. T	130. T	139. T	148. T	157. T	166. T
122. T	131. F	140. T	149. F	158. T	
123. T	132. T	141. T	150. T	159. T	
124. T	133. F	142. T	151. T	160. T	
125. T	134. F	143. T	152. T	161. T	

Fill in the Blanks

167. Aline
168. Hardwood cutting
169. China
170. Italy
171. 65 to 75
172. J & K
173. India
174. 2n=49
175. 300-500 g/ha
176. Leaf curl
177. 2x2m
178. Amrapalli

179. *J grandiflorum*
180. Square
181. 85-90%
182. Pollinizer
183. 1000-1600
184. Peach
185. Kanak Champa
186. Edge
187. Insulin
188. Bottle gourd
189. Bitter

190. Gurgaon
191. Jodhpur
192. Mango
193. Pruning
194. Seed
195. Boron
196. 685%
197. India
198. Brinjal
199. Blanching
200. Peeling